Mining in Ecologically Sensitive Landscapes

Mining in Ecologically Sensitive Landscapes

Editor

Harish Kumar

scitus
academics

Mining in Ecologically Sensitive Landscapes
Edited by **Harish Kumar**

ISBN: 978-1-68117-675-8
Library of Congress Control Number: 2016958835

SCITUS Academics LLC,
616, Corporate Way, Suite 2, 4766,
Valley Cottage, NY 10989

www.scitusacademics.com

Notice

Printed in United States of America on Acid Free Paper

Preface

Earth protection is much more important than it was and the preservation of fragile ecosystems and the safety of communities in the vicinity of mining sites can present challenges to the start of a profitable operation. The mining projects proposed vary according to the type of materials or metals to be extracted from the planet. The vast majority of proposed mining projects involve the extraction of ore deposits including copper, molybdenum, zinc, lead, silver, gold, cobalt, nickel and platinum. The subject of the publication is the environmental impact of large-scale mining projects involving these metal ores. Mining work affects the environment by revealing radioactive elements, removing topsoil, increasing the risk of contamination of nearby surface and groundwater sources and acidifying the environment. Other effects include the disruption of existing ecosystems, damaging landscapes by depletion and erosion of surrounding freshwater sources. The interaction between geology and conservation and mining in environmentally sensitive landscapes is discovered by mining. Many areas of rare geology that understand the physics of minerals also have unique ecological communities of animals and plants. Increasingly, ridiculous mineral deposits can lead to a conflict between the interests of mining and conservation in such landscapes.

This book brings contributions from conservation and mining experts around the globe to tackle this pressing problem and to work for a good effect for everyone. It will be a valuable tool for environmental science professionals, restoration ecologists, advanced environmental studies and ecology graduates, mine managers, regulators, consultants, environmental specialists in mining, and government departments concerned with this area.

Table of Contents

1

CHAPTER

Landscape-Scale Disturbance: Insights into the Complexity of Catchment Hydrology in the Mountaintop Removal Mining Region of the Eastern United States

Andrew J. Miller † and Nicolas Zégre *,†

School of Natural Resources, West Virginia University, P.O. Box 6125, Morgantown, WV 26506, USA

ABSTRACT

Few land disturbances impact watersheds at the scale and extent of mountaintop removal mining (MTM). This practice removes forests, soils and bedrock to gain access to underground coal that results in likely permanent and wholesale changes that impact catchment hydrology, geochemistry and ecosystem health. MTM is the dominant driver of land cover changes in the central Appalachian Mountains region of the United States, converting forests to mine lands and burying headwater streams. Despite its dominance on the landscape, determining the hydrological impacts of MTM is complicated by underground coal mines that significantly alter groundwater hydrology. To provide insight into how coal mining impacts headwater catchments, we compared the hydrologic responses of an MTM and forested catchment using event rainfall-runoff analysis, modeling and isotopic approaches. Despite similar rainfall characteristics, hydrology in the two catchments differed in significant ways, but both catchments demonstrated threshold-mediated hydrologic behavior that was attributed to transient storage and the release of runoff from underground mines. Results suggest that underground mines are important controls for runoff generation in both obviously disturbed and seemingly undisturbed catchments and interact in uncertain ways with disturbance from MTM. This paper summarizes our results and demonstrates the complexity of catchment hydrology in the MTM region.

Keywords: mountaintop removal mining; catchment hydrology; disturbance hydrology; streamflow generation; underground coal mining

1. INTRODUCTION

Determining the hydrological impacts of an individual disturbance from multiple, overlapping disturbances in space and time is a major unresolved issue for hydrological sciences [1], even when considering landscape-scale disturbances, such as mountaintop removal mining (MTM). MTM is a special form of surface mining adapted to mountain landscapes that removes forests, soils and overlying bedrock to gain access to underground coal seams. This practice leads to likely permanent and wholesale changes to catchment organization, structure, soils and vegetation that have important implications for catchment hydrology, geochemistry and ecosystem health [2,3,4]. The hydrological impacts of this practice are under increasing scrutiny following several devastating floods (e.g., [5]) and well-established patterns of biological impairment downstream of MTM [4,6,7] that implicate disturbance-driven changes in hydrology.

Surface mining for coal is the dominant driver of land cover changes [8,9] in the portion of the Appalachian Coalfields region within West Virginia, Kentucky, Virginia and Tennessee, impacting nearly 7% of the mostly forested 4.86 million hectare area (Figure 1), burying ~4000 km of headwater streams under valley fills [10]. Despite the scale and extent of its disturbance, impacts from MTM on catchment hydrology are poorly understood. Knowledge gaps are vast and stem primarily from a paucity of pre- *vs.* post-impact studies due to limited access to private property [6] and the lack of consistent, long-term, scale-appropriate gauged watersheds [2]. The challenges of conducting research in heavily-disturbed and ever-changing environments [11] often prevents process-based and long-term studies, and the challenges of isolating the hydrological impacts of one disturbance within the context of multiple, overlapping disturbances [1] precludes better understanding of the hydrological changes that result from MTM.

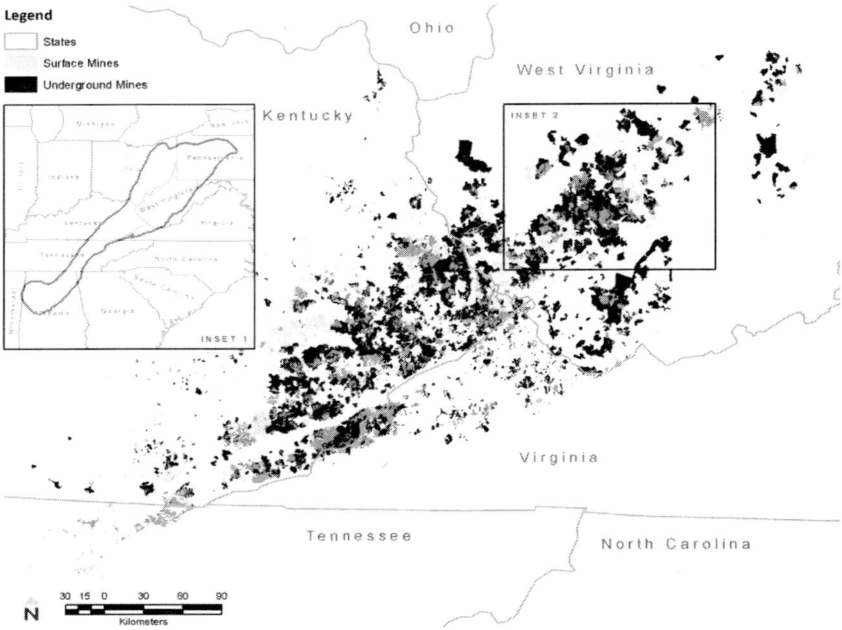

Figure 1. Extent of mountaintop removal mining and underground coal mining in the central Appalachian Mountains region. Surface mines have disturbed nearly 7% of this predominantly forested 4.86 million hectare region of West Virginia, Kentucky, Virginia and Tennessee, and underground mining has impacted 25% of the area. Inset 1: the Appalachian Coalfields region, extending from Pennsylvania to Alabama, has a long history of coal mining. Inset 2: location of the current study.

While surface mining is the dominant disturbance in the region, nearly two centuries of heavy natural resources extraction (e.g., mining, oil and gas, timber), agriculture and natural disturbances (e.g., fire, disease, insect defoliation) have resulted in a mosaic of land covers, land uses and land forms that complicate regional hydrology and make determining baseline conditions difficult. This region has a long history of underground coal mining with mines underlying much of the region where MTM is prominent (Figure 1). Underground mines pose significant challenges for understanding the hydrological impacts of surface mining [12,13,14], and most surface mining studies intentionally exclude the confounding effects of underground mining to isolate surface mining impacts on hydrology.

Given the prominence of underground mining and the overlap of surface and subsurface mining disturbances, it is important to frame MTM impacts within the context of the existing disturbance regime. This is not a trivial task, as isolating the hydrological impacts of one disturbance from multiple, overlapping disturbances is a major unresolved issue in hydrology [1] that is exacerbated by the paucity of streamflow, groundwater and underground mine discharge measurements in the MTM region, as well as uncertainty about the actual extents of centuries-old underground mining.

As a starting point for understanding the potential confounding impacts of multiple disturbances on catchment hydrology, this study combined the analysis of local and regional coal geology and mining maps with small catchment hydrologic monitoring to provide insights into how MTM impacts catchment hydrology within the context of the existing disturbance regime. To do this, we quantified the extent of underground mining in the MTM region and below two nearby headwater catchments, one forested and one impacted by MTM, and compared their hydrologic responses using event rainfall and runoff data, modeling and isotopic approaches. This paper summarizes our results from analyzing 23 storm events over a thirteen-month study period and demonstrates the complexity of catchment hydrology in heavily-disturbed landscapes and highlights the challenges for isolating the hydrological impacts of one disturbance from within an existing disturbance regime.

Overview of Coal Mining and Hydrology

Unlike traditional surface mining practices, such as contour, high wall and strip mining that impact a few hundred hectares, MTM disturbs thousands of hectares [2,15], with the largest MTM occupying ~4000 hectares (~40 km2) [16]. MTM removes as much as 200 m of overburden from ridge tops [17] generating large volumes of waste rock that precludes replacing the material back on the ridge. Overburden material is cast into adjacent stream valleys, creating valley fills that completely bury headwater streams. The placement of waste material in headwater valleys forces contact time between runoff and overburden geology. Streams downstream of valley fills are characterized by high pH, conductivity and total dissolved solids due to elevated concentrations of selenium, sulfate, calcium, magnesium and bicarbonate ions [6,7,10,18,19]. The wholesale reconfiguration of catchments from surface mining and valley fills certainly impacts important runoff generation controls, such as the flow path gradient [20] and riparian zone contributions [21], but how and to what degree is unknown.

Conceptually, the impacts of surface mining on hydrology are straight forward; the often permanent removal and conversion of forests to exotic grasses [22,23] alters the water balance by reducing evapotranspiration (ET) [24], and the removal of soils and severe soil compaction [25] from heavy machinery use in mining and reclamation activities alters dominant flow paths and catchment storage. Runoff tends to be dominated by infiltration-excess overland flow [26,27], and some studies have found that streamflow downstream of surface mines tends to be flashy with large peak flows and storm flow volumes (e.g., [14,28,29]). However, MTM results in a two-part system with contradictory controls on hydrology [2]. Valley fills act like headwater aquifers that temporarily store and release water overtime [30], increasing catchment storage and potentially offsetting runoff increases from impervious surfaces. Streamflow, base flow and flow durations downstream of valley fills are frequently elevated, shifting the hydrologic regime of historically ephemeral headwater streams to perennial [31,32]. The effects of valley fills on storm flows are less well understood; both larger and smaller peak flows have been observed downstream of valley fills [12,23,28]. Differential responses of streamflow to surface mining are primarily due to different mining and reclamation practices, different scales of disturbance [2] and, to an uncertain extent, different disturbance histories.

Our previous work in the MTM region quantified the expansion of MTM in the Big Coal River watershed and the nature of hydrologic impacts across spatial and temporal scales [33,34]. At the basin scale (>103 km2) and over the lifetime of this practice (1969–2012), maximum streamflow and streamflow variability decreased while the base flow ratio increased. At the headwater scale (101 km2), we have found a rapid translation of rainfall to event runoff based on transfer function modeling that showed little variation between response curves, high runoff ratios and small proportions of quickflow for four dormant seasons storms in White Oak Creek (also the focus of the current study). Based on the extensive underground mining, it was uncertain if increasing base flows, decreasing variability and the high runoff ratios were attributed to surface mining and valley fills or from unaccounted for water from underground coal mines that underlie much of the region.

Groundwater flow in un-mined areas of the Coalfields region is primarily controlled by fractures that formed as a result of erosional stress relief [12]. Underground mining drastically alters the subsurface structure [35] and increases hydrologic connectivity between the surface and subsurface and between water-bearing subsurface geologic units [36,37]. The effects of underground mining on groundwater flow will largely depend on the connection of mines to fracture systems [12] that efficiently connect groundwater to surface streams. At multiple spatial scales, substantial volumes of water can be transferred between basins and between the subsurface and surface that can diminish, obfuscate or exacerbate surface mining-driven changes in hydrology [12,13,14].

2. METHODS

2.1. Study Area

This study was conducted in two headwater catchments of the Clear Fork River of the Big Coal River located in southern West Virginia, USA (Figure 2). The Big Coal River is in the Appalachian Plateau physiographic region and is characterized as mountainous, deeply-incised terrain. Ridges are narrow and winding with a dendritic drainage networks dominated by ephemeral and perennial first-order streams that drain convergent hillslopes. Hillslopes are steep, and low slope areas are limited to ridge tops and valley bottoms. The Big Coal River watershed is underlain by sedimentary rocks primarily of the Pennsylvanian Age and, specifically, the Kanawha Formation of the Pottsville Group [38]. The Kanawha formation consists of massive beds of sandstone separated by thinner beds of shale, siltstone and coal [39]. Soils in this region are thin and shallow (<1 m) and underlain by the impermeable sedimentary rocks. Limited soil storage and steep slopes lead to rapid hydrograph response even in forested catchments [40]. Groundwater movement occurs primarily along horizontal and vertical stress relief fractures and responds quickly to surface moisture conditions [39,41,42].

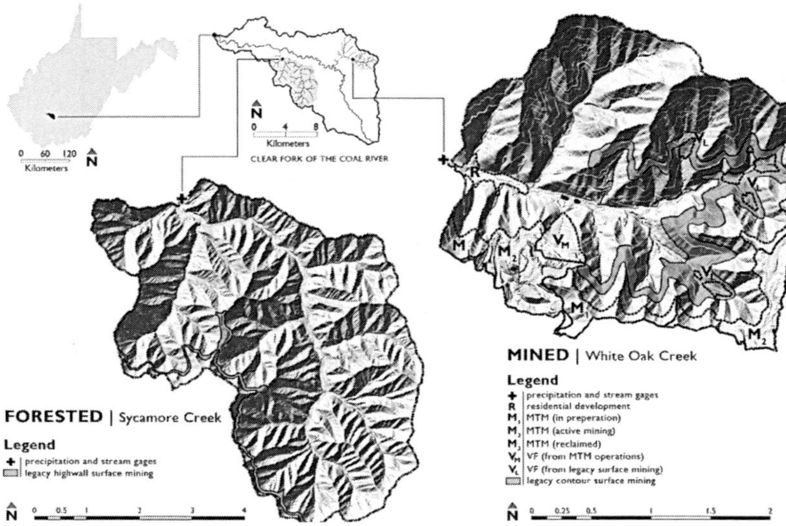

Figure 2. Location of the mountaintop mined White Oak Creek catchment and the forested Sycamore Creek catchment in the southern West Virginia Coalfields region. Hillshade and catchment boundaries were derived from a 1-m LiDAR DEM flown and processed by West Virginia University Natural Resource Analysis Center in April of 2010. Land use boundaries were digitized from 1-m aerial imagery from the 2011 National Agriculture Imagery Program (NAIP).

This region experiences a humid continental climate with warm summers and cold winters. Average temperature during the warmest month (July) is 24 ◦C and average temperature during the coldest month (January) is 0.5 ◦C. Average annual precipitation from 1973–2010 measured nearby at Madison, WV, is 1224 mm [34]. Precipitation is heavily influenced by prevailing wind direction and surface topography and is generally derived from frontal or tropical storm systems. Intense rainfalls frequently exceed 100 mm in a 24-h period [39]. Catastrophic flooding in small catchments (<1000 km2) in this region is often driven by orographic and convective thunderstorms [43].

White Oak Creek (herein referred to as mined) is a 6.5-km2 catchment that drains primarily east to west where it confluences with the Clear Fork River near Artie, WV (Figure 2). Elevations range from 475–991 m; hillslope slopes average 55.6%; the total stream length is 21 km; and the drainage density is 3.3 km km−2 (Table 1). While second and third growth deciduous forest cover dominate the catchment, 17% (1.2 km2) is impacted by recent MTM mining and by a large, partially reclaimed valley fill that occupies ~2% (0.11 km2) of the catchment area. The MTM area in this catchment can be classified into three distinct stages: preparation ~4%), active mining (~10%) and reclaimed (~3%). Areas in preparation are characterized by the excavation of terraces around the section of ridges to be excavated; the initial removal of vegetation; and the development of temporary road networks. Actively-mined areas are completely devoid of vegetation and undergoing active excavation, and reclaimed areas have been regraded and revegetated with herbaceous grasses and minimal woody vegetation. In addition to MTM, traditional contour mining on the mid-slopes of the catchment occurred during the mid-1990s, impacting ~12% of the catchment area, all of which has been revegetated using herbaceous cover and reclaimed in accordance with the Surface Mining Control and Reclamation Act (SMCRA) [44]. As part of this operation, seven small valley fills were created on the incipient drainage network of White Oak Creek, although these structures are significantly smaller than valley fills associated with contemporary MTM practices. Typical of surface mining operations, two small retention ponds were constructed on the main stem White Oak Creek to control sedimentation originating from the surface mines. Outside of MTM areas, no timber extraction occurred in the catchment during or leading up to the study period, although it has been periodically harvested over the last several decades. Residential development in the catchment is limited to the valley floor with a small amount (0.07 km2; ~1%) of light residential development located upstream of our stream gauge.

Sycamore Creek (forested) is a 25.5-km2 forested catchment that drains primarily south to north until it confluences with the Clear Fork near the town of Colcord, WV (Figure 2). Elevations range from 333–1013 m; hillslope slopes average 60.4%; the total stream length is 70 km; and drainage density is 2.75 km km−2 (Table 1). Second and third growth deciduous forest dominate the catchment (~98%). Similar to many headwater catchments in the Big Coal River basin, traditional contour mining extracted coal from this catchment during the 1970s, but analysis of aerial photography shows that this area has been reclaimed back to forest. MTM is absent from this catchment. Several small gas wells are located in the central valley, and a gravel access road parallels Sycamore Creek. In addition, two right-of-ways covered in grasses and herbaceous cover cross the catchment: a 50 m-wide natural gas pipeline in the southern half of the catchment and a 30 m-wide electric transmission line that runs north-south, paralleling Sycamore Creek for much of its length. Sycamore Creek has been timbered throughout history, but no active forest harvesting occurred leading up to or during the study period. An extensive network of logging roads from previous timbering and on-going gas operations follows the valley upstream. Light

Table 1. Catchment characteristics of the mined White Oak Creek catchment and the forested Sycamore Creek catchment.

Catchment	A.							Surface Mining				Underground Mining			
								B.	C.	D.	E.	F.	G.	H.	I.
	Station Location	Station Elevation	Catchment Area	Elevation Change	Mean Slope	Stream Length	Drainage Density	Legacy [1]	MTM [2]	Vally Fill Area / % Catchment Area [3]	Total Surface Mining Disturbance [4]	Catchment Area Undermined [5]	Underground Mines Contributing Area [6]	Catchment Area + Underground Mines Contributing Area [7]	Catchmen Area Difference
	(Lat/Long)	(m)	(km^2)	(m)	[° (Std Dev)]	(km)	($km\ km^{-2}$)	[% (km^2)]	[% (km^2)]	[% (km^2)]	[% (km^2)]	[% (km^2)]	(km^2)	(km^2)	(%)
White Oak Creek (Mined)	37°56.47′ −81°19.93′	475	6.5	516	55.6 (±19.2)	21	3.3	10.4 (0.8)	17.3 (1.1)	2.9 (0.2)	31 (2.1)	84 (5.4)	15.5	22	238
Sycamore Creek (Forested)	37°56.47′ −81°26.01′	333	25.5	680	60.4 (±13.7)	70	2.8	2.1 (0.5)	0	0	0.5 (2.0)	20 (5.2)	20	45.5	80

[1] Legacy surface mining consists of traditional surface mining practices that include highwall mining and contour mining. In White Oak Creek contour surface mining took place during the 1990s. This area was reclaimed and is currently covered by herbacous grasses and some woody vegetations. In Sycamore Creek, highwall mining took place on the western ridge during the 1970s. The highwall and mining bench were not reclaimed, but natural succession of deciduous forest has occurred. [2] MTM in White Oak Creek consists of areas "in preparation" (0.3 km^2), "active surface mining" (0.7 km^2), and "reclaimed" (0.2 km^2). [3] Valley fills in White Oak Creek consists of one large contemporary valley fill (0.1 km^2) and several fills (0.1 km^2). [4] Calculated as B + C + D. [5] Underground mining within topographically delineated catchment area. [6] Area of underground mining updip of stream gaging station. Represents subsurface contributing drainage area. [7] Calculated as A + G. [8] Estimated differences between surface delineated catchment area and contributing area (surface + underground mining). Calculated as (H-A)/A.

residential development in the broad floodplain is located downstream of our stream gauging station.

Underground coal mining throughout the Appalachian Coalfields region began in late 1800s and in the 1930s in the Big Coal River basin. To quantify the extent of underground mining throughout the region and to assess the potential for confounding impacts of underground mine disturbance and inter-basin transfer, coal geology and mining activity maps were acquired from the West Virginia University Natural Analysis Resources Center and analyzed in GIS to determine the dip direction of coal seams and to estimate the regional and local extent of underground mining. The subsurface contributing recharge area for each catchment was estimated by calculating the area of individual and overlapping up-dip, underground coal mines. Coal seams in this region often overlap and merge [45,46], and the actual extent, integrity and connectivity of underground mines is highly uncertain without detailed field reconnaissance. As a result, subsurface recharge areas are gross estimates. Digital coal geology and mining activity data are made publicly available from the West Virginia Department of Environmental Protection TAGIS group (http://tagis.dep.wv.gov/home/).

2.2. Instrumentation, Data Collection and Analyses

Precipitation and streamflow were measured upstream of catchment outlets from 1 September 2011–30 September 2012. Precipitation was measured using TE525-L tipping bucket rain gauges installed in clearings located adjacent to each stream gauging station (Figure 2) and recorded using Campbell Scientific (Logan, UT, USA) CR 800 data loggers at 10-minute intervals. Stream stage was measured using Campbell Scientific CS450 pressure transducers installed along stream reaches and also recorded at 10-min intervals using CR 800 data loggers. The recorded stream stage was converted to streamflow using stage-streamflow rating curves developed for each location. Streamflow was measured using the velocity area methodology [47] using a SonTek Handheld Acoustic Doppler Velocimeter and verified using the salt tracer dilution method [48]. The rating curve for the mined catchment was developed using eight measurements ranging from 0.02–1.21 mm h−1($r2$= 0.99) and using 10 measurements ranging from 0.001–0.84 mm h−1 ($r2$ = 0.99) for the forested catchment. Seven months into the study period, the stream gauging station in the forested catchment was moved 5 m downstream to a bedrock controlled pool to minimize measurement error associated with sedimentation. The original pressure transducer elevation was surveyed and marked using a 1-m re-bar pin, and a stage at the new location was adjusted using a correction factor.

Rainfall duration, depth and intensity and total unit runoff, peak flow, time-to-peak and the runoff ratio (Q/P) were characterized for each event. In addition, events were separated into quickflow [49], and hydrologic response times were modeled using a runoff transfer function approach [50,51] (see below). Events were defined by one hour prior to the onset of rain and until the stream stage returned to pre-storm base flow levels or when stream recession was interrupted by a second storm event.

Events-based isotope samples were collected by interfacing Isco automated water samplers (Model 3700, Teledyne ISCO, Inc., Lincoln, NE, USA) with precipitation and stream stage data loggers. Precipitation samplers consisted of 200-mm funnels connected to tubing that bypassed the Isco pump and fed directly into the Isco distribution arm to distribute samples to 1000-mL polyethylene bottles. After the onset of precipitation, rainfall water samples were collected incrementally every hour and/or after exceeding a 7.0-mm threshold. This enabled capturing isotopic variation during events [52,53] and allowed for event precipitation to be weighted for mixing models [54]. Stream water was collected once daily during base flow conditions and between 3- and 6-h intervals during event rising, peak and falling limbs. All Isco bottles were lined with 2-oz Nasco sample bags pre-treated with mineral oil to prevent fractionation from evaporation. Samples were separated from mineral oil by puncturing the bottom of the sample bag and draining water into a 25-mL cone capped scintillation vial for storage until processing. Isotope samples were collected for all storms during the study period, except from December 2011–March 2012, when precipitation fell as snow or frozen rain.

Water samples were analyzed for deuterium (D) and oxygen-18 (18O) using a laser-based liquid water isotope analyzer (DLT-100 Version 2, Los Gator Research, Inc., San Jose, CA, USA) located at the West Virginia University Mountain Hydrology Laboratory. In-house standards were developed by calibrating Hawaiian spring water enriched with heavy isotopes and Colorado spring water depleted of heavy isotopes to the Vienna-Standard Mean Ocean Water (VSMOW) issued by the International Atomic Energy Association (IAEA). Isotope values are reported in per mil (‰) [55] relative to a standard as δD or δ 18O. The precision reported was 0.58(‰) and 0.11(‰) for δD and δ18O, respectively.

2.3. Hydrograph Separation and Response Time Modeling

Streamflow was separated into event and pre-event components using two isotope hydrograph separation methods: a traditional two-component mixing model [56,57] and the transfer function hydrograph separation (TRANSEP) model [58]. Hydrograph separation using these approaches is based on the assumption that pre-event water composition is constant in space and time [57] and that streamflow can be separated into event and prevent components:

$$Q = Q_p + Q_e \tag{1}$$

$$QC = Q_p C_p + Q_e C_e \tag{2}$$

where Q is streamflow; Qp and Qe are contributions from pre-event and event water; and C, Cp, Ce are D or 18O concentrations in streamflow, pre-event and event water [58].

It is well accepted that the isotopic composition of rain water changes through time [52,53], and therefore, the rainfall composition used in the traditional two-component mixing model was incrementally weighted by rainfall intensity [54]:

$$\delta D \; or \; \delta^{18}O = \sum_{i=1}^{n} Ii\delta i / \sum_{i=1}^{n} Ii \tag{3}$$

where Ii is the average rainfall intensity in mm h−1; δi is the measured rainfall isotopic composition during the increment.

While this weighting approach accounts for temporal variability and mass tracer allocation [58], it assumes an instantaneous transfer of event water to the stream and does not incorporate travel time, which is an important descriptor of catchment hydrology. The TRANSEP model, on the other hand, integrates isotope hydrograph separation and an instantaneous unit hydrograph approach to model rainfall-runoff dynamics and travel times. We only briefly describe TRANSEP below; more details can be found in [58] and [59].

The TRANSEP model uses a simple rainfall-runoff model consisting of a nonlinear module that converts precipitation into effective precipitation [50] and a linear model that represents the transformation of effective rainfall into streamflow [58]. TRANSEP requires the a priori selection of transfer functions for both the runoff and the transport transfer function [60]. Based on our previous work in these catchments [34] and successful application in other studies [58,59,61], we used the two parallel linear reservoirs (TPLR) [58] transfer function. The runoff transfer function was optimized to measured streamflow using ant colony optimization (AOC) [58,62] and then used to constrain the optimization of the transport transfer function on the basis of observed isotope compositions and effective precipitation [60]. One hundred thousand Monte Carlo simulations were used to explore parameter sensitivity and to calculate median hydrologic response times from the cumulative distribution function for the top 1% performing models based on the Nash–Sutcliffe efficiency (NSE) objective function [63].

3. RESULTS

3.1. Extent of Underground Mining

Based on our analysis, underground coal mining underlies approximately 1.2 million hectares or 25% of the MTM region (Figure 1), and underground mining below the study catchments is extensive (Figure 3). While no underground mines are located directly under the main stems of either of the creeks, underground coal mining has occurred under headwater tributaries of both catchments.

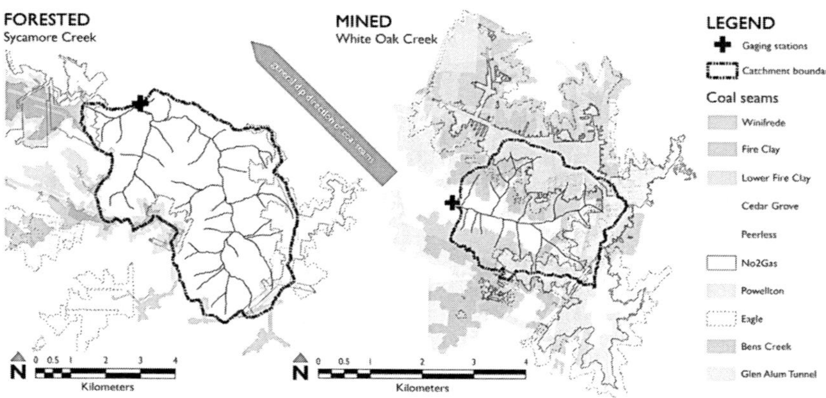

Figure 3. Underground mines below the mountaintop mined White Oak Creek and the forested Sycamore Creek catchments. Coal seams are listed in stratigraphic order, and the dip direction is northwest. Mean dip in the mined catchment is 0.92° and 0.74° in the forested catchment. Approximately 85% of the mined catchment area and 20% of the forested catchment area are directly underlain by underground mines. When including up-dip underground mine areas, the contributing drainage area increases over the topographically-delineated area by ~240% in the mined catchment and by >80% in the forested catchment.

Seven coal seams of the Kanawha formation were mined under the eastern, southern and western hillslopes of the MTM mined catchment (Figure 3) with ~85% of the catchment having been directly undermined (Table 1). Underground mined areas up-dip of the stream gauging station, or the underground recharge area, were approximately 15.5 km2, increasing the effective drainage area to 22 km2. Underground mining has occurred as recently as 2008 and extends deeper into the Kanawha formation than under the forested catchment, mining the Ben's Creek and Glen Alum Tunnel coal seams. Six coal seams were mined under the eastern, southern and western hillslopes of the forested catchment (Figure 3) with ~20% of the catchment having been directly undermined (Table 1). Underground mine contributing area in this catchment was approximately 20 km2, increasing the effective drainage area to ~46 km2. The deepest seam mined below this catchment was the Eagle seam. The dip direction of the coal seams in the area runs southeast to northwest at a slope of ~0.91° in the mined catchment and ~0.74°.

3.2. Rainfall-Runoff Relationships

Rainfall over the thirteen-month study period totaled 1470 mm in the mined catchment and 1511 mm in the forested catchments. Runoff averaged 914 mm (62% of P) and 322 mm (21% of P), respectively. Rainfall was mostly evenly distributed throughout the year with slightly more skew towards spring and early summer. Flow duration curves for the two catchments were markedly different for all but the largest flows (Figure 4).

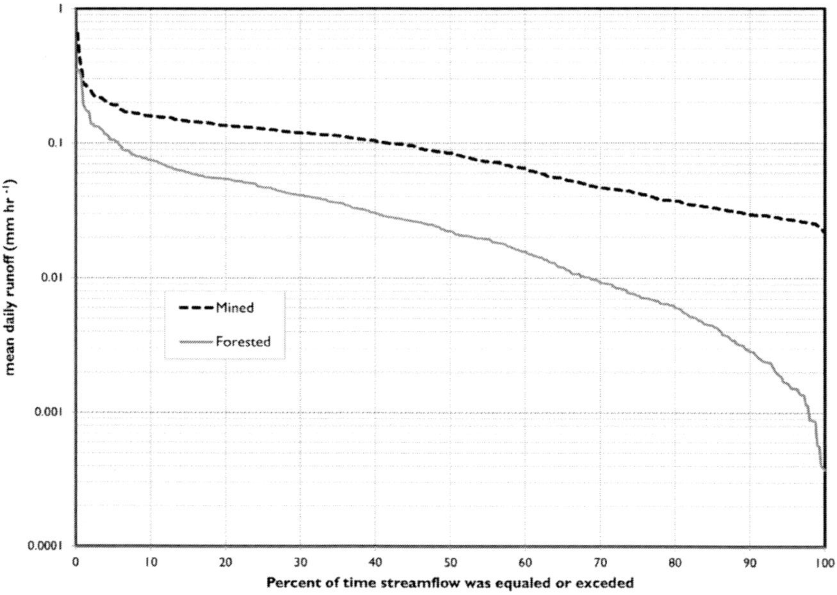

Figure 4. Flow duration curves for the mined and forested catchments from 1 September 2011–30 September 2012 showing the different flow regimes; runoff in the mined catchment is consistently greater than the forested catchment during all times of the study period, particularly during base flow conditions when flow is sustained by valley fills.

The twenty-three events that were used to characterize rainfall-runoff and hydrologic responses of our two study catchments are shown in Figure 5 and summarized in Table 2. Several events occurring in late-December–January and in February were excluded from the analysis because precipitation either fell as snow or as freezing rain and was under-measured by tipping bucket rain gauges. Event rainfall duration, total rainfall, rainfall intensity and maximum rainfall intensity were similar for the two catchments based on the Wilcoxon signed rank test (Figure 6). Events that occurred during the dormant season were generally lower in intensity, longer in duration and had more total rainfall than growing season events (Table 2).

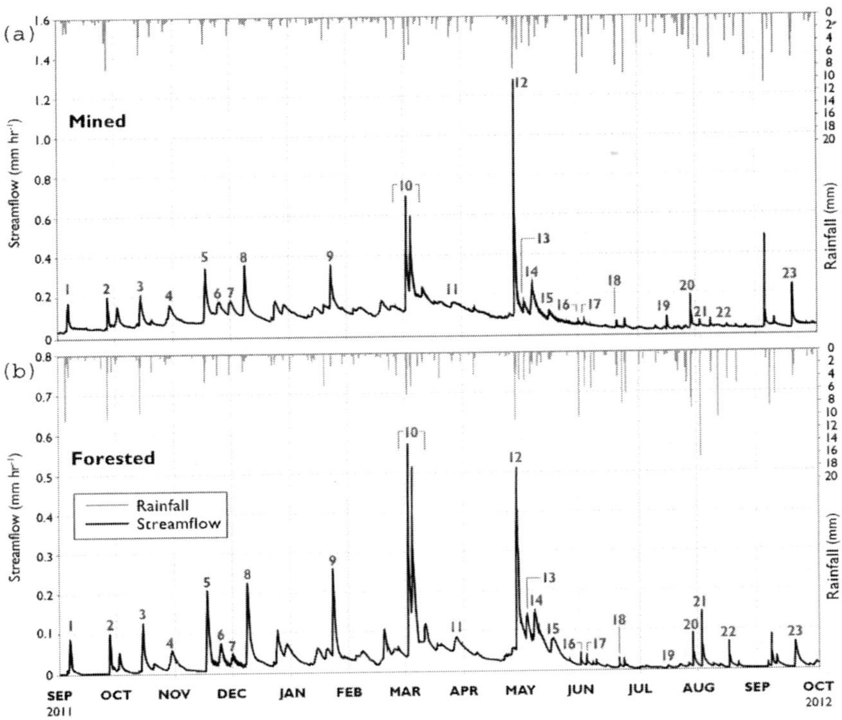

Figure 5. Storm event hydrographs and hyetographs for the (**a**) mined and (**b**) forested catchments for the study period from September 2011–October 2012. The storm events included in the analyses are noted sequentially. The large storm in the mined catchment in late September 2012 was excluded from analysis because the vent tube to compensate for atmospheric pressure was incorrectly assembled after exchanging desiccant.

Runoff characteristics for the 23 events are shown by catchment in Figure 6 and summarized in Table 2. Total unit streamflow, peak flow and runoff ratio were generally larger in the mined catchment, while the time-to-peak and quickflow ratio were greater in the forested catchment irrespective of season. Time-to-peak was faster during the growing season in the mined catchment, but similar to the forested catchment during the dormant season.

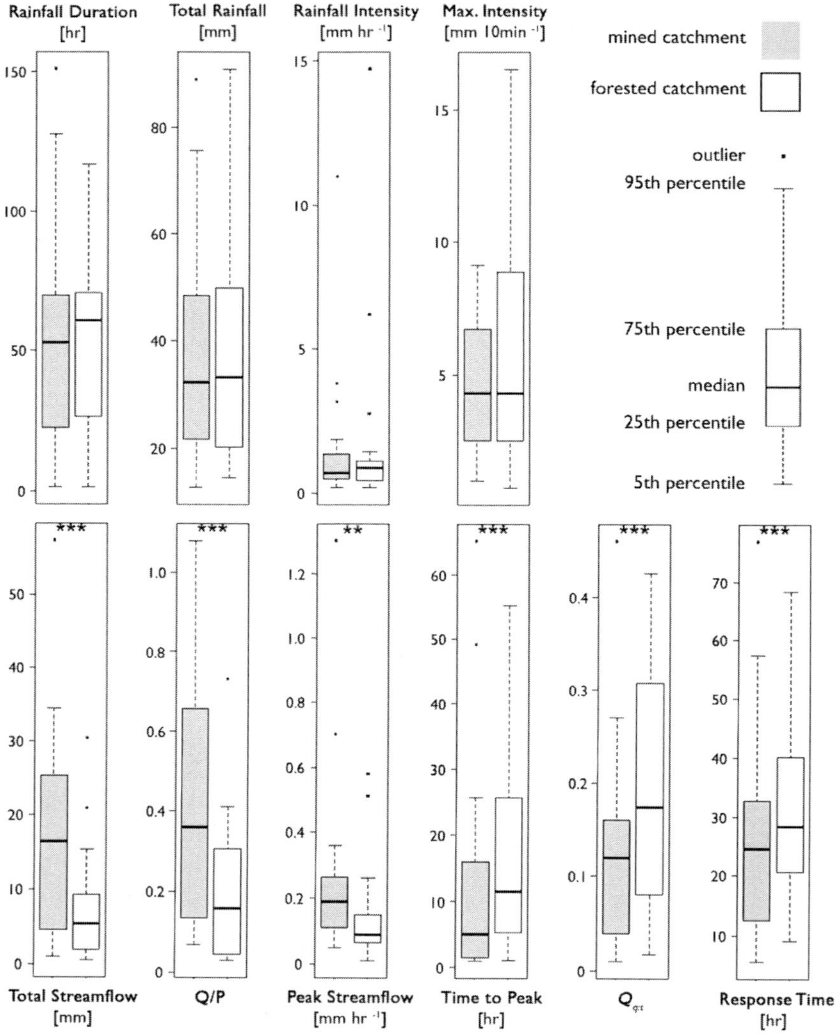

Figure 6. Event rainfall and runoff boxplots for the mined and forested catchments over 23 storm events that occurred during September 2011–October 2012. *** and ** indicate statistically-significant differences at the 99% and 95% confidence level based on the Wilcoxon signed rank test. Statistically-significant differences in rainfall characteristics were not detected. Q/P represents the runoff ratio; Qqis the fraction of quickflow [49]. Median response times were calculated from the cumulative distribution function of the top 1000 performing transfer function models.

Table 2. Rainfall and runoff storm characteristics and hydrologic response times for 23 events measured in the mined and forested catchment over the 13-month study period. Dormant season events are shaded.

Event	Event Start[a]		Mined Catchment									Forested Catchment								
			Rainfall			Runoff						Rainfall			Runoff					
			Duration	Depth	Intensity	Total Streamflow	Q/P	Peak Flow	Time-to-Peak[c]	Lag[d]	QF/Q[e]	Duration	Depth	Intensity	Total Streamflow	Q/P	Peak Flow	Time-to-Peak[c]	Lag[d]	QF/Q[e,f]
	Date	Time[b]	(hr)	(mm)	(mm hr^{-1})	(mm)	-	(mm hr^{-1})	(hr)	(hr)	-	(hr)	(mm)	(mm hr^{-1})	(mm)	-	(mm hr^{-1})	(hr)	(hr)	-
1	04 September 2011	1440	41.7	65.0	1.6	16.4	0.25	0.18	41.7	1.5	0.08	78.5	90.9	1.2	2.7	0.03	0.09	42.0	41.5	-
2	26 September 2011	1510	12.7	48.0	3.8	5.2	0.11	0.21	4.7	1.2	0.27	10.0	62.0	6.2	2.1	0.03	0.10	10.3	6.7	0.42
3	11 October 2011	1930	55.8	51.8	0.9	17.3	0.33	0.22	57.5	15.5	0.13	55.7	49.5	0.9	5.5	0.11	0.13	61.7	20.0	0.17
4	26 October 2011	1130	68.3	34.8	0.5	34.4	0.99	0.17	65.2	65.2	0.02	65.8	30.7	0.5	6.9	0.23	0.06	73.5	55.3	-
5	15 November 2011	0500	52.7	49.0	0.9	20.8	0.42	0.35	37.7	10.8	0.21	52.3	49.5	0.9	7.7	0.16	0.21	37.7	11.3	0.35
6	20 November 2011	1010	71.7	23.4	0.3	25.3	1.08	0.18	84.0	25.7	0.02	71.0	23.1	0.3	6.0	0.26	0.08	86.5	28.3	-
7	29 November 2011	0020	34.8	22.4	0.6	20.1	0.90	0.19	20.8	18.8	0.03	34.5	16.8	0.5	3.7	0.22	0.05	24.3	23.0	-
8	05 December 2011	1950	63.7	46.5	0.7	30.7	0.66	0.36	43.3	5.2	0.09	63.2	50.6	0.8	15.4	0.31	0.22	47.5	37.7	0.19
9	20 January 2012	1720	63.5	32.3	0.5	26.0	0.80	0.36	16.7	5.3	0.12	63.3	33.3	0.5	10.8	0.33	0.26	17.7	7.0	0.31
10	29 Feburary 2012	0450	151.0	88.9	0.6	57.5	0.65	0.70	17.8	5.0	0.23	70.5	73.9	1.0	30.4	0.41	0.58	17.5	7.3	0.43
11	23 March 2012	1320	127.7	27.2	0.2	25.5	0.94	0.17	49.2	49.2	0.00	50.8	17.8	0.3	13.0	0.73	0.09	53.7	34.8	-
12	25 April 2012	1520	78.7	75.7	1.0	34.3	0.45	1.30	19.7	9.0	0.46	78.0	86.9	1.1	20.9	0.24	0.51	19.8	9.7	0.37
13	30 April 2012	2340	26.3	19.6	0.7	9.8	0.50	0.19	3.3	3.2	0.04	22.0	23.4	1.1	7.5	0.32	0.14	16.7	16.3	0.08
14	04 May 2012	2330	127.3	47.8	0.4	21.4	0.45	0.28	35.5	17.0	0.12	117.2	37.3	0.3	13.5	0.36	0.15	23.3	15.3	0.04
15	13 May 2012	1700	38.3	26.7	0.7	9.6	0.36	0.13	20.5	16.3	0.02	60.7	26.2	0.4	7.9	0.30	0.08	41.3	32.0	0.02
16	29 May 2012	1300	1.5	16.5	11.0	1.1	0.07	0.08	1.5	1.5	0.05	1.3	19.6	14.7	0.8	0.04	0.05	1.2	1.0	0.10
17	01 June 2012	0300	10.8	20.3	1.9	2.7	0.13	0.09	10.5	1.5	0.02	11.8	17.3	1.5	1.5	0.08	0.04	13.0	3.7	0.05
18	17 June 2012	2000	14.3	18.8	1.3	2.2	0.12	0.07	14.5	1.3	0.04	14.2	15.2	1.1	0.7	0.04	0.03	15.2	2.7	0.07
19	12 July 2012	2110	67.2	24.4	0.4	4.2	0.17	0.08	45.2	1.2	0.04	92.7	21.1	0.2	0.6	0.03	0.01	17.8	14.3	-
20	26 July 2012	2050	95.7	39.9	0.4	5.8	0.14	0.20	2.3	1.3	0.16	15.0	41.4	2.8	1.9	0.05	0.09	4.2	2.5	0.21
21	31 July 2012	1500	6.7	21.1	3.2	1.8	0.08	0.07	5.5	2.7	0.04	114.2	68.8	0.6	3.1	0.04	0.15	1.8	1.7	0.29
22	14 August 2012	1500	18.8	12.7	0.7	1.4	0.11	0.05	2.8	0.8	0.01	70.3	14.5	0.2	1.4	0.10	0.07	4.2	2.0	0.15
23	17 September 2012	1410	41.7	59.7	1.4	11.7	0.20	0.25	13.7	0.8	0.12	31.0	41.9	1.4	3.5	0.08	0.07	15.7	8.0	0.13
	Average		55.3	37.9	1.5	16.7	0.43	0.26	26.7	11.3	0.10	54.1	39.6	1.7	7.3	0.20	0.14	28.1	16.6	0.20
	Std. Dev.		40.7	20.4	2.3	14.2	0.33	0.27	22.8	16.3	0.11	32.3	23.4	3.1	7.4	0.17	0.14	23.4	15.1	0.14

[a] Start times defined as start of precipitation in the forested catchment. [b] Time listed in Eastern Standard Time (EST). [c] Time to peak defined as time from start of precipitation to peak discharge. [d] Time lag defined as time from maximum instantaneous storm intensity to peak discharge. [e] *QF/Q is* the ratio of quickflow to total flow based on *Hewlett and Hippert* (1967). [f] Hydrograph separation unsuccessful for events 1,4,6,7,11,19 due to the separation slope being greater than the rising limb of the hydrograph.

Quickflow ratios for both catchments were generally greater during the dormant season (Table 2). In addition, we found that quickflow was strongly correlated with gross precipitation in both catchments (mined: $r2 = 0.78$, p-value < 0.0001; forested: $r2 = 0.68$, p-value $= 0.003$). When quickflow and precipitation were compared across growing and dormant seasons, a threshold pattern emerged (Figure 7). Below a threshold of around 30 mm, the relationship between precipitation and quickflow was significant and strongly correlated in both catchments during the growing season (mined: $r2 = 0.76$, p-value < 0.01; forested: $r2 = 0.99$, p-value < 0.01). Above this value, precipitation and quickflow were moderately correlated in the mined catchment ($r2 = 0.32$, p-value $= 0.49$) and strongly correlated in the forested catchment ($r2 = 0.76$, p-value $= 0.08$). During the dormant season, the inverse was true in the mined catchment; precipitation and quickflow were significant and correlated during the dormant season ($r2 = 0.92$, p-value $= 0.02$). Of the four dormant season events in the forested catchment, only one below-threshold event was successfully separated into quickflow [49] and, hence, inference below the threshold was limited.

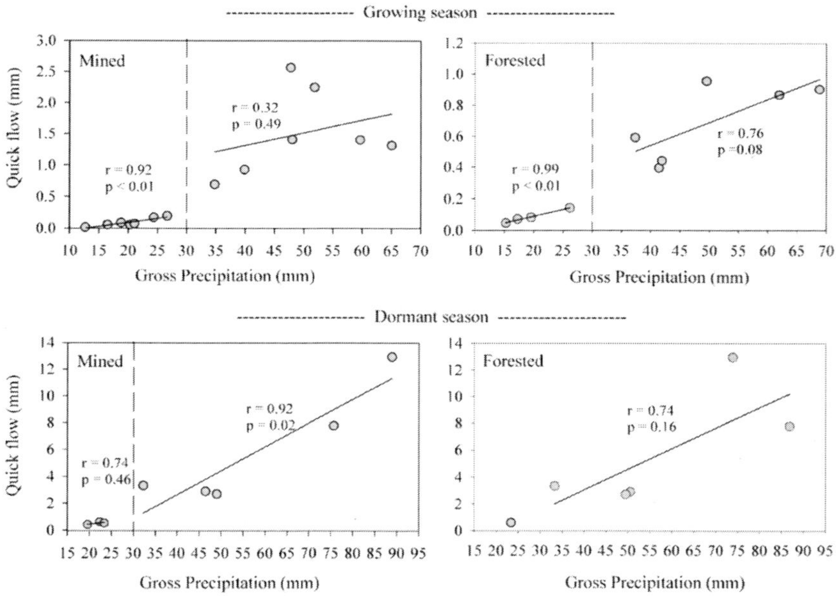

Figure 7. Gross precipitation and quickflow for all events based on the growing season and the dormant season for the mined and forested catchment. The threshold value is ~30 mm. Quickflow response is non-linear and generally more variable above threshold values. $r2$ is the coefficient of determination based on regression analysis

3.3. Runoff Modeling and Isotope Hydrograph Separation

Storm event response times determined from the top 1% performing models (NSE > 0.97) are shown in Table 2, and box plots are shown in Figure 6. Model parameters were generally poorly identified (dotty plots not shown), a common problem with rainfall-runoff models [64]. Response times were generally faster (26.3 ± 17.1 h) in the mined catchment than the forested catchment (31.0 ± 15.2 h) and consistently faster and less variable during the dormant season than the growing season for both catchments. The shape of the response curves (Figure 8) reflects the TPLR model structure where effective precipitation is partitioned into fast and slow draining reservoirs. The steeply sloped recession in the first 20–30 h represents the initial period when the fast draining reservoir is active, whereas the longer, more gradual recession beyond 30 h is the depletion of the slow draining reservoir. In general, response curves for the mined catchment were more variable than the forested catchment and more variable for both catchments early and later on during storms.

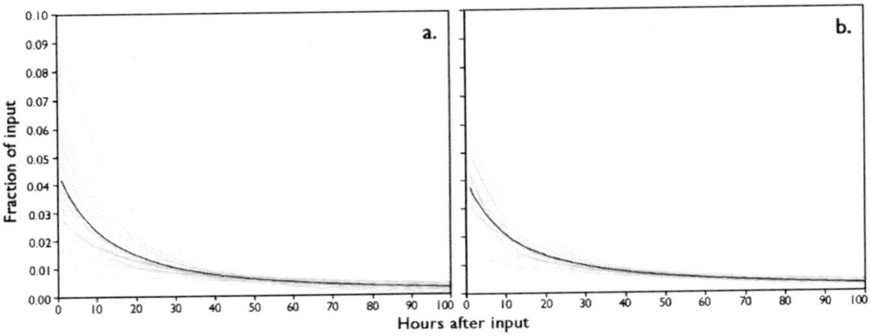

Figure 8. Hydrologic response curves for the 23 storm events in the (**a**) mined and (**b**) forested catchments. Response curves for individual events are shown in gray; black line shows the average of response curves for each catchment.

Of the 23 events analyzed in this study, only nine had complete isotope records, and only one of these events (Event 20) was separated into pre-event and event water contributions using both isotopes and mixing models. In both catchments, runoff was dominated by pre-event water (72%–79%) (Table 3). Event 20 was also separated at peak flow that resulted in higher proportions of event water in the forested catchment (35%–43%) than the mined catchment (14%–26%) (Table 3). For the eight remaining events with complete records, streamflow was separated by only one method, by one isotope or in only one catchment. Hydrograph separations for these events mostly agreed with higher proportions of pre-event water, but the results are inconclusive given the large variability (Table 3). In many cases, we were not able to separate streamflow because: (1) the fraction of pre-event or event water was greater than total streamflow; (2) the poor performance (NSE < 0.50) of runoff and/or tracer models; and/or (3) event water isotopic signatures were not distinguishable in sequential samples of streamflow (example shown for deuterium, Figure 9). The incomplete records of the remaining 14 events were primarily attributed to failure of the automated rainfall and streamflow sampling systems due to fouling or power loss.

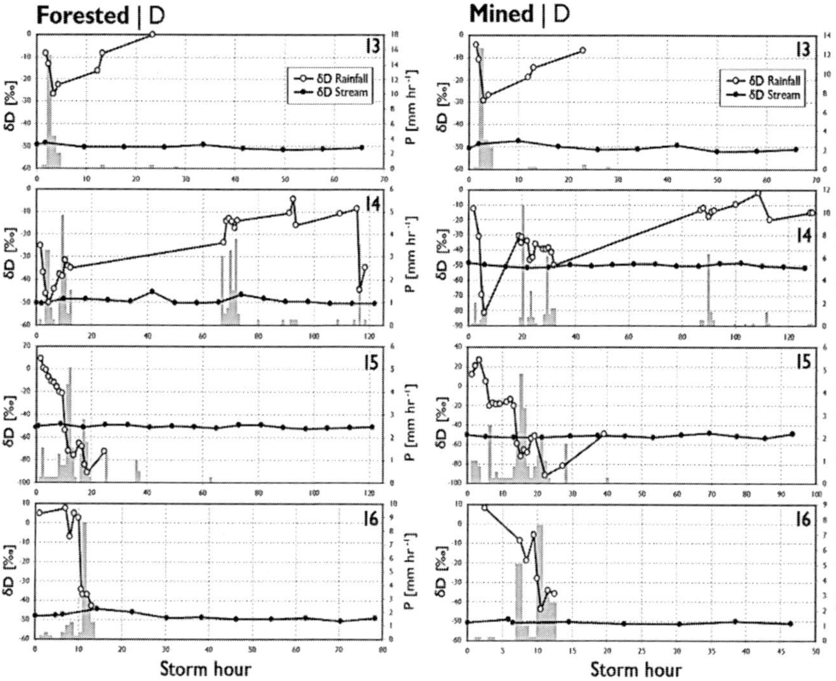

Figure 9. Plots of δD of rainfall and stream for selected events (13–16) for the mined and forested catchments. δ18O showed similar patterns for the same events. Rainfall in mm h−1 shown in gray. Despite variations in rainfall isotopic compositions throughout storm events that in come cases crossed over, the stream isotopic composition showed little response to rainfall input, making hydrograph separation impossible for all but one event.

Table 3. Storm event and pre-event water compositions based on isotope hydrograph separation using δD and δ18O in a traditional two-component mixing model and the transfer function hydrograph separation (TRANSEP) model. Of the 23 storm events measured in the study, only nine had complete isotope records, and only Event 20 was successfully separated into components, both isotopes and separation models. (-) indicates failure to separate the hydrograph, while the superscript indicates the source of error.

Event	Mined Catchment				Forested Catchment			
	Two-Component (%/%)		TRANSEP (%/%)		Two-Component (%/%)		TRANSEP (%/%)	
	δD	$\delta^{18}O$	δD	$\delta^{18}O$	δD	$\delta^{18}O$	δD	$\delta^{18}O$
2	39/61	-[a]	-[b,c]	-[b,c]	-[a]	68/32	-[b,c]	-[b,c]
5	-[a]	-[a]	-[b]	-[b]	19/81	-[a]	-[b]	-[b]
13	-[a]	-[a]	-[b]	-[b]	-[a]	-[a]	-[b]	-[b]
14	-[a]	-[a]	-[b]	-[b]	-[a]	-[a]	-[b]	-[b]
15	-[a]	-[a]	-[b]	-[b]	-[a]	-[a]	-[b]	-[b]
17	-[a]	-[a]	-[b]	-[b]	-[a]	-[a]	-[b]	-[b]
20	21/79	23/77	21/79	28/72	24/76	23/77	23/77	22/78
[d] peak	22/88	17/83	26/74	23/77	36/64	36/64	43/57	35/65
21	6/94	5/95	-[b]	-[b]	27/73	21/79	-[b]	-[b]
23	-[a]	30/70	-[b]	-[b]	-[b]	-[b]	16/84	-[b]

[a] Fraction of preevent or event water greater than total streamflow. [b] Poor (NSE < 0.50) performance of event tracer model. [c] Poor (NSE < 0.50) performance of preevent tracer model. [d] Runoff separated at peak.

4. DISCUSSION

4.1. The Not-So-Clear Impacts of Mountaintop Removal Mining

Despite significant differences in disturbance intensity between the two catchments, direct comparison and inference about the drivers of hydrologic change are complicated by scale differences [14] and potentially unaccounted for controls on catchment hydrology. While reconciling hydrologic responses across spatial scales remains an unresolved issue in hydrology [21], it is generally accepted that hydrologic responses are different between small and large watersheds [65]. Size disparity between the topographically-delineated contributing areas of the MTM-mined and forested catchment could explain hydrologic difference, since metrics, such as peak flow, quickflow, area-normalized streamflow and time-to-peak, tend to be negatively correlated with catchment size [66,67]. Based on subsurface recharge, the size of effective contributing areas is reversed. Hence, it could be expected that the hydrologic responses also differ. Given the large degree of uncertainty in the actual extent and connectivity of underground coal mines and coal seams, reanalysis of hydrometric, isotopic and response time modeling using the updated catchment areas was beyond the scope of the current study, but should be considered in future analyses.

While similarly-sized paired catchments could minimize hydrologic differences between a reference and a disturbed catchment [68], finding appropriate and accessible catchments in the region is nearly impossible given the long disturbance history, the mosaic of land cover, private property ownership and the ubiquity of surface and underground mining. A promising approach would be to examine the hydrology of a large number of catchments across scales and disturbance gradients instead of focusing on just two headwater catchments. Nevertheless, our study provides important insights into the complexity of catchment hydrology in the central Appalachian Mountains region that have important implications for flooding and water quality.

Given the propensity of surface mining to alter water balance components and flow paths (i.e., [69]), many of our results were expected and similar to previous studies. Despite similar event rainfall characteristics, the hydrology in the two catchments differed in significant ways; times-to-peak were shorter; hydrologic response times were faster; and total streamflow, peak flow, base flow and runoff ratios were greater in the mined catchment than the forested catchment. With 10% of the catchment actively being mined and completely devoid of vegetation at the time of our study and another 20% classified as grasses instead of native hardwood forests, greater streamflow (Figure 10) could plausibly be explained by water balance changes in ET from forest losses that exceed the 20% disturbance thresholds shown in the forest hydrology literature (e.g., [70,71]). Elevated base flow and flow duration downstream in the MTM-mined catchment are also consistent with previous studies that hypothesize that valley fills regulate runoff by temporarily storing and releasing water overtime [2,24,31,32,72,73].

However, our results and interpretation also differed in important ways, particularly with respect to the sources of more water; the unexpectedly smaller proportions of quickflow in the mined catchment; and the failure to separate runoff into event and pre-event components. While the streamflow and runoff ratio were significantly larger in the mined catchment, we do not believe that the larger volumes of water were solely attributed to reduced ET, but also to transient storage in valley fills and, potentially more important, water from abandoned underground coal mines and mine pools that underlie much of the catchment [34,46]. For changes in ET to be the dominant source of more water, we would expect the greatest differences between catchments during the growing season when plant water demand is highest. Instead, differences were greater during the dormant season when ET is minimal and antecedent soil moisture is highest.

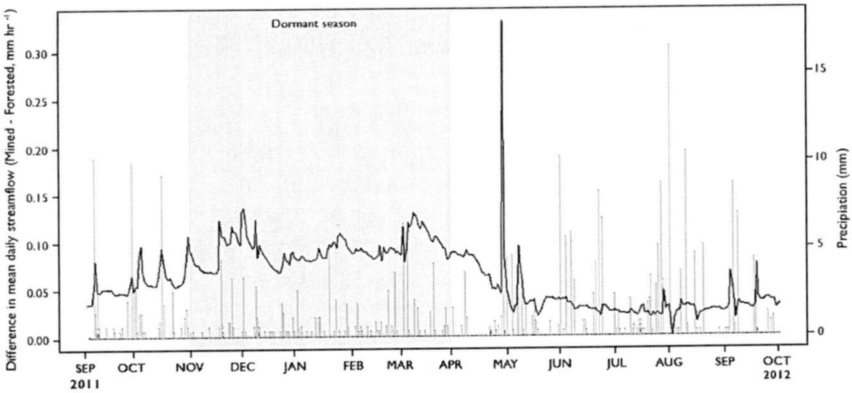

Figure 10. Runoff differences between the mined and forested catchment over the 13-month study period. The dormant season (November–April) is shaded. The greatest differences between catchments were during the dormant season, likely due to unaccounted for water from underground mine pools and inter-basin transfer.

4.2. Uncertain Role Valley Fills and Underground Mines

4.2.1. Valley Fills

The placement of fill material in headwater valleys substantially increases catchment storage over pre-disturbance conditions. Ross et al. [74] estimated that all of the valley fills located throughout the MTM region have a theoretical capacity to store >1.3 km3 (or a depth of 7 m of water spread over the 180 km2 of land area occupied by valley fills) that is approximately equivalent to one year's worth of regional precipitation [74].

While increased catchment storage from valley fills has the clear effect of elevating streamflow, base flow and flow duration, its impact on storm flows in less well understood. Conceptually, valley fills should decrease storm flows by transiently storing and releasing runoff overtime (i.e., elevated flow duration), but some studies [5,24,73], including this one, observed larger peak flow downstream of valley fills. The paradox of elevated streamflow, base flow and peak flow downstream has been explored by [2], who conceptualized runoff generation in valley fills by the activation of preferential flow paths. During low flow conditions, runoff is dominated by steady-state matrix flow through torturous flow paths that sustain streamflow. During sufficiently-sized storms, the matrix and larger air-filled voids created by a large rock fragment within a valley fill become saturated, activating large preferential flow paths that could be responsible for large peak flows downstream.

4.2.2. Underground Mines

Groundwater flow directions in coal seams are primarily controlled by dip direction [35,75], and streams located down-dip and stratigraphically below coal seams can gain significant volumes of water [12]. Underground mines alter the subsurface structure [35] by creating large voids and subsidence fractures that increase hydraulic conductivity (e.g., 1.5 m d^{-1} [45]), as well as connectivity between the surface and subsurface and between water-bearing geologic units [36,37,45]. Rainfall can be rapidly translated to streamflow through near surface-stress relief fractures and underground mine pools that efficiently connect groundwater to surface streams [76]. This was observed in the nearby Elkhorn Creek watershed; dissolved oxygen, temperature and specific conductivity data collected at the outlet of a mined coal seam indicated rapid response to rainfall [45].

Large voids created by underground mining are major conduits for water movement and storage. Mine pools located below drainage often are completely flooded, where as mine pools located above drainage can be partially flooded or completely flooded during periods of high antecedent moisture and recharge. Connectivity to other mined and overlapping coal seams and proximity to perched aquifers also play important roles on the hydrologic condition and storage capacity of mine pools [46]. According to the West Virginia Mine Pool Atlas [46], mines pools in our study area are predominantly above drainage and partially flooded, but below drainage and completely flooded pools also exist. The maximum potential storage of partially-flooded and completely-flooded mine pools of the seven coal seams located in our study area totals ~1 km^3 [46], which, like valley fills, is approximately equivalent to mean annual rainfall for the region [74]. Hence, the transient storage potential of mine pools is vast and certainly plays a critical, albeit uncertain role in streamflow generation in our catchments.

Inter-basin transfer resulting from underground mining can move large volumes of water between catchments, further obfuscating dominant streamflow generation controls and hydrologic response to surface disturbance. Inter-basin transfer was an important source of streamflow in the aforementioned Elkhorn Creek watershed. Streamflow in two headwater catchments of Elkhorn Creek had anomalously large streamflow when compared to their topographic delineated drainage areas [45]. It was determined that nearly 2× and 9× larger drainage areas that extended ~5 km beyond the topographic watershed divides were necessary to produce observed streamflow at the gauging stations. While detailed delineation of the underground recharge areas is beyond the scope of this study, the large number and estimated extent of up-dip underground mines below our catchments increases effective drainage areas by 2× or greater in the mined catchment and by as much as half in the forested catchment (Figure 3).

With streamflow in our study comprised of large contributions of water from unaccounted for water sources, isotope rainfall event signals would be indistinguishable in streamflow, making hydrograph separation impossible for all but one event. Why this one event was successful for both catchments was uncertain, but could be the result of similar pre-storm storage conditions of mine pools that resulted in similar proportions of pre-event water that dominated hydrographs. The general absence of isotopic variations in storm flow (Figure 9) in both catchments could be attributed to thorough mixing of rainfall with water from mine pools and by large recharge areas that incorporate water from up-dip underground mines that extend beyond the surface drainage area [45]. In a sense our failure to separate runoff into pre-event and event components using two-component mixing models supports this idea. Another possibility is the transfer of rainwater outside of our catchments through inter-basin transfer, but based on dip direction, we believe the former was more likely. More research is necessary to understand the significance of underground mine pools for runoff generation, particularly with respect to MTM-driven changes in hydrology.

Despite large uncertainties about the actual extents of effective recharge area, underground mine pools and inter-basin transfer likely play important roles in runoff generation, but how, and to what degree, remains uncertain. By simplifying catchment storage into a single reservoir [56,77,78], waters originating from sources other than rainfall are not accounted for. Clearly, a two component mixing model is not appropriate for these systems. While geochemistry and multiple component mixing models could provide important insights into catchment processes, these sorts of detailed studies are complicated by restricted access [34], the ever-changing configuration of surface-mined catchments [11] and uncertainty in the extent and hydrologic conditions (partially or completely flooded) of mine pools [46].

4.2.3. A Case for Threshold-Mediated Storm Responses

The large variations in response curves early and later on during storms and the threshold behavior of storm flow in our catchments are consistent with threshold-mediated, connectivity-controlled runoff responses [79] that have been observed across different land forms, land cover and scales (e.g., [80,81]). While previous studies identified dominant roles of soil and hillslope storage on controlling runoff responses, we believe that streamflow in our catchments is primarily controlled by the filling, mixing and release of water from the transiently-connected and dynamic storage conditions of partially-flooded and completely-flooded mine pools. Our working hypothesis for streamflow generation in catchments that contain up-dip underground mine pools is that below-threshold storm responses likely reflect a mine pool storage threshold below which precipitation inputs mix with older water to recharge partially-flooded mine pools, and above which precipitation inputs activate a larger network of partially- to completely-flooded mine pools that results in non-linear storm flow responses with larger variability (Figure 7). This is not dissimilar to Miller and Zégre's [2] conceptualization of valley fill hydrology, although we believe underground mines play a more dominant role in streamflow generation in these catchments.

Threshold hydrologic responses in our study are consistent with findings from the Elkhorn Creek watershed study that showed a rapid response of mine discharge to precipitation despite the absence of variations in isotopic composition. It was determined that meteoric water of relatively recent origin (pre-event) derived from near surface-stress relief fractures and an older component of slow recharge of water from overlying perched aquifers and phreatic flow paths associated with stress relief, bedding plane and mining-induced fractures were dominant sources of recharge for streamflow in Elkhorn Creek [45]. Further, it was determined that recharge primarily occurred during warmer periods of the spring or fall, coincident with periods of high flow [45]. High streamflow in our catchments during the growing season (Figure 5) is consistent with seasonal recharge and could explain the larger variability and weaker correlation between precipitation and quickflow in the above-threshold storms that resulted from the activation of a large mine pool network beyond catchment boundaries.

5. IMPLICATIONS AND CONCLUSIONS

MTM and underground mining significantly alter catchment hydrology by creating voids that increase catchment storage and runoff processes. The threshold behavior of runoff is likely the result of filling, mixing and releasing of water from partially- and completely-flooded mine pools and, to a lesser degree, valley fills. Fill and release cycles have important implications for water chemistry, since oxygen and water are unlimited under these conditions. The quality of water draining partially-flooded, above-drainage mines is often poorer for longer periods than completely flooded mines [76]. Unfortunately, water chemistry sampling and analysis was beyond the limits of this study and should be the focus of future work.

Catchment hydrology in the Appalachian Coalfields region is complex, driven in part by obvious landscape-scale disturbances, such as mountaintop removal mining, but also by underground coal mining disturbance. The nearly two centuries of resource extraction has resulted in a mosaic of land covers, land uses and land forms that present significant challenges for determining baseline conditions and for understanding the extent and significance of MTM-driven impacts. The challenges for understanding the hydrological impacts of MTM are particularly evident given the paucity of catchment studies despite the nearly five decade-long use of this practice.

Nevertheless, our findings provide important insight into the complexity of catchment hydrology in this region and other areas with multiple, overlapping disturbances. Results suggest that underground mines likely are important controls for runoff generation in both obviously disturbed and seemingly undisturbed catchments having important implications for water chemistry and interact in uncertain ways with MTM-driven hydrological change. Future research should aim to isolate the respective impacts of MTM and underground mining to determine the nature of their interaction.

ACKNOWLEDGMENTS

Funding for this research was provided by National Science Foundation Awards No. 1042683 and 1458952; the Oak Ridge Associated Universities (ORAU) Ralph E. Powe Junior Faculty Enhancement Award; the West Virginia University Faculty Senate Research Award R-14-025; and the Hatch Act Formula Funds. We are indebted to Bacon Brown for his support in this project. We thank Shannon Alex Shy for GIS support.

AUTHOR CONTRIBUTIONS

Zégre conceived of and designed the experiments. Miller performed the experiments. Miller and Zégre analyzed the data and wrote the paper.

CONFLICTS OF INTEREST

The authors declare no conflict of interest.

REFERENCES

1. Ebel, B.A.; Mirus, B.B. Disturbance hydrology: Challenges and opportunities. *Hydrol. Process.* **2014**, *28*, 5140–5148.

2. Miller, A.; Zégre, N. Mountaintop removal mining and catchment hydrology. *Water* **2014**, *6*, 472–499.

3. Griffith, M.B.; Norton, S.B.; Alexander, L.C.; Pollard, A.I.; LeDuc, S.D. The effects of mountaintop mines and valley fills on the physicochemical quality of stream ecosystems in the central Appalachians: A review. *Sci. Total Environ.***2012**, *417–418*, 1–12.

4. Merriam, E.R.; Petty, J.T.; Strager, M.P.; Maxwell, A.E.; Ziemkiewicz, P.F. Landscape-based cumulative effects models for predicting stream response to mountaintop mining in multistressor Appalachian watersheds. *Freshw. Sci.* **2015**, *34*, 1006–1019.

5. Wiley, J.B.; Brogan, F.D. *Comparison of Peak Discharges Among Sites With and Without Valley Fills for the 8–9 July 2001, Flood in the Headwaters of Clear Fork, Coal River Basin, Mountaintop Coal-Mining Region, Southern West Virginia*; Technical Report 03-133; U.S. Geological Survey: Reston, VA, USA, 2003.

6. Lindberg, T.T.; Bernhardt, E.S.; Bier, R.; Helton, A.M.; Merola, R.B.; Vengosh, A.; Di Giulio, R.T. Cumulative impacts of mountaintop mining on an Appalachian watershed. *Proc. Natl. Acad. Sci. USA* **2011**, *108*, 20929–20934.

7. Bernhardt, E.S.; Lutz, B.D.; King, R.S.; Fay, J.P.; Carter, C.E.; Helton, A.M.; Campagna, D.; Amos, J. How many mountains can we mine? Assessing the regional degradation of central Appalachian rivers by surface coal mining.*Environ. Sci. Technol.* **2012**, *46*, 8115–8122.

8. Saylor, K.L. *Land Cover Trends: Central Appalachians*; U.S. Geological Survey: Reston, VA, USA, 2008.

9. Townsend, P.A.; Helmers, D.P.; Kingdon, C.C.; McNeil, B.E.; Beurs, K.M.D.; Eshleman, K.N. Changes in the extent of surface mining and reclamation in the Central Appalachians detected using a 1976–2006 Landsat time series. *Remote Sens. Environ.* **2009**, *113*, 62–72.

10. EPA. *The Effects of Mountaintop Mines and Valley Fills on Aquatic Ecosystems of the Central Appalachian Coalfield*; Technical Report

EPA/600/R-09/138F; US Environmental Protection Agency: Washington, DC, USA, 2011.

11. Bonta, J. Challenges in conducting hydrologic and water quality research in drastically disturbed watersheds. *J. Soil Water Conserv.* **2005**, *60*, 121–133.

12. Borchers, J.; Ehlke, T.; Mathes, M.; Downs, S. *The Effects of Coal Mining on the Hydrologic Environmnet of Selected Stream Basins in Southern West Virginia*; U.S. Geological Survey: Reston, VA, USA, 1991.

13. McCormick, B.C.; Eshleman, K.N.; Griffith, J.L.; Townsend, P.A. Detection of flooding responses at the river basin scale enhanced by land use change. *Water Resour. Res.* **2009**, *45*.

14. Negley, T.L.; Eshleman, K.N. Comparison of stormflow responses of surface-mined and forested watersheds in the Appalachian Mountains, USA. *Hydrol. Process.* **2006**, *20*.

15. Phillips, J.D. Impacts of surface mine valley fills on headwater floods in eastern Kentucky. *Environ. Geol.* **2004**.

16. Keene, T.; Skousen, J. Mine spoil reclamation with switchgrass for biofuel production. In Proceedings of the 2010 National Meeting of the American Society of Mining and Reclamation, Pittsburgh, PA, USA, 5–11 June 2010.

17. Peng, S.S. Mountaintop removal controversy slows West Virginia Coal Mining. *Min. Eng.* **2000**, *52*, 53–58.

18. Hartman, K.; Kaller, M.; Howell, J.; Sweka, J. How much do valley fills influence headwater streams? *Hydrobiologia***2005**, *532*, 91–102.

19. Merriam, E.R.; Petty, J.T.; Merovich, G.T.; Fulton, J.B.; Strager, M.P. Additive effects of mining and residential development on stream conditions in a central Appalachian watershed. *J. N. Am. Benthol. Soc.* **2011**, *30*, 399–418.

20. McGuire, K.J.; McDonnell, J.J. Hydrological connectivity of hillslopes and streams: Characteristic time scales and nonlinearities. *Water Resour. Res.* **2010**, *46*, W10543.

21. McGlynn, B.; McDonnell, J.; Seibert, J.; Kendall, C. Scale effects on headwater catchment runoff timing, flow sources, and groundwater-streamflow relations. *Water Resour. Res.* **2004**, *40*.

22. Zipper, C.; Burger, J.; Skousen, J.; Angel, P.; Barton, C.; Davis, V.; Franklin, J. Restoring forests and associated ecosystem services

on Appalachian coal surface mines. *Environ. Manag.* **2011**, *47*, 751–765.

23. Taylor, T.J.; Agouridis, C.T.; Warner, R.C.; Barton, C.D.; Angel, P.N. Hydrologic characteristics of Appalachian loose-dumped spoil in the Cumberland Plateau of eastern Kentucky. *Hydrol. Process.* **2009**, *23*.

24. Messinger, T. *Comparison of Storm Response of Streams in Small, Unmined and Valley-Filled Watersheds, 1999–2001, Ballard Fork, West Virginia*; Technical Report 02-4303; USGS: Reston, VA, USA, 2003.

25. Bussler, B.H.; Byrnes, W.R.; Pope, P.E.; Chaney, W.R. Properties of minesoil reclaimed for forest land use. *Soil Sci. Soc. Am. J.* **1984**, *48*, 178–184.

26. Guebert, M.D.; Gardner, T.W. Macropore flow on a reclaimed surface mine: Infiltration and hillslope hydrology.*Geomorphology* **2001**, *39*, 151–169.

27. Chong, S.K.; Cowsert, P.T. Infiltration in reclaimed mined land ameliorated with deep tillage treatments. *Soil Tillage Res.* **1997**, *44*, 255–264.

28. Bonta, J.V.; Amerman, C.R.; Harlukowicz, T.J.; Dick, W.A. Impact of coal-surface mining on three Ohio watersheds - surface-water hydrology. *J. Am. Water Resour. Assoc.* **1997**, *33*, 907–917.

29. Ferrari, J.R.; Lookingbill, T.R.; McCormick, B.; Townsend, P.A.; Eshleman, K.N. Surface mining and reclamation effects on flood response of watersheds in the central Appalachian Plateau region. *Water Resour. Res.* **2009**, *45*.

30. Dickens, P.S.; Minear, R.A.; Tschantz, B.A. Hydrologic alteration of mountain watersheds from surface mining. *J. Water Pollut. Control Fed.* **1989**, *61*, 1249–1260.

31. Wiley, J.B.; Evaldi, R.D.; Eychaner, J.H.; Chambers, D.B. *Reconnaissance of Stream Geomorphology, Low Streamflow, and Stream Temperature in the Mountaintop Coal-Mining Region, Southern West Virginia, 1999–2000*; U.S. Geological Survey: Reston, VA, USA, 2001.

32. Green, J.; Passmore, M.; Childers, H. *A Survey of the Condition of Streams in the Primary Region Of Mountaintop Mining/Valley Fill Coal Mining*; US EPA: Washington, DC, USA, 2000.

33. Zégre, N.P.; Maxwell, A.; Lamont, S. Characterizing streamflow response of a mountaintop-mined watershed to changing land use. *Appl. Geogr.* **2013**, *39*, 5–15.

34. Zégre, N.P.; Miller, A.J.; Maxwell, A.; Lamont, S.J. Multiscale analysis of hydrology in a mountaintop mine-impacted watershed. *JAWRA J. Am. Water Resour. Assoc.* **2014**, *50*, 1257–1272.

35. Puente, C.; Atkins, J.T. *Simulation of Rainfall-Runoff Response in Mined and Unmined Watersheds in Coal Areas of West Virginia*; Technical Report Water Supply Paper 2298; US Geological Survey: Reston, VA, USA, 1989.

36. Hobba, W.J. *Effects of Underground Mining and Mine Collapse on the Hydrology Of Selected Basins in West Virginia*; Technical Report 4900012; US Geological Survey: Reston, VA, USA, 1981.

37. Hawkins, J.W.; Dunn, M. Hydrologic characteristics of a 35-year-old underground mine pool. *Mine Water Environ.* **2007**, *26*, 150–159.

38. Cardwell, D.; Erwin, R.; Woodward, H. *Geologic Map of West Virginia*; 1:250,000; West Virginia Geological and Economic Survey: Morgantown, WV, USA, 1968.

39. Ehlke, T.; Runner, G.; Downs, S. *Hydrology of Area 9, Eastern Coal Province, West Virginia*; Open-File Report 81-803; US Geological Survey: Reston, VA, USA, 1982.

40. Adams, M.B.; Edwards, P.J.; Ford, W.M.; Schuler, T.M.; Thomas-Van Gundy, M.; Wood, F. *Fernow Experimental Forest: Research History and Opportunities*; USDA Forest Service: Washington, DC, USA, 2012.

41. Kipp, J.A.; Dinger, J.S. Stress-relief fracture control of ground-water movement in the Appalachian plateaus. In Proceedings of the Fourth Annual Eastern Regional Groun Water Conference, Burlington, VT, USA, 14–16 July 1987; pp. 423–438.

42. Wyrick, G.G.; Borchers, J.W. *Hydrologic Effects of Stress-Relief Fracturing in an Appalachian Valley [West Virginia]*; US Geological Survey: Reston, VA, USA, 1981.

43. Smith, J.A.; Baeck, M.L.; Ntelekos, A.A.; Villarini, G.; Steiner, M. Extreme rainfall and flooding from orographic thunderstorms in the central Appalachians. *Water Resour. Res.* **2011**, *47*, W04514.

44. US Congress. *Surface Mine Control and Reclamation Act*; US Congress: Washington, DC, USA, 1977.

45. Kozar, M.D.; McCoy, K.J.; Britton, J.Q.; Bascombe M. Blake, J. *Hydrogeology, Groundwater Flow, and Groundwater Quality of an Abandoned Underground Coal-Mine Aquifer, Elkhorn Area, West Virginia*; Technical Report Bulletin B-46; USGS: Reston, VA, USA, 2012.

46. McColloch, J.S.; Richard D. Binns, J.; Bascombe M. Blake, J.; Clifford, M.T.; Gooding, S.E. *West Virginia Mine Pool Atlas*; Technical Report 036; West Virginia Department of Environmental Protection: Charleston, WV, USA, 2012.

47. Turnipseed, D.; Sauer, V. *Discharge Measurements at Gaging Stations*; U.S. Geological Survey: Reston, VA, USA, 2010.

48. Hudson, R.; Fraser, J. Introduction to Salt Dilution Gauging for Streamflow Measurement Part IV: The Mass Balance (or Dry Injection) Method. *Streamline* **2005**, *9*, 1–7.

49. Hewlett, J.D.; Hibbert, A.R. *Factors Affecting The Response Of Small Watersheds To Precipitation in Humid Areas*; Pergamon Press: Oxford, UK, 1967.

50. Jakeman, A.J.; Hornberger, G.M. How much complexity is warranted in a rainfall-runoff model? *Water Resour. Res.* **1993**, *29*, 2637–2649.

51. Nippgen, F.; McGlynn, B.L.; Marshall, L.A.; Emanuel, R.E. Landscape structure and climate influences on hydrologic response. *Water Resour. Res.* **2011**, *47*, W12528.

52. Heathcote, J.A.; Lloyd, J.W. Factors affecting the isotopic composition of daily rainfall at Driby, Lincolnshire. *J. Climatol.* **1986**, *6*, 97–106.

53. Matsuo, S.; Friedman, I. Deuterium content in fractionally collected rainwater. *J. Geophys. Res.* **1967**, *72*, 6374–6376.

54. McDonnell, J.J.; Bonell, M.; Stewart, M.K.; Pearce, A.J. Deuterium variations in storm rainfall: Implications for stream hydrograph separation. *Water Resour. Res.* **1990**, *26*, 455–458.

55. Craig, H. Isotopic variations in meteoric waters. *Science* **1961**, *133*, 1702–1703.

56. Sklash, M.G.; Farvolden, R.N. The role of groundwater in storm runoff. *J. Hydrol.* **1979**, *43*, 45–65.

57. Buttle, J. Isotope hydrograph separations and rapid delivery of pre-event water from drainage basins. *Prog. Phys. Geogr.* **1994**, *18*, 16–41.

58. Weiler, M.; McGlynn, B.L.; McGuire, K.J.; McDonnell, J.J. How does rainfall become runoff? A combined tracer and runoff transfer function approach. *Water Resour. Res.* **2003**, *39*.

59. Roa-García, M.C.; Weiler, M. Integrated response and transit time distributions of watersheds by combining hydrograph separation and long-term transit time modeling. *Hydrol. Earth Syst. Sci.* **2010**, *14*, 1537–1549.

60. Lyon, S.W.; Desilets, S.L.E.; Troch, P.A. Characterizing the response of a catchment to an extreme rainfall event using hydrometric and isotopic data. *Water Resour. Res.* **2008**, *44*, W06413.

61. McGuire, K.J.; McDonnell, J.J.; Weiler, M.; Kendall, C.; McGlynn, B.L.; Welker, J.M.; Seibert, J. The role of topography on catchment-scale water residence time. *Water Resour. Res.* **2005**, *41*.

62. Abbaspour, K.C.; Schulin, R.; van Genuchten, M.T. Estimating unsaturated soil hydraulic parameters using ant colony optimization. *Adv. Water Resour.* **2001**, *24*, 827–841.

63. Nash, J.; Sutcliffe, J. River flow forecasting through conceptual models. Part I: A discussion of principles. *J. Hydrol.* **1970**, *10*, 282–290.

64. Beven, K. Parameter Estimation and Predictive Uncertainty. In *Rainfall-Runoff Modelling: The Primer*; John Wiley and Sons, Ltd.: Chichester, UK, 2001.

65. Pilgrim, D.H.; Cordery, I.; Baron, B.C. Effects of catchment size on runoff relationships. *J. Hydrol.* **1982**, *58*, 205–221.

66. Heerdegen, R.G.; Reich, B.M. Unit hydrographs for catchments of different sizes and dissimilar regions. *J. Hydrol.* **1974**, *22*, 143–153.

67. Post, D.A.; Jakeman, A.J. Relationships between catchment attributes and hydrological response characteristics in small Australian mountain ash catchments. *Hydrol. Process.* **1996**, *10*, 877–892.

68. Hewlett, J.D.; Pienaar, L. Design and analysis of the catchment experiment. In Proceedings of the Symposium on Use of Small Watersheds in Determining Effects of Forest Land Use on Water Quality, Lexington, KY, USA, 22–23 May 1973.

69. Ritter, J.B.; Gardner, T.W. Hydrologic evolution of drainage basins disturbed by surface mining, central Pennsylvania.*GSA Bull.* **1993**, *105*, 101–115.

70. Bosch, J.M.; Hewlett, J.D. A review of catchment experiments to determine the effect of vegetation changes on water yield and evapotranspiration. *J. Hydrol.* **1982**, *55*, 3–23.

71. Stednick, J.D. Monitoring the effects of timber harvest on annual water yield. *J. Hydrol.* **1996**, *176*, 79–95.

72. Paybins, K.S. *Flow Origin, Drainage Area, and Hydrologic Characteristics for Headwater Streams in the Mountaintop Coal-Mining Region of Southern West Virginia, 2000-01*; Water-Resources Investigations Report 02-4300; U.S. Geological Survey: Reston, VA, USA, 2003.

73. Messinger, T.; Paybins, K. *Relations between Precipitation and Daily and Monthly Mean Flows in Gaged, Unmined and Valley-Filled Watersheds, Ballard Fork, West Virginia, 1999–2001*; Technical Report WRi no.2003-4113; USGS: Reston, VA, USA, 2003.

74. Ross, M.R.V.; McGlynn, B.L.; Bernhardt, E.S. Deep impact: Effects of mountaintop mining on surface topography, bedrock structure, and downstream waters. *Environ. Sci. Technol.* **2016**, *50*, 2064–2074.

75. Rozell, D.J.; Reaven, S.J. Water pollution risk associated with natural gas extraction from the marcellus shale. *Risk Anal.* **2012**, *32*, 1382–1393.

76. Demchak, J.; Skousen, J.; McDonald, L.M. Longevity of acid discharges from underground mines located above the regional water table. *J. Environ. Qual.* **2004**, *33*, 656–668.

77. Sklash, M.G. Environmental isotope studies of storm and snowmelt runoff generation. In *Process Studies in Hillslope Hydrology*; Anderson, M.G., Burt, T.P., Eds.; John Wiley and Sons: Chichester, UK, 1990; pp. 401–435.

78. Sklash, M.G.; Farvolden, R.N. The use of environmental isotopes in the study of high-runoff episodes in streams. In*Isotope Studies of Hydrologic Processes*; Perry, E.C., Jr., Montgomery, C.W., Eds.; Northern Illinois University Press: Dekald, IL, USA, 1982; pp. 65–73.

79. Spence, C. A paradigm shift in hydrology: Storage thresholds across scales influence catchment runoff generation.*Geogr. Compass* **2010**, *4*, 819–833.

80. Tromp-vanMeerveld, H.J.; McDonnell, J.J. Threshold relations in subsurface stormflow: 2. The fill and spill hypothesis.*Water Resour. Res.* **2006**, *42*.

81. Detty, J.M.; McGuire, K.J. Threshold changes in storm runoff generation at a till-mantled headwater catchment. *Water Resour. Res.* **2010**, *46*.

2

CHAPTER

Eliciting Drivers of Community Perceptions of Mining Projects through Effective Community Engagement

Liang Wang [1,*], Kwame Awuah-Offei [2], Sisi Que [3] and Wei Yang [3]

[1]State Key Laboratory of Coal Mine Disaster Dynamics and Control, Chongqing University, Chongqing 400014, China

[2]Department of Mining Engineering, Missouri University of Science & Technology, Rolla, MO 65409, USA

[3]Key Laboratory of Hydraulic and Waterway Engineering of the Ministry of Education, College of River and Ocean Engineering, Chongqing Jiaotong University, Chongqing 400074, China

ABSTRACT

Sustainable mining has received much attention in recent years as a consequence of the negative impacts of mining and public awareness. The aim of this paper is to provide mining companies guidance on improving the sustainability of their sites through effective community engagement based on recent advances in the literature. It begins with a review of the literature on sustainable development and its relationship to stakeholder engagement. It then uses the literature to determine the dominant factors that affect community perceptions of mining projects. These factors are classified into five categories: environmental, economic, social, governance and demographic factors. Then, we propose a new two-stage method based on discrete choice theory and the classification that can improve stakeholder engagement and be cost-effective. Further work is required to validate the proposed method, although it shows potential to overcome some of the challenges plaguing current approaches.

Keywords: sustainability; mining; local community perspective; community acceptance

1. INTRODUCTION

In the past decade, concerns over how to ensure sustainable development (the ability of current generations to meet their needs without compromising the ability of future generations to meet their own needs) have increased over the world [1,2,3,4,5,6,7,8,9,10]. Mining companies have been forced to adapt their operating practices to fit this paradigm under public pressure. While there is no doubt that mineral and metal products have made a significant contribution to ensuring sustainable development, the juxtaposed adverse impacts cannot be ignored. The negative environmental and social impacts of mining operations have attracted attention from governments, non-governmental organizations, the public and other stakeholders. How mining can contribute to sustainable development has become a key challenge for the industry.

Currently, most of the major mining companies produce audited, annual sustainability reports that document their sustainability impacts [11]. The mining industry has progressed from environmental compliance (and associated standards, like ISO 14001), to corporate social responsibility (CSR) programs, to social license to operate (SLO) and now to sustainability reporting with standards like the Global Reporting Initiative (GRI) [12,13,14,15,16,17,18]. Additionally, there are calls for mines and mining businesses, like their counterparts in other sectors, to operate in a way that creates shared value for all stakeholders [19]. Furthermore, modern mine or project management involves obtaining, and maintaining, a social license to operate [12] and free, prior and informed consent, which is a related concept [20].Social license to operate and informed consent require mines to actively engage their stakeholders. Stakeholder engagement has, thus, become a key component of managing mines or projects for sustainable outcomes [21]. Several organizations, including the International Finance Corporation (IFC) and the International Council on Mining & Metals (ICMM), have proposed guidelines for adequate stakeholder engagement [22,23,24,25,26,27,28,29,30,31]. The peer-reviewed literature also contains many contributions in this area [12,32,33,34,35,36,37,38]. All of these show that mines and mining businesses have a role to play in the sustainable development of their host communities and the world, at large.

However, there are still numerous mining projects that have been postponed, interrupted and even shut down due to poor community engagement [12,13,33,37,39,40]. Stakeholder-related risk has been shown to be one of the major non-technical risks responsible for delays in mining projects [33]. From a mine operator's standpoint, community engagement is the best way to mitigate these community-related risks and achieve sustainable outcomes (e.g., informed consent and social license to operate). Community engagement should elicit the dominant factors or issues affecting community members' views of a mining project and how these factors affect perceptions of a project [41]. While previous research has discussed the factors that affect community perceptions of mining projects (individually or a few at a time) [42,43,44,45,46,47,48,49], the literature lacks a systematic review that synthesizes this knowledge to inform further research and industrial practice. Furthermore, an approach is needed to use our current understanding of these factors and how they drive community perceptions to facilitate effective engagement.

To bridge this gap, we conducted a comprehensive review of the literature to establish the dominant factors affecting community members' views of a mining project. This research divides the factors found in the literature into five categories: environmental, economic, social, governance and demographic aspects. This classification is based on previous research by the authors [40,41]. The authors also propose a novel two-stage approach, based on this classification and discrete choice theory, for effective stakeholder engagement. This research could provide mine operators with guidance on how to improve social license to operate, obtain free, prior and informed consent and, thus, improve the sustainability of their operations.

2. METHODS

To identify the dominant factors that drive community perceptions of mining, a systematic review of published mining community engagement studies was conducted to identify English language studies available in print or online between 1990 and 2016. Peer-reviewed publications were identified through searches in major abstract databases (e.g., Web of Science, Scopus, Compendex, Google Scholar). Search terms included: "sustainability(able) mining", "mining community(ies)", "community engagement", "community acceptance", "stakeholder analysis", "stakeholder engagement", "discrete choice experiment(s)" and "discrete choice model(l)ing". Studies were included if they were relevant and published as full text articles (not just abstracts).

The authors screened search results for relevance by reviewing titles and abstracts. The review mostly focused on peer-reviewed journal publications since the intention was to rely on rigorous research to address the stated objectives. In a few cases, relevant papers in peer-reviewed conference proceedings were included in the list of reviewed papers. Furthermore, some technical reports from reputable government agencies, non-governmental and industry groups were included in the list of references for review.

We relied on the results of the literature review and our own previous research [41] to classify the factors. Subsequently, we used the classification to formulate a classification of the factors and formulate our proposed approach for effective community engagement. The literature review results are presented in Section 3 and Section 4. Section 5 contains the classification and the proposed approach.

3. SUSTAINABILITY, STAKEHOLDER ENGAGEMENT AND SLO

Sustainable development is defined as the ability of current generations to meet their needs without compromising the ability of future generations to meet their own needs [50]. In order to make it possible to apply this definition to decision making, several refining definitions have been offered by various authors. For example, sustainable development has been defined to include social, economic and environmental impacts, which has been widely referred to as the triple bottom line [51]. Furthermore, it has been defined in relation to social, natural, human, physical and financial capital (the five capitals) [52]. These definitions provide various frameworks for assessing development to determine whether it is sustainable or not [40].

Recently, concerns about corporate sustainability have increased over the world [1,2,3,4,5,6,7,8,9,10]. Besides bad publicity due to negative environmental impacts and the resulting stricter government legislation and public pressures, poor sustainability performance affects the long-term profitability of a business. Thus, businesses have both an interest and a responsibility to incorporate sustainable development concepts into their long-term business strategy [5,40,53,54,55,56].

Sustainable development can only be given real meaning by investigating the ideas through a multi-stakeholder approach [8] (a stakeholder is any group or individual who can affect or is affected by the achievement of the organization's objectives [57]). The Institute of Social and Ethical Accountability [58] defines stakeholder engagement as "the process of seeking stakeholder views on their relationship with an organization in a way that may realistically be expected to elicit them." A mining project and its stakeholders are interdependent. This relationship is confirmed by Rotheroe [8], who indicates that industry has to engage stakeholders in the decision making process and throughout the whole project to achieve sustainable development [59].

In recent years, mining has witnessed an increasing demand for sustainable development from the public and regulators, as well as internal advocates who cite the sector's own long-term benefit [60]. Many mining companies realize the important role of other stakeholders and emphasize stakeholder engagement in the process of mine planning and design, permitting, operation and closure. For the mining sector, ICMM [21] defines stakeholders as a comprehensive list of people and groups who may be affected by, can affect or have an interest in a project. Examples include the local and indigenous groups, employees and contractors, labor unions, suppliers, governments and regulators, media, non-governmental organizations and investors [61,62]. In mining industry terms, the community is generally defined as the inhabitants of the immediate and surrounding areas who are affected by a company's activities [63]. Actually, local communities are the first stakeholder on the ICMM Checklist of possible stakeholders [21]. However, in practice, it is difficult to define the limits of the "local" community, and communities can be excluded from consultation when they should be included. This challenge is exacerbated by the fact that it is not easy to predict how communities may be affected by the secondary impacts of mining; particularly, social and economic impacts.

It is increasingly evident that mining community engagement is important for the success of mining operations (indeed, for all industrial activity). The examples of mining projects that have been disrupted due to lack of community support, cited earlier, are proof of this [12,13,33,37,39]. Community engagement is critical for obtaining permits prior to commencing mining; because without community engagement prior to commencing mining, it is not possible to obtain free, prior and informed consent, which is a best practice for modern mining projects, especially when mines are located on lands that historical belong to indigenous peoples [20]. In fact, community acceptance (i.e., informed consent) is a requirement for the permitting process in some jurisdictions (e.g., Peru passed the Law on the Right of Consultation of Indigenous Peoples in 2011 in accordance with various international conventions they had ratified). In the USA, the local community's acceptance is not necessarily a requirement for granting a permit. However, public participation is required during environmental impact assessment [64].

This concept of community approval of mining operations and its relationship to socio-political risk has been formalized as the social license to operate, in the last decade [12]. The social license to operate (SLO) is defined as a community's perceptions of the acceptability of a company and its local operations [12]. SLO is inversely proportional to the level of socio-political risk faced by a mining operation. Others, however, point out that SLO is nebulous and question whether it is useful, as a practical matter. For example, Owen and Kemp [65] contend that corporate goals to "obtain" or "retain" SLO assume that it can indeed be granted by communities in a manner similar to legally-mandated permits, which have specific permit conditions and carry specific consequences if the conditions are violated by the company. Regardless of its operational usefulness, SLO, conceptually, is a measure of community-related socio-political risk and is used in this work in that sense. Community-related risk is one of the major non-technical risks responsible for delays in mining project execution [33]. For a mining project, the cost of delays can be significant. Davis and Franks [33] estimated the cost of such delays to be approximately US$ 10,000/day, during the exploration stage of a new mine. Good community engagement that leads to acquiring social license to operate is the best way to mitigate these community-related risks.

Mine managers are gradually coming to understand the special importance of the host community and are attempting to address this issue. It is common to see mine managers referring to local communities as "primary" or "key" stakeholders [26]. However, even with increased effort, the mines and mining businesses still struggle to avoid conflicts with host community members. In fact, there appears to be a rise in the number of conflicts in the face of increased community engagement from mines [60].

The host community's perception of a mining project is often different from the perception of the mine's management (or proponents) and other stakeholders [66]. In order to reduce the perception dissonance between the mine's management and the host community, it is important for management to understand the factors that shape a community's perception of a project. Without this view, management may perceive a project to be acceptable to the community although the community may have a totally different perception. Such differences in perception are likely to lead to discontent and possible conflict between the members of the community and the mine's management. This affects the mine's social license to operate.

A community consists of individuals, who interact with each other more frequently than with those outside the community [67]. The community's perceptions emerge from individuals' choices and strategic interactions [68]. Individuals' choices are the basis for understanding the strategic interactions and the community's perception. For example, McGee [69] shows that community responses were characterized by private rather than public responses and individual rather than collective actions. There are other examples of community groups that facilitated a collective response to environmental contamination [70]. Hence, the community's perception of a mining project is an aggregation of the individual perceptions of the community members. In this study, we focus on the factors affecting individuals' choices and how to acquire good information on individual preferences to explain community perceptions.

There is no one unique formula for mining companies to understand the host communities' perceptions of a project. Take the stakeholder analysis matrix in Table 1, for example. It is the approach suggested by ICMM [26] and the most widely-used method for stakeholder analysis in the minerals sector. This method requires the analyst(s) to evaluate each stakeholder's view of the project (positive, neutral, negative), how influential they are (high, medium, low) and how they will be impacted by the project (high, medium, low). The analyst then fills a stakeholder analysis matrix (Table 1) with stakeholder information and then classifies the stakeholders into three groups: highly influential supporter of the project, neutral about the project and highly influential opponent of the project.

Table 1. Stakeholder analysis matrix [26].

Name/Group of Stakeholders	View of Project			Influence			Impact		
	Positive	Neutral	Negative	High	Middle	Low	High	Middle	Low
X	√	□	□	√	□	□	□	□	√
Y		√			√				√
Z	□	□	√	√	□	□	√	□	□

With this method, an individual in the local community may consider the characteristics of the project, its effect on him or her and the general costs and benefits when he or she is asked "impWhat is your view of the project?" Then, he or she can make a decision on whether his or her view of the project is positive or negative. This sort of instrument that elicits responses affected by many factors is prone to confounding effects. What is more, the results obtained from this approach are difficult to act upon. For example, the company even after obtaining data using this result still does not know how to bridge the gap between the stakeholder's perceptions and the company's aspirations with only data on positive or negative views of the project.

This method only elicits the community's view of the project as a whole without information on which aspects of the project are acceptable or unacceptable to community members. For effective community engagement, the approach should inform why and how the local community prefers one project over another. The approach proposed in Section 5 provides such information.

4. FACTORS AFFECTING INDIVIDUAL PERCEPTION OF MINING

There are many factors that can affect an individual's perception of a mining project. It is important to understand these factors because they drive the community's perception, which is a summation of the individual perceptions. The community's perception of the project directly affects the mine's social license to operate.

There is a significant amount of work in the literature on the factors that affect an individual's perception of mining. Generally, the factors that affect community acceptance are mining impacts, governance and community demographics.

4.1. Mining Impacts

4.1.1. Positive Impacts

Mining operations can result in three positive impacts on the host community: job opportunities, income increase and infrastructure improvement.

The impact of job opportunities and related economic impacts (income increases) in the United States (U.S.), for example, are summarized in Table 2. In 2012, U.S. mines provided more than 634,000 jobs directly and 1.27 million jobs indirectly or induced [71]. ICMM describes job opportunities as the first issue and claims that the most frequently-asked question by members of local communities is, "how many jobs will go to their community members", when they hear that a mine may be developed in their community [26]. In some instances, individuals may be disillusioned when the new mining jobs are taken by migrants into the community and may have a negative view of job opportunities. However, the literature suggests that community members, generally, consider increased job opportunities a positive impact [41,70].

Table 2. Economic contribution of U.S. mining.

Item	Direct	Indirect and Induced	Total
Employment	634,600	1,268,800	1,903,440
Labor Income (billions of dollars)	$46.2	$71.0	$118.2
Contribution to GDP (billions of dollars)	$102.1	$123.0	$225.1
Taxes Paid (billions of dollars)	$18.9	$26.9	$45.8

Source: IMPLAN (IMpact analysis for PLANning) modeling system (2012 database) [71].

Income increases due to higher paying jobs and/or the unemployed joining the mine's supply chain is another important impact of mining [26,72]. For example, the direct labor income created by U.S. mining in 2012 was over $46 billion with the total (direct, indirect and induced) exceeding $118 billion [71]. Petkova et al. indicate that the relatively high incomes of people working in the mining and allied industry were seen, by the local community, to generate positive impacts on all towns [72].

Infrastructure improvement is another obvious positive impact of mining. The host community infrastructure that often receives improvements due to mining include health services, educational institutions, power and water supply, sewerage and sanitation and transportation infrastructure [26]. Some of this investment in infrastructure is for business purposes (for instance, a quarry needs to improve roads so their product can be transported efficiently to market). However, a significant portion also comes through corporate social responsibility programs that invest in the host community. For example, BHP Billiton Ltd. says it is committed to invest 1% of pre-tax profits, which amounts to US$ 241.7 million, in community programs [61].

4.1.2. Negative Impacts

However, mining also has juxtaposed adverse impacts, including environmental pollution, increases in housing costs, labor shortages for other businesses and traffic and crime increase. The environmental issue is the main issue raised by those concerned with the impacts of global mining and the first reason often cited by residents for rejecting mining [37]. The environmental impacts include water use and pollution, air, land and noise pollution.

Mining affects water resources through the use of large amounts and contamination of water. In the U.S. state of Nevada, for example, the United States Geological Survey (USGS) estimates that the water table surrounding open-pit mines has dropped by 300 meters [73]. There are many sources of contaminants at a mine site that a can pollute water bodies nearby. These include sediments from exposed soil, diesel fuel (large amounts of fuel are stored on site) and process chemicals (e.g., cyanide for gold processing and sulfuric acid for copper processing). Acid mine drainage (AMD) is recognized as one of the more serious environmental problems in the mining industry due to the number of watersheds affected and the costs incurred for remediation [74]. It is an acidic leachate with high concentrations of heavy metals and sulfate that result from the oxidation of sulfidic minerals. AMD can severely contaminate surface and ground-water, as well as soils [74,75,76]. For example, at the Summitville gold mine in Colorado, AMD polluted seventeen miles of the Alamosa River, severely impacting aquatic life. The place was designated as a Federal Superfund site and cost the U.S. Environmental Protection Agency (US EPA) $30,000 a day to rehabilitate [77]. Opponents of mining are concerned about potential environmental impacts, in particular, possible water contamination [11].

Mining activity can potentially impact terrestrial ecosystems. For example, contaminated water can impact terrestrial ecosystems, including accumulation of toxic elements in soil, soil acidification, damage to soil biota, loss of soil fertility, plant contamination, plant toxicity and food chain contamination [78]. Solid waste is another big issue, since mining products are, mostly, a small fraction of total excavated mass. In surface gold mining, for example, one ton of ore is likely to yield less than one gram of gold, with the rest ending up as tailings. In addition, several tonnes of barren rock may be mined to expose the ore. The amount of solid waste tends to increase with time since improved mining technology makes it possible to exploit lower-grade deposits with time.

Air pollution is another important impact. The major area of concern is dust, from excavation and transportation, causing air quality degradation [25]. In addition, the processing (including refining) of material produces pollutants (e.g., oxides of nitrogen and sulfur) that pollute the air. Worldwide, smelters add 142 million tons of sulfur dioxide to the atmosphere every year, 13 percent of global emissions [77]. Furthermore, air pollutants result from fuel use to power equipment at mine sites.

Noise pollution results from traffic, blasting and operating heavy machinery [25]. Noise pollution is the single largest type of community complaint [23]. For instance, BHP Billiton reports that out of 536 complaints in 2008 at all BHP sites, 200 were related to noise [79]. Ivanova and Rolfe [80] also identified noise impacts, together with vibration and dust, as a significant factor (90% confidence) in explaining community members' preferences for mining developments.

All of these environmental issues affect how community members perceive a particular mining project. If members of the community perceive that a particular mine (e.g., due to its reputation for environmental violations) has a reputation for poor environmental performance, they are less likely to accept the mine and, thus, grant SLO [37,81].

Beside the environmental issues, increases in housing costs and labor market shortages are negative impacts of mining projects that affect individual's perceptions of mining. Petkova et al. [72] observed significant increases in housing costs in Bowen Basin in Queensland, Australia, following the boom in coal prices between 2003 and 2008 (Table 3). The growth rates of median weekly rents from 1998 to 2008 were all at least 160% for the five studied communities with reported data. Unsurprisingly, they found accommodation in short supply in all six surveyed communities. Furthermore, Ivanova and Rolfe [80] found that "housing and rental prices" were significant, at the 5% level, for explaining the preferences for mine development options. Mining can lead to labor shortages, especially for other businesses in the local community that cannot compete with large mines for talent. Petkova et al. found that businesses in five of the six studied communities complained of difficulty recruiting from the labor pool [72]. Intuitively, one would expect that individual's in the local community incur other costs as their living expenses rise with the higher living costs. However, housing costs tend to be the bulk of household costs incurred as a result of new mines.

Table 3. Accommodation and staff shortages [40,72].

Median Weekly Rents	Moranbah	Nebo	Rolleston	Blackwater	Springsure	Coppabella
1998	137	117	80	145	100	N/A
2003	235	220	85	140	137	N/A
2008	680	450	220	380	260	N/A
Growth rate 1998 to 2008	393%	283%	175%	162%	160%	N/A

Traffic and crime have been observed to increase in host communities with the onset of large-scale mining. For example, two social impact assessment (SIA) studies of Central Queensland's Coppabella coal mine report that residents perceive that crime risk, general anti-social behavior and crimes against property were increasing in the community [82]. The police confirmed this observation in absolute terms, although they observed that the criminal activity increase was proportional to population growth from 2003 to 2006. This link between criminal activity and mining is supported by other research. Hajkowics et al. observes that indicators of crime and domestic violence reflect serious social problems in mining communities [83].

Traffic increase has also been observed in the SIA studies [82]. For instance, residents near the Coppabella coal mine in Australia believed that traffic volumes and accidents have increased, including the large trailers and mining equipment. Road use statistics indicated that traffic volumes had increased. The additional traffic can be associated with miners traveling between their residence and employment site. Other studies (environmental impact assessments (EIAs)) in Bowen Basin, Australia, also observe an increase in road traffic and traffic incidents [82,84]. A possible reason for the increased traffic incidents is the higher occurrence of drivers travelling home while fatigued after the end of their shift.

Taken together, the negative perceptions of a mining project due to these negative impacts drive community perceptions, which in turn affect the community's willingness to grant SLO.

4.1.3. Other Impacts

Besides the obvious positive and negative impacts, mining projects can also cause population increases and cultural impacts. Furthermore, two additional attributes of the mine can affect the communities' perception of the intensity and duration of impacts: mine buffer (how far the mine is from the community) and life (duration of mining operation). These impacts are difficult to characterize as negative or positive. For example, whereas one person may think a mine with a long life is a good thing (e.g., because the jobs will last longer), another person in the same community might think otherwise (e.g., because the impacts will last longer).

A consequence of a boom in mining is the associated population growth, especially in small communities without enough skilled labor [85]. Resource exploitation can be directly linked to local population changes, as there is often population growth from migrants looking for job opportunities. This is shown by the population census of four mining communities in Bowen Basin, Queensland, Australia, in Table 4 [72]. The mining boom started in 2001, and the population growth is apparent in four of six studied communities. The population growth from 2001 to 2006 varied from +2.4% to +18.5%.

Table 4. Description of the case study communities [72].

No. Permanent Residents	Blackwater	Moranbah	Nebo	Springsure
2001 *	4913	6124	238	770
2006 **	5031	7133	282	829
Growth rate 2001 to 2006	+2.4%	+16.5%	+18.5%	+7.7%

Australian Bureau of Statistics (ABS) * [86]; ** [87].

A new mining company and migrants looking for job opportunities have impacts on the local community's lifestyle and traditions (culture). Indigenous populations' ways of life can be particularly affected in such situations. The diverse cultural backgrounds of the mining communities and management styles of the mining companies are a factor in determining the extent of this impact [88]. ICCM specifically identifies cultural (heritage) impacts as a factor in community engagement [26]. Cultural impacts include any effects on the cultural norms and practices, which include effects on intangible and tangible cultural heritage, and access to and the vibrancy of cultural facilities. This will be of critical importance when indigenous peoples are present within the area of impact for the mining project.

Besides the above two factors, mine buffer and life affect the mining community's perceptions of the impacts of the mine. Community opposition to a mining operation is an all too familiar picture, and this phenomenon has been called the "not in my backyard" ("NIMBY") syndrome or NIMBY-ism [89]. The key to NIMBY opposition is the location of the proposed project. For an increasing number of people today, their backyard is so vast that projects located relatively far away are still affected by this phenomenon. Ivanova and Rolfe (2011) found "buffer for mine impacts" to be a significant (at the 5% level) factor that explains community preferences for mine developments. The mine life is a measure of the persistence of all impacts (positive and negative). Therefore, it determines how long the job opportunities and noise impacts, for example, will last. It is a measure of the "length contract", which has been found to be a significant factor (at the 1% level) for explaining local acceptability of renewable energy adoption in an ageing population [90].

4.2. Governance

In most jurisdictions, mineral rights are held by government on behalf of the public. Even in cases where mineral rights are held by private individuals or entities, land use and other business restrictions require that a mining company obtain a permit from government prior to initiating a mine. During mining, all aspects of mining activity are regulated by the government. Consequently, mineral resources are seen as something of a public trust that is given to a mining company to exploit with the consent of the public. During permitting, the mine owner proposes a development plan for the project, which is approved, often with modification, when a permit is granted. The permit conditions and the general regulatory framework for mining activity become compliance obligations that the mine must meet to keep the public trust. The government and its representatives, as representatives of the people, make decisions on whether to approve a permit application, what conditions to impose and how to regulate an active mine. The way these decisions are made have a significant impact on the community's perception of the mine's owners and government. Perceptions that government officials are not acting in the best interest of the community (whether independently or under the influence of mine management) can undermine the legitimacy of the mine. Hence, a key factor that affects the community's perception (and hence, social license to operate) is governance, including the mechanism for making permit decisions and the availability of transparent information.

The decision making mechanism and availability of independent and transparent information complement each other. Local communities need to have the right to be engaged in the decision making mechanism first for independent and transparent information to be useful for them. Additionally, available independent and transparent information is the foundation for them to make meaningful contributions to the decision making process. The decision making mechanism describes how decisions are made when disagreements arise on the outcome of the permitting process or other regulatory process. The information refers to all information relevant to the decision to permit a mine or evaluate regulatory compliance, including reports on mining impacts and baseline studies, such as those contained in an environmental impact assessment (EIA), for example. These decision making mechanisms vary from purely legal (i.e., the mining company meets the regulatory requirements) to those that take cognizance of the SLO and seek legitimacy [91]. In most recent cases, the information is provided by the mining company and/or government. The local community often does not trust the available information on the potential impacts from both sources [26]. Community members are more likely to trust the information if they perceive it to be independent and transparent, provided (or reviewed) by multiple groups with technical expertise but no commercial stake in the industry. The information should cover both the broad industry and also relate to specific proposals, which can facilitate local community participation in the decision making and help the community develop.

4.3. Community Demographics

Compared to the mining project's impacts and characteristics, there is much less in the literature that discusses demographic factors that affect an individual's likelihood to support a (proposed) mining project in their community. Age, gender, income and number of children were found to be significant, at the 1% or 5% significant level, in explaining respondents preferences for various mining projects in a choice experiment in a mining community [80] (Table 5). The positive coefficients of female (gender), number of children and age implies that the individuals who are female, older or that have more children were more likely to prefer a mining project than individuals who are male, younger or that have fewer children.

Table 5. Demographic factors [80,92].

Factors	Coefficient	Standard Error
Female [1]	1.243 **	0.259
Number of children [1]	0.261 **	0.098
Income [1]	0.000 *	0.000
Age [1]	0.037 *	0.015
Education [2]	−0.422 *	−2.293

** Significant at the 1% level; * significant at the 5% level; [1] Ivanova and Rolfe 2011; [2] Dimitropoulos and Kontoleon 2009.

Dimitropoulos and Kontoleon [92] observed that the level of education was significant for local acceptability of wind-farm investment at the 5% level. The level of education may be important for mining decisions, as well. The negative coefficient of education, in Table 5, implies that there is a higher probability that people with a higher education level will be opponents of the mining project than people with a lower education level.

In an attempt to understand the local community after a massive demonstration and violent conflict, Muradian et al. [91] uses gender, level of education, age and job field as important background characteristics. Other research indicates job field may be a useful factor [93].

Nonetheless, it seems reasonable to assume that demographic factors have an impact on community perceptions of a project. The community is not one homogenous block, and views vary depending on several factors that may include gender, age, income, level of education, number of children and job field.

5. CLASSIFYING FACTORS FOR EFFECTIVE COMMUNITY ENGAGEMENT LEADING TO SLO

5.1. Factors' Classification

Effective community engagement requires that stakeholders engage members of the community on the issues that matter to the community. In order to do this, mine managers need to be able to use all tools available to them for community consultation (e.g., surveys, focus group discussions, town hall meetings) [26,67,94]. However, the limitations of these approaches mean that the many issues that need addressing may not be adequately addressed individually. Some sort of classification is necessary to ensure management elicits relevant information efficiently (e.g., reduce the cost of surveys or the number of focus group discussions required to acquire the same information). It is only when this efficiency (i.e., balance between costs and time and eliciting relevant information) is achieved that the engagement program will be effective and lead to social license to operate.

There are many subjective ways to classify the factors that affect community support. In this work, the authors attempted to classify the factors equally into five main groups: environmental, economic, social, governance and miscellaneous others and demographics factors. The classification is designed to facilitate efficient communication between the mining company and the local community. This classification is also subjective and not a general one; but it has proven to be appropriate for the preliminary community engagement research [95]. The classification is shown in Figure 1. Four of the classes deal with issues that community engagement needs to elicit opinions on while the last class groups demographic factors that may play a role in explaining the diversity of opinions within the host community. Such classification is important because it can be used to facilitate efficient and effective community engagement.

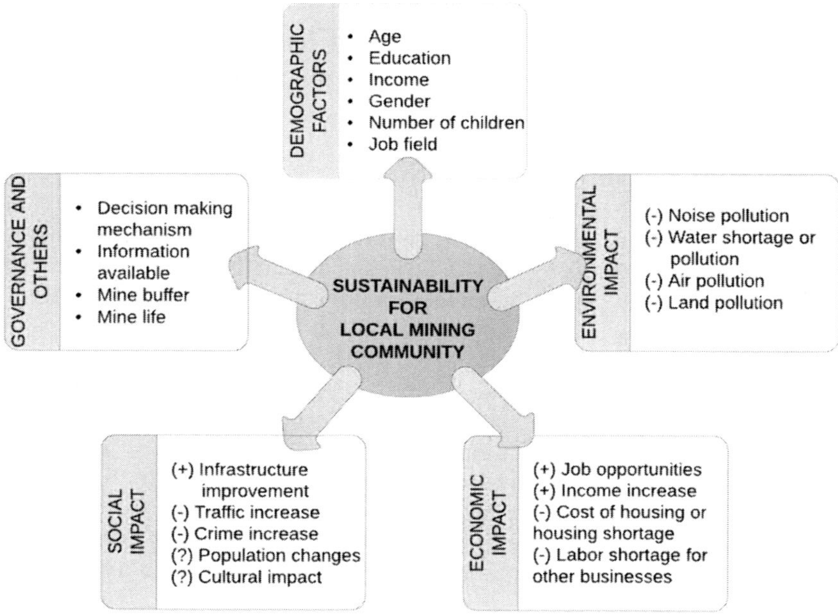

Figure 1. Classification of factors for effective community engagement. (+) and (−) are used to indicate factors that have positive or negative impacts, respectively.

5.2. Two-Stage Framework

Beside the ICMM approach involving the stakeholder analysis matrix (Table 1), discrete choice theory can also be used for mining community engagement using the classified factors. Discrete choice theory [96] has been used successfully to understand community preferences for mining development [80,84]. Compared to the stakeholder analysis matrix method, this method provides a more quantitative way to evaluate the effect of the characteristics of mining projects and demographic factors on community perceptions. Discrete choice experiments (a survey where respondents are asked to select their preferred alternative when presented with multiple hypothetical or real alternatives) and modeling (statistical modeling of discrete choice experiment results using one of several theoretical models) are the two main components of applying discrete choice theory.

A sample discrete choice experiment (DCE) choice set is shown in Table 6. This choice set is designed using four factors, which are the four upper level impacts (Figure 1). Each factor has three levels (high, medium/status quo and low). The combinations of the different levels of the factors are used to generate alternative mine development scenarios for respondents drawn from the local community. Respondents have to be drawn with valid statistical sampling techniques to ensure the sample is representative and adequate to support the conclusions [97]. In the survey, respondents drawn from the local community will be shown several choice sets, but with different alternatives. Each time, respondents will be asked to choose the alternative they prefer, if a new mine were to open near them (their residence).

Table 6. Sample Stage 1 discrete choice set.

	Environmental Impact	ECONOMIC Impact	Social Impact	Government Approval Process and Other Ancillary Factors
Option A	Same as a similar mine in the area	Same as a similar mine in the area	Same as a similar mine in the area	Same as a similar mine in the area
Option B	20% worse than a similar mine in the area	Same as a similar mine in the area	20% better than a similar mine in the area	20% better than a similar mine in the area
Option C	20% better than a similar mine in the area	20% better than a similar mine in the area	Same as a similar mine in the area	20% worse than a similar mine in the area

In this method, the respondent is asked to choose one of the alternatives (A, B or C), compared to the stakeholder matrix approach, where he or she is only asked to indicate whether his or her view of the project is "positive", "negative" or "neutral". The difference is that the proponent of the project will get more information about what drives the community's preferences: environmental impacts, economic impacts, social impacts or the government approval process and other ancillary factors. The additional information collected by the DCE approach can help the mining company to better understand the community's perception of the mine, as an initial step. DCE has been used to understand mining community perceptions of mine developments [81,92].

We are proposing a new framework where DCE can be used in a two-stage approach for effective community engagement. In this approach, Stage 1 uses DCE to elicit views from respondents based on the four broad factors: environmental, social and economic impacts and governance and others. Stage 2 elicits views on an expanded list of factors from any of the broad factors that are found to be of particular interest to the respondents in Stage 1. A key part of discrete choice experiments and modeling is identifying and classifying the factors that affect an individual's preferences. This first step (identifying and classifying factors) can be accomplished with the factors and classification presented in this study. Using the four broad factors based on the classification in this work, a Stage 1 discrete choice experiment (DCE) can be conducted and the resulting data used for discrete choice modeling.

If necessary, a Stage 2 survey can be designed using a hybrid of the upper level factors (the four categories) and lower level factors of the upper level factors that are found to be significant. For example, assume the Stage 1 survey shows that only economic impacts are important (either positive or negative) to the community and that respondents are neutral on the other impacts. The Stage 2 survey can then have the following factors: governance, environmental impacts, social impacts, job opportunities, income increases, housing costs and labor shortage for other business. The proposed survey can collect much more useful information without being unduly complex or costly when compared to the traditional approach of applying DCE to stakeholder analysis [81,92]. This advantage exists in cases where the number of relevant factors considered in the DCE is high (e.g., Que [92] considers 16 factors).

What is more, the demographics can be tracked in both stages. As the literature suggests, demographic factors have an impact on community perceptions of a project. The community is not one homogenous block, and views vary depending on several factors that may include gender, age, income, level of education, number of children and job field. The data from any survey can be analyzed to evaluate whether demographics play a role in explaining diversity in perceptions.

We believe this two-stage approach is advantageous because it is cost-effective and time-flexible. Compared to the stakeholder matrix method (Table 1), the Stage 1 survey is likely to cost the same and take as much time, but result in more detailed information for assessing community preferences. The Stage 2 survey, if found necessary, will provide managers and researchers more information by which to examine the broader concepts in detail. Additionally, the cost of this second stage will only be incurred when found necessary.

However, there are two main limitations of the proposed approach. First, the approach requires a classification (grouping) of factors prior to any input from the community. Since the remainder of the community engagement depends on the validity of the classification, invalid classification can lead to misleading results. The classification in Figure 1 has been validated in the U.S. context [41]. There is no reason, however, to believe that this classification is universally valid without further research. Second, by using a two-stage approach where only some of the primary factors are included in the Stage 2 surveys, some confounding effects between the primary factors can be lost in the data.

This could be an issue in situations where the effects of a particular factor may be heightened in the presence of another factor.

Overall, we believe that this two-stage approach would be beneficial. However, further work is required to examine whether it is valid and under what conditions it is valid. Such research can be done by conducting surveys with all relevant factors and in two stages, as proposed. The results of the two kinds of surveys can be compared to see if the results are consistent.

6. CONCLUSIONS

This study presents 17 factors that affect community members' views of a mining project, which impacts a mine's social license to operate. This is done through a comprehensive review of the literature. The authors classify the 17 factors found in the literature into five categories: environmental, economic, social, governance and demographic aspects. The work also proposes a two-stage community engagement approach based on the classification of the dominant factors presented in this work and discrete choice theory. The approach is cost-effective and less time consuming than traditional community engagement. Further work is required to validate the proposed approach. The aim of this paper is to provide mining companies guidance to improve their chance of obtaining a social license to operate for their operations through better community engagement.

ACKNOWLEDGMENTS

This work is financially supported by grants from the National Natural Science Foundation of China (Grant No. 51504045) and the National Key Research Program of China (2016YFC0402103).

AUTHOR CONTRIBUTIONS

All authors conceived of and designed the research. Sisi Que and Liang Wang wrote the initial draft of the manuscript. Kwame Awuah-Offei reviewed the draft manuscript, making significant revisions. Wei Yang helped with the major revision.

CONFLICTS OF INTEREST

The authors declare no conflict of interest.

REFERENCES AND NOTES

1. Dechant, K.; Altman, B.; Downing, R.M.; Keeney, T. Environmental leadership: From compliance to competitive advantage. *Acad. Manag. Exec.* **1994**, *8*, 7–28.
2. Epstein, M.J.; Roy, M.-J. Making the business case for sustainability: Linking social and environmental actions to financial performance. *J. Corp. Citizsh.* **2003**, *9*, 79–96.
3. Freeman, R.E.; Gilbert, D.R. *Corporate Strategy and the Search for Ethics*; Prentice Hall: Englewood Cliffs, NJ, USA, 1988.
4. Friedman, A.L.; Miles, S. Social responsible investment and corporate social and environmental reporting in the UK: An exploratory study. *Br. Acc. Rev.* **2001**, *33*, 523–548.
5. Gao, S.S.; Zhang, J.J. Stakeholder engagement, social auditing and corporate sustainability. *Bus. Process. Manag. J.* **2006**,*12*, 722–740.
6. Mathews, M.R. Twenty-five years of social and environmental accounting research: Is there a silver jubilee to celebrate? *Acc. Audit. Account. J.* **1997**, *10*, 481–531.
7. Rowe, J.; Enticott, R. The role of local authorities in improving the environmental management of SMEs: Some observations from partnership programmes in the West of England. *J. Eco-Manag. Audit.* **1998**, *5*, 75–78.
8. Rotheroe, N.; Keenlyside, M.; Coates, L. Local agenda 21; articulating the meaning of sustainable development at the level of the individual enterprise. *J. Clean. Prod.* **2003**, *11*, 537–548.
9. Schaefer, A. Corporate sustainability—Integrating environmental and social concerns? *Corp. Soc. Environ. Manag.* **2004**,*11*, 179–187.
10. Shrivastava, P. The role of corporations in achieving ecological sustainability. *Acad. Manag. Rev.* **1995**, *20*, 936–960.
11. Fonseca, A. How Credible are Mining Corporations' Sustainability Reports? A Critical Analysis of External Assurance under the

Requirements of the International Council on Mining and Metals. *Corp. Soc. Responsib. Environ. Manag.* **2010**, *17*, 355–370.

12. Thomson, I.; Boutilier, R.G. Social license to operate. In *SME Mining Engineering Handbook*; Darling, P., Ed.; SME: Englewood, CO, USA, 2011; pp. 1779–1796.

13. Browne, A.L.; Stehlik, D.; Buckley, A. Social licences to operate: For better not for worse; for richer not for poorer? The impacts of unplanned mining closure for "fence line" residential communities. *Local Environ.* **2011**, *16*, 707–725.

14. Hedberg, C.; Malmborg, F. The global reporting initiative and corporate sustainability reporting in Swedish companies. *Corp. Soc. Responsib. Environ. Manag.* **2003**, *10*, 153–164.

15. Brown, H.S.; de Jong, M.; Lessidrenska, T. The rise of the Global Reporting Initiative: A case of institutional entrepreneurship. *Environ. Politics* **2009**, *18*, 182–200.

16. Willis, A. The Role of the Global Reporting Initiative's Sustainability Reporting Guidelines in the Social Screening of Investments. *J. Bus. Ethics* **2003**, *43*, 233–237.

17. Wood, D.J. Cororate social performance revisited. *Acad. Manag. Rev.* **1991**, *16*, 691–718.

18. Wood, D.J. Measuring Corporate Social Performance: A Review. *Int. J. Manag. Rev.* **2010**, *12*, 50–84.

19. Porter, M.E.; Kramer, M.R. Creating shared value. *Harv. Bus. Rev.* **2011**, *89*, 62–77.

20. Esteves, A.M.; Franks, D.; Vanclay, F. Social impact assessment: The state of the art. *Impact Assess. Proj. Apprais.* **2012**,*30*, 34–42.

21. International Council on Mining and Metals. *Planning for Integrated Mine Closure: Toolkit*; ICMM: London, UK, 2008.

22. International Council on Mining and Metals. *Human Rights in the Mining & Metals Inductry: Handling and Resolving Local Level Concerns & Grievances*; ICMM: London, UK, 2009.

23. International Council on Mining and Metals. *Good Practice Guide: Indigenous Peoples and Mining*; ICMM: London, UK, 2010.

24. International Finance Corporation. *Projects and People: A Handbook for Addressing Project-Induced in-Migration*; IFC: Washington, DC, USA, 2009.

25. International Council on Mining and Metals, International Committee of the Red Cross, IFC. *Voluntary Principles on Security and Human Rights: Implementation Guidance Tools*; ICMM: London, UK, 2011.

26. International Council on Mining and Metals. *Community Development Toolkit*; ICMM: London, UK, 2012.

27. IFC. *Doing Better Business through Effective Public Consultation and Disclosure*; IFC: Washington, DC, USA, 1998.

28. IFC. *Stakeholder Engagement: A Good Practice Handbook for Companies Doing Business in Emerging Markets*; IFC: Washington, DC, USA, 2007.

29. IFC. *Good Practice Note: Addressing Grievances from Project-Affected Communities*; IFC: Washington, DC, USA, 2009.

30. IFC. *Guide to Human Rights Impact Assessment and Management (HRIAM)*; IFC: Washington, DC, USA, 2010.

31. IFC. *Strategic Community Investment: A Good Practice Handbook for Companies Doing Business in Emerging Markets*; IFC: Washington, DC, USA, 2010.

32. Azapagic, A. Developing a framework for sustainable development indicators for the mining and minerals industry. *J. Clean. Prod.* **2004**, *12*, 639–662.

33. Davis, R.; Franks, D.M. The costs of conflict with local communities in the extractive industry. In Proceedings of the First International Seminar on Social Responsibility in Mininig, Sanitiaco, Chile, 19–21 October 2011; pp. 1–13.

34. Gunningham, N.; Sinclair, D. Regulation and the Role of Trust: Reflections from the Mining Industry. *J. Law Soc.* **2009**,*36*, 167–194.

35. Jenkins, H.; Yakovleva, N. Corporate social responsibility in the mining industry: Exploring trends in social and environmental disclosure. *J. Clean. Prod.* **2006**, *14*, 271–284.

36. Kempe, J. Review of water pollution problems and control strategies in the South African mining industry. *Water Sci. Technol.* **1983**, *15*, 27–58.

37. Moffat, K.; Zhang, A. The paths to social licence to operate: An integrative model explaining community acceptance of mining. *Resour. Policy* **2014**, *39*, 61–70.

38. O'Faircheallaigh, C. Community Development Agreements in the Mining Industry: An Emerging Global Phenomena.*Community Dev.* **2013**, *44*, 222–238.

39. Prno, J.; Scott Slocombe, D. Exploring the origins of "social license to operate" in the mining sector: Perspectives from governance and sustainability theories. *Resour. Policy* **2012**, *37*, 346–357.

40. Que, S.; Awuah-Offei, K.; Samaranayake, V.A. Classifying critical factors that influence community acceptance of mining projects for discrete choice experiments in the United States. *J. Clean. Prod.* **2015**, *87*, 489–500.

41. Luyet, V.; Schlaepfer, R.; Parlange, M.B.; Buttler, A. A framework to implement Stakeholder participation in environmental projects. *J. Environ. Manag.* **2012**, *111*, 213–219.

42. Parent, M.; Deephouse, D. A case study of stakeholder identification and prioritization by managers. *J. Bus. Ethics***2007**, *75*, 1–23.

43. Tullberg, J. Stakeholder theory: Some revisionist suggestions. *J. Sociol. Econ.* **2013**, *42*, 127–135.

44. Bryson, J.M. Stakeholder Identification and Analysis Techniques. *Public Manag. Rev.* **2004**, *6*, 21–53.

45. Storey, K. Fly-in/Fly-out and Fly-over: Mining and regional development in Western Australia. *Aust. Geogr.* **2001**, *32*, 133–148.

46. Hilson, G. An overview of land use conflicts in mining communities. *Land Use Policy* **2002**, *19*, 65–73.

47. Viveros, H. Examining Stakeholders' Perceptions of Mining Impacts and Corporate Social Responsibility. *Corp. Soc. Responsib. Environ. Manag.* **2016**, *23*, 50–64.

48. Szablowski, D. Mining, Displacement and the World Bank: A Case Analysis of Compania Minera Antamina's Operations in Peru. *J. Bus. Ethics* **2002**, *39*, 247–273.

49. Cheney, H.; Lovel, R.; Solomon, F. *People, Power, Participation: A Study of Mining-Community Relationships*; Mining, Minerals and Sustainable Development (MMSD): London, UK, 2002.

50. Brundtland Commission, World Commission on Environment and Development (WCED). 1987.

51. Munashinge, M.; Shearer, W. *Defining and Measuring Sustainability*; The World Bank: Washington, DC, USA, 1995.

52. Goodwin, N.R. Five kinds of capital: Useful concepts for sustainable. In Proceedings of the The American Association of Legal Scholars Annual Meeting, Medford, MA, USA, September 2003.

53. Elkington, J. *Cannibals with Forks: The Triple Bottom Line of 21st Century Business*; Capstone Publishing: Oxford, UK, 1997.

54. Grant, R.W. *Contemporary Strategy Analysis: Concepts, Techniques, Applications*; Blackwell Business Books: Oxford, UK, 1997.

55. Johnson, G.; Scholes, K. *Exploring Corporate Strategy: Text and Cases*; Prentice-Hall: Hemel Hempstead, UK, 1993.

56. Russo, M.; Fouts, P. A resource-based perspective on corporate environmental performance and profitability. *Acad. Manag. J.* **1997**, *40*, 534–559.

57. Freeman, R.E. *Strategic Management: A Stakeholder Approach*; Prentice-Hall: Englewood Cliffs, NJ, USA, 1984.

58. Institute of Social and Ethical AccountAbility. *ISEA AccountAbility 1000 (AA1000): Standard, Guidelines and Professional Qualification*; ISEA: London, UK, 1999.

59. Cheney, G.; Christensen, L.T. Organisational communication: Linkages between internal and external communication. In *The New Handbook of Organisational Communication: Advances in Theory, Research, and Methods*; Sage: Thousand Oaks, CA, USA, 2001; pp. 231–269.

60. Hodge, R.A. Mining company performance and community conflict: Moving beyond a seeming paradox. *J. Clean. Prod.* **2014**, *84*, 27–33.

61. BHP Billiton. *Sustainability Reporting Navigator*; BHP Billiton: Melbourne, Australia, 2014.

62. Rio Tinto. *Kennecott Utah Copper Sustainable Development Report*; Rio Tinto: Melbourne, Australia, 2012.

63. MCMPR. *Principles for Engagement with Communities and Stakeholders*; MCMPR: Canberra, Australia, 2005.

64. Environmental Protection Agency. *Principles of Environmental Impact Assessment*; EPA: Washington, DC, USA, 1998.

65. Owen, J.R.; Kemp, D. Social licence and mining: A critical perspective. *Resour. Policy* **2013**, *38*, 29–35.

66. Solomon, F.; Katz, E.; Lovel, R. Social dimensions of mining: Research, policy and practice challenges for the minerals industry in Australia. *Resour. Policy* **2008**, *33*, 142–149.

67. Umadevi, V. Community Mining in Co-authorship Network. In Proceedings of the International Conference on Information Communication and Embedded Systems (ICICES), Chennai, India, 27–28 February 2014.

68. Keane, A.; Jones, J.P.G.; Milner-Gulland, E.J. Modelling the effect of individual strategic behaviour on community-level outcomes of conservation interventions. *Environ. Conserv.* **2012**, *39*, 305–315.

69. McGee, T.K. Private responses and individual action. Community responses to chronic environmental lead contamination. *Environ. Behav.* **1999**, *31*, 66–83.

70. Edelstein, M.R. *Contaminated Communities: The Social and Psychological Impacts of Residential Toxic Exposure*; Westview: Boulder, CO, USA, 1988.

71. National Mining Association. *The Economic Contributions of U.S. Mining*; National Mining Association: Washington, DC, USA, 2014.

72. Petkova, V.; Lockie, S.; Rolfe, J.; Ivanova, G. Mining developments and social impacts on communities: Bowen Basin case studies. *Rural Soc.* **2009**, *19*, 211–228.

73. Rockwell, B.W. *The Goldfield Mining District, Nevada: An Acid-Sufate Bonanza Gold Deposit*; United States Geological Survey: Denver, CO, USA, 2000.

74. Akcil, A.; Koldas, S. Acid Mine Drainage (AMD): Causes, treatment and case studies. *J. Clean. Prod.* **2006**, *14*, 1139–1145.

75. Valente, T.M.; Leal Gomes, C. Occurrence, properties and pollution potential of environmental minerals in acid mine drainage. *Sci. Total Environ.* **2009**, *407*, 1135–1152.

76. Peppas, A.; Komnitsas, K.; Halikia, I. Use of organic covers for acid mine drainage control. *Miner. Eng.* **2000**, *13*, 563–574.

77. Earthworks and Oxfam America. *Dirty Metals: Mining, Communities and the Environment*; Earthworks and Oxfam America: Washington, DC, USA, 2004.

78. Dudka, S.; Adriano, D.C. Environmental Impacts of Metal Ore Mining and Processing: A Review. *J. Environ. Qual.* **1997**, *26*, 590.

79. BHP. *Resourcing the Future: Sustainability Report 2008*; BHP: Melbourne, Australia, 2008.

80. Ivanova, G.; Rolfe, J. Assessing development options in mining communities using stated preference techniques.*Resour. Policy* **2011**, *36*, 255–264.

81. Walsh, P.R. A license to operate? An empirical examination of the influence of environmental and social performance on the financial performance of mining sector firms. *Int. J. Innov. Sustain. Dev.* **2014**, *8*, 190–206.

82. Lockie, S.; Franettovich, M.; Petkova-Timmer, V.; Rolfe, J.; Ivanova, G. Coal mining and the resource community cycle: A longitudinal assessment of the social impacts of the Coppabella coal mine. *Environ. Impact Assess. Rev.* **2009**, *29*, 330–339.

83. Hajkowicz, S.A.; Heyenga, S.; Moffat, K. The relationship between mining and socio-economic well being in Australia's regions. *Resour. Policy* **2011**, *36*, 30–38.

84. Ivanova, G.; Rolfe, J.; Lockie, S.; Timmer, V. Assessing social and economic impacts associated with changes in the coal mining industry in the Bowen Basin, Queensland, Australia. *Manag. Environ. Qual.* **2007**, *18*, 211–228.

85. Lockie, S.; Franetovich, M.; Sharma, S.; Rolfe, J. Democratisation versus engagement? Social and economic impact assessment and community participation in the coal mining industry of the Bowen Basin, Australia. *Impact Assess. Proj. Apprais.* **2008**, *26*, 177–187.

86. ABS. *Australian Bureau of Statistics. Census Data. Canberra: Australian Government*; ABS: Houston, TX, USA, 2001.

87. ABS. *Australian Bureau of Statistics. Census Data. Canberra: Australian Government*; ABS: Houston, TX, USA, 2006.

88. Sassoon, M. Social Benefits and Costs: Workshop Report. In *Mining and the Comrmunity: Results of the Quito Conference*; McMahon, G., Ed.; The World Bank: Washington, DC, USA, 1998.

89. MPE. Mining, People and the Environment. Available online: http://comhlamh.org/wp-content/uploads/2013/09/Southern-Alternatives-Policy-report.pdf (accessed on 4 July 2016).

90. Willis, K.; Scarpa, R.; Gilroy, R.; Hamza, N. Renewable energy adoption in an ageing population: Heterogeneity in preferences for micro-generation technology adoption. *Energy Policy* **2011**, *39*, 6021–6029.

91. Muradian, R.; Martinez-Alier, J.; Correa, H. International capital versus local population: The environmental conflict of the Tambogrande mining project, Peru. *Soc. Nat.* **2003**, *16*, 775–792.

92. Dimitropoulos, A.; Kontoleon, A. Assessing the determinants of local acceptability of wind-farm investment: A choice experiment in the Greek Aegean Islands. *Energy Policy* **2009**, *37*, 1842–1854.

93. Mason, C.; Paxton, G.; Parr, J.; Boughen, N. Charting the territory: Exploring stakeholder reactions to the prospect of seafloor exploration and mining in Australia. *Mar. Policy* **2010**, *34*, 1374–1380.

94. Basu, P.K.; Hicks, J.; Krivokapic-Skoko, B.; Sherley, C. Mining operations and corporate social responsibility: A case study of a large gold mine in regional Australia. *Extr. Ind. Soc.* **2015**, *2*, 531–539.

95. Que, S.; Awuah-Offei, K.; Samaranayake, V.A.; Weidner, N. Using Discrete Choice Theory in Mining Stakeholder Analysis: A Case Study in the United States. *J. Clean. Prod.* **2015**. in press.

96. McFadden, D. Coditional logit analysis of qualitative choice behavior. In *Frontiers in Econometrics*; Academic Press: New York, NY, USA, 1974; pp. 105–142.

97. Hensher, D.; Rose, J.; Greene, W. *Applied Choice Analysis: A Primer*; Cambridge University: Cambridge, UK, 2005.

3
CHAPTER

Integrating Industrial Ecology Thinking into the Management of Mining Waste

Éléonore Lèbre * and Glen Corder

Center for Social Responsibility in Mining, The Sustainable Minerals Institute, The University of Queensland, St Lucia QLD 4072, Australia

ABSTRACT

Mining legacies are often dominated by large waste facilities and their associated environmental impacts. The most serious environmental problem associated with mine waste is heavy metals and acid leakage through a phenomenon called acid mine drainage (AMD). Interestingly, the toxicity of this leakage is partly due to the presence of valuable metals in the waste stream as a result of a diversity of factors influencing mining operations. A more preventive and recovery-oriented approach to waste management, integrated into mine planning and operations, could be both economically attractive and environmentally beneficial since it would: mitigate environmental impacts related to mine waste disposal (and consequently reduce the remediation costs); and increase the resource recovery at the mine site level. The authors argue that eco-efficiency and resilience (and the resulting increase in a mine's lifetime) are both critical—yet overlooked—characteristics of sustainable mining operations. Based on these arguments, this paper proposes a framework to assist with identification of opportunities for improvement and to measure this improvement in terms of its contribution to a mine's sustainability performance.

Keywords: mining; metals; eco-efficiency; resilience; mine waste; sustainable development; Material Flow Analysis; industrial ecology; sustainable resource management

1. INTRODUCTION

In our finite world the consequences of the continuously growing global demand for metals are becoming more and more concerning. It is common thinking that metal mining is by definition an unsustainable activity because of the non-renewability of its feedstock and many authors have raised the issue of resource depletion. Some have estimated the remaining years of supply for some of the commonly used metals [1,2,3]; others have talked about an upcoming “peak minerals” [4,5]; and global trends in ore grade decline have been observed [6].

Even though a mining operation cannot inherently be sustainable it is nonetheless crucial to identify areas of improvement in order to stimulate a positive change. In this context this paper raises the key question of what defines a “sustainable” mining operation, and the associated questions: what is currently unsustainable in mining practices and what would a more sustainable mining operation look like? In the first part of this paper the authors provide key elements of the answers by stating the most critical problems the mining industry is facing: the consequences of ore grade decline, a lack of economic viability, and increasing mining legacies.

The authors propose to view these legacies—in particular the giant mine waste deposits—as opportunities rather than an environmental and economic burden. In the second part the question is therefore addressed from a waste management angle and the paper investigates how a new approach to mine waste management could potentially improve a mine’s sustainability performance. The authors argue that prioritising resource recovery and viewing waste as a future resource could potentially address all three critical problems stated previously.

In the third and final part the authors propose a framework based on the principles of industrial ecology. This framework aims at testing and assessing the hypothesis enounced in the second part, and eventually:

- Quantify the benefits of a recovery-oriented waste management system; and
- Identify external and internal incentives to facilitate a desirable change in mining practices.

2. THE CHALLENGE OF SUSTAINABLE MINING

The sustainability challenge is multifaceted and multidimensional and the present paper does not intend to provide a comprehensive picture of such a complex problem. The aim is rather to give attention to a critical aspect of mining that has been mostly overlooked in the literature, namely the responsibility of the mine's operator to manage the finite resource in a sustainable way. The following section presents three interrelated facets to this challenge.

2.1. Economic Resource Depletion

The Club of Rome published "Limits to growth" in 1972, warning decision-makers about the consequences of exponential growth in a finite world. When the initial report was updated 30 years later, the system-dynamic analysis performed showed that abiotic resource depletion and the consequent increase in resource prices would become a major problem during the first half of the twenty first century [7]. Other studies support these conclusions [1,8].

Despite these results there is an ongoing debate about whether abiotic resource depletion rate is becoming critical. Gordon *et al.* [9] and other authors evaluate that there is still no immediate concern, although Gordon *et al.* [9] admit that human and natural limitations are likely to arise in the future. Hence, even though the question of physical depletion is not considered as crucial, the authors still acknowledge the growing constraints of resource use. This aspect was well-stated by the European Commission [10]: "at present the environmental impacts of using non-renewable resources like metals, minerals and fossil fuels are of greater concern than their possible scarcity".

This will become eventually a matter of physical *versus* economic depletion. Prior *et al.* [6] make this distinction, arguing that "physical depletion is not the primary determinant of a mineral's availability". Economic depletion, namely the moment when the costs of resource extraction compensate for its benefits, is a real issue that is likely to come sooner than physical depletion. Interestingly, economic depletion and physical depletion are not necessarily connected: according to Prior *et al.*[6] the social and environmental impacts of extraction "may stimulate a resource production peak that is completely unrelated to resource stocks".

Indeed, beyond the discussion on whether the forecasts for the remaining years of supply are correct, the reality of abiotic resources gradual exhaustion is already impacting the mining industry: the decline in ore grade and quality is the "visible symptom" [11] and results in increased energy demand [3], water use [12] and waste generation [13].

A number of studies show global trends in ore grade decline as well as a trend for more complex and finer-grained ore deposits [6,14,15]. As a result minerals are more difficult to extract and require more energy. The energy increase as a function of ore grade decline is even exponential [8,16]. Finally, a lower ore grade leads to lower mineral processing recovery rates, which implies higher metal losses to tailings [17].

The real problem of resource depletion is therefore the reduced accessibility of the resource, which results in increased environmental impacts and costs. Technological improvements have allowed for exploiting poorer and more complex deposits while keeping metal prices stable [9], but this might not be sufficient as the problem will tend to increase exponentially in the future [11]. Mining companies have been exploiting easily accessible, high-grade reserves in priority, often wasting lower-grade material (see the next section). This is not a sustainable strategy considering the global trends that require a long-term resource management strategy.

2.2. A resilience Problem

Adding to the resource depletion discussion is a seemingly conflicting trend of a lack of resilience in mining operations. Laurence [18] analysed one thousand mine closures that occurred in a 30-year period, and observed that only 25% of them closed after complete extraction of the ore. The remaining 75% were unplanned closures that resulted in valuable resource being left behind. This means that 75% of mining projects are not resilient and consequently not sustainable in the original sense of the word: not able to sustain their operations in time.

Having resilient operations is essential for sustainability, as the extractive industries cannot contribute to sustainable development if they cannot succeed economically or even survive. The Mining, Minerals and Sustainable Development research project [19] highlighted the "viability of the minerals industry" as the first of nine main challenges the mining industry is facing.

Moran and Kunz [20] proposed a maturity framework for sustainable mining operations and in that framework the final, most sustainable stage of maturity is "adaptable and resilient". Walker *et al.* [21] define resilience as "the capacity of a system to absorb disturbance and reorganize while undergoing change so as to still retain essentially the same function, structure, identity, and feedbacks". Adaptability is the ability of the system's actors to influence resilience. A resilient mining project is therefore able respond to external changes, one of the most common being a change in commodity prices [18]. Laurence's study shows that the mining industry is far from reaching that final stage of maturity.

Premature closures have negative consequences at all levels. At the social level, employees lose their jobs, and the subsidence of the local community is affected. The financial actors lose their investments. Unplanned closures lead to safety issues, poor rehabilitation and consequently unmitigated environmental impacts. And, above all, a premature closure does not consider either alternative land uses for the waste or the possibility to re-mine it in the future. What is left of the resource is most likely sterilised, that is to say rendered non-economic to extract.

This resilience problem shows how crucial the time perspective is in evaluating the sustainability of a mining project. Due to the non-renewability of its feedstock, it is physically impossible to prolong a mining activity forever. Hence, compared to many other industrial activities mine sites inherently have a shorter lifetime. However, mining operations can be prolonged and Laurence's results show that there is a great potential for improvement. A more resilient mine site would consequently have a longer lifetime, and would allow the complete extraction of the resource. This would potentially be beneficial at all levels: a stable local community, a well-planned rehabilitation and economic gains resulting from a more careful handling of the exhaustible resource.

2.3. Mining Legacies

According to Laurence's study [19] significant amounts of valuable material can be found on three out of four closed mine sites. Hence, in a global context of resource depletion, the mineral resource at the local level is not being recovered as efficiently as possible. This leads to the third aspect of the sustainable mining challenge: mine legacies. The unprocessed resource and all remaining rock deposits become waste and contribute to the environmental legacy after closure. However, mining legacies are present on every mine site and even resilient ones still generate significant amounts of waste. Defining mining legacies becomes even more difficult considering that the distinction between waste and valuable material is often blurry, as this section will show.

Hence, what are mining legacies? What kind of material is left behind on mine sites and is there a better way to make use of it? While mining operations are inherently limited in time, the environmental legacies they generate can last for hundreds of years or effectively in perpetuity. Yet, mine waste still contains significant amounts of valuable material, which may be economically recoverable now or in the future with new or emerging technologies.

This section investigates the content of mining legacies, focusing on the most important and problematic waste streams. It shows the main sources of metal losses at the mine site level, discussing how mine waste is generated and what factors define its quantity and composition. Understanding these processes is the first step in evaluating the potential for resource recovery from mine waste. Section 2.3.2 also shows a barrier for improvement at this level, pointing out the lack of available data surrounding mining legacies.

2.3.1. How is Mine Waste Generated?

There are three main processes in metal mining that generate waste: mining itself, which is the extraction of the ore from the ground; mineral processing, which produces a mineral concentrate out of the ore; and metallurgical processing which generates a refined metal out of the mineral concentrate (see Figure 1).

Mining is the first stage of the exploitation of a mineral resource. It can be defined as the extraction of material from the ground in order to recover the ore, which is the material of economic value.

The aim of mineral processing, also called beneficiation or concentration, is to separate the valuable minerals contained in the ore from the gangue, which is the name given to the worthless material surrounding the minerals in the ore. This includes processes such as: crushing, grinding, flotation and gravity, magnetic or electrostatic separation.

Metallurgical processing is generally done in two different ways: hydrometallurgy and pyro-metallurgy, although electrometallurgy is another less common option [22]. Hydrometallurgy (or leaching) relies on the use of solvents to dissolve the wanted metals while pyro-metallurgy (or smelting) uses heat to break down the crystalline structure of the ore mineral. In both cases the chemical combination of the minerals is destroyed to release the metal in its pure form.

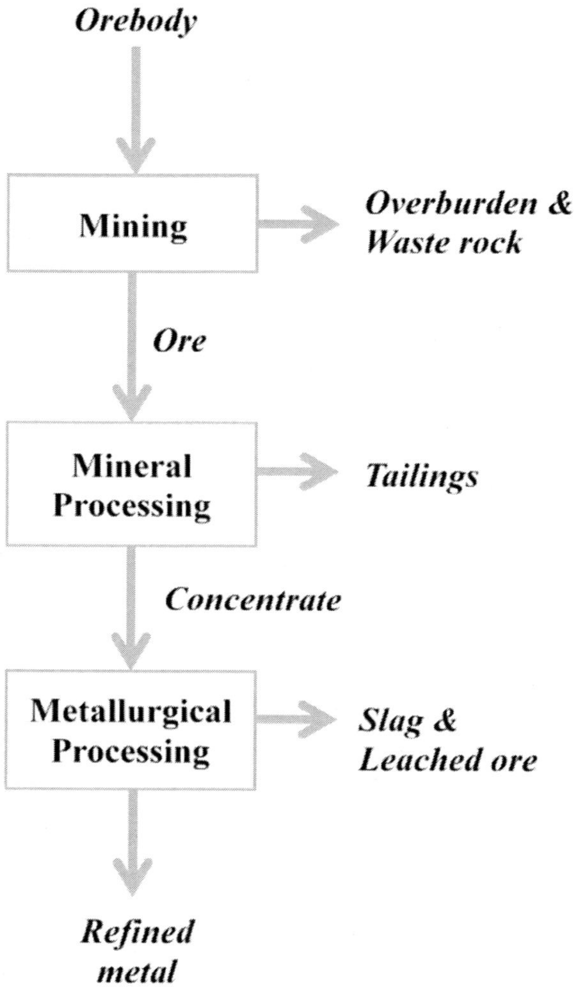

Figure 1. Schematic product and waste streams at a metal mine.

While hydrometallurgical processes are nowadays usually happening on the mine site, electro and pyro-metallurgical processes are mainly situated off-site [22]. The reason is that electro and pyro-metallurgical processes require much more energy and mine sites are typically located in remote areas where access to energy is expensive. However, there are other factors that determine where a metallurgical processing operation will be held. For example pyro-metallurgical processing generates sulfuric acid as a by-product while hydro-metallurgical processing is a net consumer of it. Therefore there is an incentive for these two operations to be combined [15].

It is also worth noting that in some cases of very high grades the ore can be directly sent to the refining stage without previous mineral processing, or with a very basic processing stage. This is sometimes the case for iron ore for example that can contain on average 60% iron [23].

Therefore, depending on the site, a mining company may perform only the mining stage, or it may include a mineral processing stage as well, and sometimes it can also have a refining plant. In a majority of cases however, a mine site performs the mining and mineral processing operations [22].

Three main types of waste are generated by these three activities (see Figure 1):

Overburden or Waste Rock

It is the material that covers or surrounds the ore body and that has to be removed in order to access the economical material. What distinguishes waste rock from ore, that is to say what is economical to extract and what is not, is the cut-off grade: under a certain mineral percentage the material will be treated as waste. The value of the cut-off grade will depend on all the factors that determine the costs of the overall mining operations. For example surface mining (in an open pit) is typically much cheaper than underground operations; as a result open cut mining can afford to have a much lower cut-off grade. Mineral extraction is scheduled and the cut-off grade is updated overtime to adapt to change in costs and metal prices to ensure an economically viable operation.

Tailings

Tailings are generated by the mineral processing of the ore. As a result of milling and flotation they are generally finely grained rocks in suspension in water. The ore grade will determine how much tailings are produced. For example, the mineral processing of 100 tonnes of copper ore with a 2% grade will roughly generate 98 tonnes of dry tailings, to which must be added the water content. The grade varies greatly from one metal to another.

Beneficiation processes are not 100% efficient in performing the separation between minerals and gangue: the mineral concentrate produced still contains a significant percentage of gangue and valuable minerals also end up in the waste stream. The mineral recovery rates will highly depend on the technology chosen and how adapted it is to the specificity of the ore (e.g., its mineralogy). However, metal recovery can and often needs to be adjusted in order to meet the desired concentrate grade. Recovery rates and concentrate grades are correlated as a more selective process will lead to a higher concentrate grade as well as a lower recovery rate, and consequently higher metal losses in tailings.

The concentrate grade is determined by the smelter contract and the smelter's benefits will depend on the metal prices. The optimum concentrate grade leading to maximum profit is typically higher in times of low metal prices, and *vice versa*. Consequently, a smelter contract designed in times of low prices would lead to more metal losses in tailings [17].

Slag and Leached Ore

Slag is the solid waste produced by pyro-metallurgical processing. Hydrometallurgical processing generates another significant waste deposit in mine sites: leached ore. After being crushed in the mill the ore is deposited as a heap leach pad and acid runs through it to extract the wanted metal. When the remaining ore's grade is considered to be uneconomical the operations end and the leached ore is left as waste.

More generally, inefficiencies in mineral characterization, extraction, processing, *etc.* generate metal losses throughout the life of the mine, from exploration to closure. In all cases several factors (mineralogical, technological, managerial, *etc.*) make the difference between ore and waste, and hence determine the amount and the composition of the mining legacies. These factors influence the costs and revenues of the operations and economic considerations eventually determine the final decision.

The extent of the problem is multiplied when one considers the complexity of mineral ores and the diversity of elements they usually contain. Indeed ore deposits usually contain several kinds of metallic minerals, and some valuable non-metallic minerals are often associated (e.g., fluorites and barytes). However, they may not be present in sufficient amounts for their extraction to be economic, or their extraction simply does not align with the company's strategy.

When these minor minerals are found in the concentrate, they are considered as impurities by metallurgists and are rejected in the slag. The same happens to a larger extent in poly-metallic ore mines where several elements are present at economic levels. In this case separate concentrates are sometimes produced for each metal extracted and "a valuable metal in the "wrong" concentrate will be considered as impurity" [17].

When metallurgical processing and mineral processing are performed by different companies, a contract between the two defines the quality of the concentrate and fixes its price. In particular, the contract sets acceptable levels of impurities and a minimum grade for the concentrate. Sometimes however, a smelter may extract more than one metal out of the concentrate and will pay for these additional metals if they are present above a certain grade, also set in the contract. This is typically the case for precious metals such as gold and silver found in copper concentrates [17].

In conclusion, even though the mineralogy of the ore deposit sets the basis for how a mining project can be executed, the difference between ore and waste, or between valuable material and impurities, is eventually determined by human factors. Mining companies adapt to the ore deposit and choose the technologies to be used, design the process flow sheet and plan the entire mine project in order to eventually maximise their profit. This makes the system also subject to human error and one can question the success of current strategies given the lack of resilience exposed earlier, and speculate on whether metal losses could partly be avoided.

2.3.2. Mine Waste and Mining Legacies

Avoiding metal losses and preventing mine waste is of critical importance. This section considers the extent of mining legacies, how essential it is to mitigate mine waste generation as well as a current barrier for improvement, namely a general lack of reporting.

Lottermoser [22] estimated that the global mine waste deposit was in the order of several hundred billion tonnes, covering an area of one hundred million hectares. Metal mine waste generation in particular would be currently averaging 15 billion tonnes per year. This is ten times larger than global municipal waste generation, which was estimated to be 1.3 billion tonnes in 2012 [24].

To put these numbers into perspective, the total amount of mine waste is the same order of magnitude as the materials moved by geological processes, such as oceanic crust formation, soil erosion, *etc.* [22]. Lottermoser concludes that "the Earth is getting increasingly shaped by mine wastes rather than by natural geological processes".

Furthermore, mine waste generation has been growing exponentially and this trend is likely to continue in the future [25], though Bringezu [26] pointed out that it could be partially offset by other factors such as the growing trend for underground mining. Indeed, underground mining generates less waste rock than open cut operations because, other than the relatively small amounts of rock excavated for access tunnels and shafts, only ore is excavated.

Mining is by far the industry that generates the most waste. Coal mining is the largest, followed by metal mining industries together with the industrial minerals industry [22]. Extraction and production of clay, sand and gravel generally produce much less waste.

Waste generation varies significantly between metals. This depends mostly on the ore grade that is directly related to the amount of tailings produced. Lottermoser [22] stated that for every tonne of metal ore extracted, at least a tonne of solid waste is generated, but often the amounts of waste are orders of magnitude greater.

Estimating the total amounts of metals lost in mine waste as well as total volumes is subject to uncertainties because of a lack of available data. Indeed, the way mine waste data is reported is mostly inconsistent and incomplete. In the "Global Reporting Initiative" (GRI), a voluntary initiative established by the United Nations in 1997, mine waste is reported under the category "EN23" [27] which encompasses the "total weight of waste by type and disposal method". The same applies to the metallurgy industry and slag is to be reported the same way. The guidelines therefore do not encourage companies to distinguish between different kinds of waste or to report their composition.

Worse, data on waste rock is mostly inexistent. Waste rock generation from mining activities is called a hidden flow [28]. Brattebø *et al.* [28] define hidden flows as material flows that do not enter our economic system. Indeed, when thinking about the global human footprint one first thinks of all the material consumed by our society: consumer goods, infrastructure, food, fossil fuels, *etc.* However, at the basis of all human activities significantly larger movements of materials are happening: the excavation of overburden in order to extract abiotic resources, but also the excavation of materials for infrastructure (roads, buildings, *etc.*), and soil erosion due to agriculture and forestry (more generally the extraction of biotic material). These movements remain mostly unquantified because they are hidden flows and do not have economic value.

Furthermore, if not managed properly, mine waste can generate significant pollution, mostly through the form of acid mine drainage (AMD). It is considered to be a major source of water pollution in countries that have historic or current mining activities [29] and one of the major environmental challenges the mining industry is facing worldwide [25]. AMD is the oxidation of sulfide minerals in the presence of water and oxygen. It results in water acidification and heavy metals solubilisation that are both highly detrimental for the surrounding environment when the contaminated water then slowly leaches from the waste dump.

The leaching rates and the long-term effects of AMD are difficult to estimate. Adding to the lack of data on mine waste amounts, mining legacies are mostly unquantified. In addition, they remain long after their corresponding mines close, which makes it difficult to penalise the people who were financially responsible.

A minimum of 50,000 recorded abandoned mines in Australia is to be added to the 550 operating mines [30]. In the case of an abandoned mine the local government becomes responsible for assessing the risks and prioritising the sites that need the most maintenance or rehabilitation. The first step in managing abandoned mine sites is to make an inventory. Sometimes this first step is not even achieved and, in Australia, one of the most resource-rich countries, the Northern Territory does not have such inventory [30]. In the other states, inventories are mostly up-to-date, but do not provide much detail on the sites except for their localisation.

Adding to this are the regular reports of tailings dams accidents. Hudson-Edwards *et al.* [25] estimated that over 70 major dam failures have occurred around the world since 1970. The International Commission on Large Dams reported about 230 accidents in the second half of the twentieth century [31] and the WISE Uranium project reported 112 accidents between 1960 and 2014 [32]. The magnitude of accidents varies significantly, from a minor seepage to a major dam collapse.

Azam and Li [33] reviewed the tailings dam failures occurring during the last hundred years and investigated both the cause of the accident and the social, economic and environmental impacts it generated. The concerning result is that for a great majority of accidents the total amounts of contaminants released remains unknown.

Finally the spatial occupancy of mine waste is an important impact to consider. Mine waste covers 100 million hectares [22] and the land occupied is rendered unusable in the long term. The Fraser Institute [34] identifies two main environmental-impact categories: the several contamination paths to the surrounding environment and the "loss of productive land". Reid *et al.* [35] argue that land use should be one of the most important impact categories to consider in environmental impact assessments applied to mining.

In conclusion, mine waste is problematic for various reasons and improvements in mine waste management are necessary. Mitigating, controlling, or even understanding mine waste generation faces a major barrier: mining companies typically do not monitor the waste they generate and the composition of mine waste thus remains largely unknown. This prevents opportunities to make a better use of the waste, extract further material from it or even reduce its generation. However, an alternative waste management oriented towards resource recovery could potentially mitigate environmental impacts, which is what the next section will explore. This paper will now develop a potentially more beneficial scenario that addresses the three sustainability challenges exposed in this first section: resource depletion, lack of resilience and mining environmental legacies.

3. A POTENTIAL SOLUTION

3.1. A Preventive and Recovery-Oriented Waste Management

According to Laurence [18] the International Council on Mining and Metals (ICMM) principles "ignore one important dimension that distinguishes mining from all other activities: a focus on the mineral resource itself". Ayres *et al.* [15] confirm that the current economic models used in designing a mine project "do not take into account energy or material resource depletion or thermodynamic constraints" . Fonseca *et al.* [36] reach the same conclusion in their assessment of five other promising sustainability frameworks: "what seems to be driving those frameworks is less the need to preserve minerals for future generations and more the need to ensure performance improvements in connection with some environmental, social and economic areas". *A priori*, there is nothing wrong about addressing the triple bottom line, except if it is done in a short-term perspective. The concerning context of resource depletion shows how important it is to adopt a long-term perspective.

Just like any other sector the mining industry requires infrastructure and labour to operate and needs to address its pollution mitigation objectives while creating positive value for a wide range of stakeholders in order to contribute to sustainable development. However, unlike most other sectors, the mining industry extracts a non-renewable resource out of the ground and the challenge is to exploit this resource "in a sustainable way" [18].

A wide range of valuable components is left behind in mine sites in the various waste deposits and, interestingly, the AMD occurring from waste deposits is partly due to the presence of these valuable components. Several authors have highlighted this fact and argued that further extraction of these elements would offer two advantages: recovering value out of the waste and mitigating environmental impacts. Lottermoser [22] states that "if innovative alternatives to current waste disposal practices are pursued and if wastes are recycled or reused, then waste disposal problems are eliminated." Dold [37] also insists on this double benefit and adds that pollution mitigation also results in cost reduction and thus both aspects can be economically desirable. Lottermoser [38] adds that mine waste recycling ideally "creates financial assets, slows consumption of natural resources, limits waste production, encourages innovation and local industries, creates jobs and teaches responsibility for the environment shared by all".

For Wills [17] mine wastes and in particular tailings still contain significant amounts of valuable metals and constitute a potential future resource. The author adds that the cost of tailings reprocessing can sometimes be lower than that of processing ore, because tailings have already been extracted and ground. However, extracting minerals from mine waste may also require additional inputs and related costs due to the low concentration of materials and possibly a more complex mineralogy [8]. Resource recovery from low-grade ore or very low-grade ore may not be economically or environmentally desirable.

In spite of these concerns, the global trend of ore grade decline may require the mining industry to adapt to lower-grade material and it is worth investigating how viewing mine waste as a potential future resource and managing it accordingly would be beneficial. Dold [37] provides a good visual description on how handling the waste as low grade ore or very-low grade ore in a recovery-oriented waste management would contribute in optimising resource extraction and mitigating AMD (Figure 2). In Figure 2, the low-grade stockpile can be processed after the higher-grade material, and ultimately the very-low grade material when the economics become favourable. Meanwhile an appropriate storage facility collects the leachate and prevents it from contaminating the surrounding environment. Pit lake water may also be treated to recover metals and prevent further contamination.

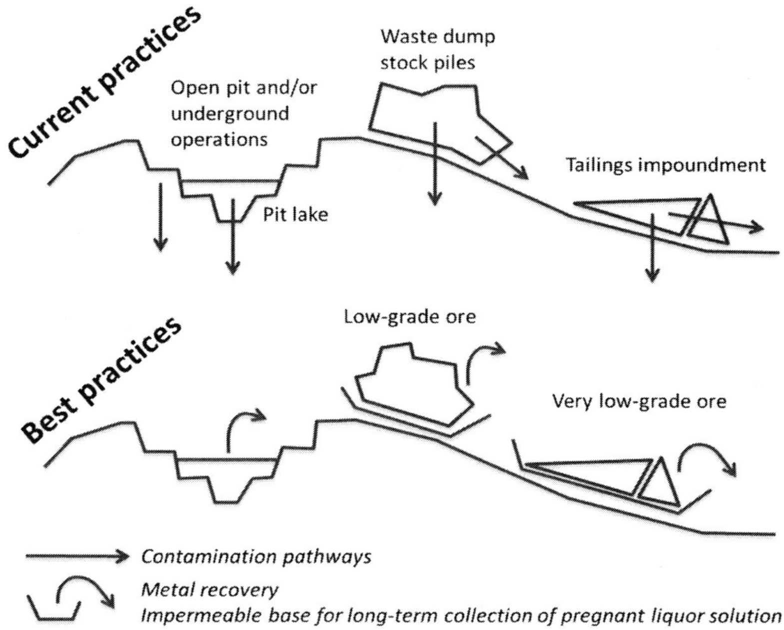

Figure 2. Current mine waste management practices versus a metal recovery oriented waste management. Adapted from Dold [37].

Such a mine site would handle and process an increased amount of material and consequently extend its lifetime. It could also potentially increase its resilience, as Folke *et al.* [39] insist on the link between resilience and diversity. A recovery-oriented mine waste management system would diversify the operations on site by processing material from various deposits, using different technologies and extracting different types of minerals. For instance, a mine extracting several valuable elements will be less sensitive to a drop in prices of only one element. However, such a hypothesis needs to be tested.

3.2. A New Pyramid of Priorities for Mine Waste Management

Dold argues that mine waste should be viewed as a potential future resource and consequently proposes an alternative way to handle mine waste. This alternative approach can be illustrated by building a new hierarchy of priorities for mine waste management. The waste management hierarchy "Reduce-Reuse-Recycle-Dispose" sets general guidelines for waste management that prioritise waste prevention, *i.e.*, waste reduction, over waste reuse and recycling, final disposal in landfills being the least desirable option. This hierarchy has been used for mine waste in the 2012 waste management plan for the Rasp mine in Broken Hill, Australia [40]. However, this hierarchy was primarily developed for municipal solid waste as guidelines for consumer behaviours. Some authors have pointed out its limitations when applied to other types of waste [41,42]. In particular, mine waste differs from post-consumer waste in many ways and the pyramid presented in Figure 3 is more adapted.

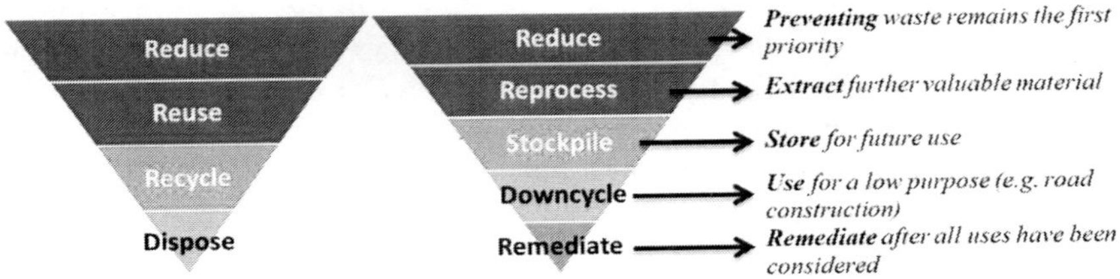

Figure 3. General waste management hierarchy (**left**); new hierarchy for mine waste management (**right**).

The proposed mine waste management pyramid still places waste prevention at the top of the priority list. It is however then followed by the reprocessing option, which allows recovering part of the valuable materials left in the waste. Stockpiling can occur when there is a need to wait for more favourable economics and until reprocessing becomes a profitable activity. When the valuable elements left in the waste are considered unrecoverable then the waste may be used for another purpose, such as on site void backfilling, or sold as road construction material. Depending on the purpose the waste may have to be proven inert and non-hazardous. This option is rated lower on the priority list as it results in the definitive loss of all remaining valuable components. Finally, remediation and site rehabilitation may happen after all other options have been considered.

However, these five options do not exclude each other and perhaps the use of an inverted pyramid can be misleading. For instance, remediation is a necessary step and needs to be addressed in all mine sites. However, if done too early it could result in sterilising the remaining resource. The new hierarchy places the priority on resource recovery while simultaneously making use of the waste. All five options may be combined in order to optimise the economic gains and mitigate the environmental impacts. This optimisation also needs to take into account the external inputs (energy, chemicals, machinery, *etc.*) required to perform the different actions.

Such a mine waste management approach requires a long-term perspective, especially because of the stockpiling possibility. Ayres [15] acknowledges that in many cases part of the waste rock deposits is low grade ore that has been stored intentionally in anticipation of future rise in prices. Stockpiling is already a common practice. However, the mine may cease its operations before the prices rise, which relates to the issue of resilience in an economically unstable environment.

The previous section showed how, in current mining operations, many factors determine whether a material is treated as waste or not. It can be unfavourable economics in the form of low prices or high costs, inefficient processing due to technological limitations and mineralogical factors [22], or more broadly personal management decisions [37] and the overall corporate strategy (e.g., high-grading). Policies and regulations also have an influence in providing incentives or discouraging further resource recovery. Some of these factors, if not all, may have prevented the optimisation of extraction at the time of mining but they may change and become favourable in the future.

Two questions arise from the previous discussions: is this alternative waste management approach as attractive as it looks? That is to say, does it: reduce the environmental legacies, increase resource recovery and favour economic resilience? And if it is, why has it not been more broadly applied in practice and what incentives could stimulate a positive change? The next section describes a framework aiming at answering these two questions.

4. THE FRAMEWORK

4.1. Three Scenarios to Compare

A recovery-oriented waste management approach at the mine site level could potentially provide an adequate answer to the challenges presented in the first section: optimise resource recovery while mitigating environmental legacies and increasing resilience. However, this hypothesis needs to be tested and verified, and a framework needs to be developed to assess the benefits of such management in a comprehensive way, as well as identifying the factors that could trigger change. Before presenting the proposed framework we will first explain the situation by distinguishing three cases (Figure 4):

- Scenario A: a traditional mine produces a certain amount of mineral concentrate P1 while generating an environmental impact E1 during its entire life cycle. After closure the mine legacies generate an additional environmental impact E1′.
- Scenario B: a traditional mine, a certain time after its initial closure is re-opened. The new mine operator decides to re-mine the waste deposits. This second operation leads to an additional production P2 and its related environmental impact E2. In between the two operations, the waste is deposited in the traditional way without planning for future extraction. Examples of case B can be found in practice, such as the large Australian mine sites Kalgoorlie and Mount Morgan.
- Scenario C: an innovative mine site is operated with a planning for future recovery as described earlier. It operates during a certain period, longer than in scenario A and without discontinuation as in scenario B, to produce a final amount of P3 while generating a certain environmental impact E3.

In this study the ratio P/E is defined as the eco-efficiency of the mining project. This production-environmental burden ratio reflects direct environmental impacts generated on site (pollution due to toxic waste generation, greenhouse gas emissions due to fuel consumption, *etc.*) and indirect environmental impacts embedded in energy and material use (e.g., the materials used to build the process plant were extracted from another mine, transportation in and out of the mine site also generates greenhouse gas emissions, *etc.*). Thus eco-efficiency is also linked to the economic aspect; the cost of inputs to the operations. It is a relevant indicator that ensures resource recovery (*i.e.*, production) is not favoured at the expense of the environmental footprint and indirect resource use.

Scenario B provides an enhanced resource recovery compared to scenario A. It however requires additional inputs and generates an additional environmental impact while in parallel treating the legacy from project 1, *i.e.*, preventing E1′. To compare these two scenarios the advantages and disadvantages need to be evaluated in a quantitative manner.

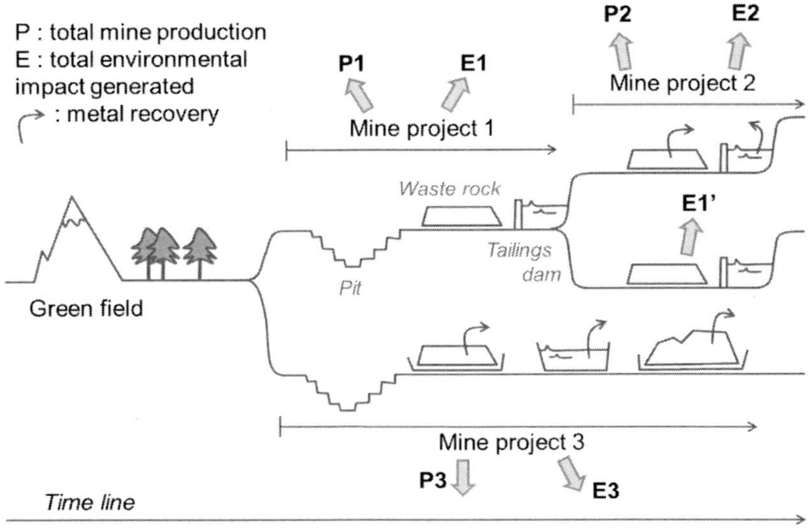

Figure 4. Distinguishing three scenarios to be compared. Notes: Scenario **A**: traditional mining; scenario **B**: re-mining; scenario **C**: "sustainable" mining.

Comparing scenario B and scenario C is more complex. E3 could potentially be smaller than E1 + E2—the total environmental impact generated by scenario B—as better planning and the continued activity allowed for a cleaner disposal of the low-grade material. Furthermore, the discontinuation between mine projects 1 and 2 may result in abandoned infrastructure that could have been reused if it had been properly maintained. The new infrastructure carries embedded environmental impacts that contribute to E2. P3 could also potentially be higher than P1 + P2 as a consequence of this better planning and the resulting efficiency improvements and cost savings. Also, project 2 initially needs a prospection stage in order to determine the composition of mine waste, while project 3 has monitored its low-grade stockpiles.

How can the three scenarios be compared and to what purpose? Scenarios A and B are both real cases and represent current practices. The eco-efficiency of project 2 can be assessed using a real case study and compared to the alternative case of "doing nothing" as in scenario A. A quantitative analysis based on onsite data collection can be performed. In particular, this analysis can determine two aspects of the problem stated above: the increased resource recovery (production over total resource in the ground) and the environmental legacies, *i.e.*, how the legacies from project 1 are addressed and what new legacies project 2 generates (see Section 4.2).

However, scenario C is a hypothetical scenario and cannot be the subject of a quantitative analysis. Furthermore, there are no conventional approaches to measure resilience [43]. Scenario C can however be studied in a qualitative manner and be used to investigate the two other parts of the research questions: the factors that influence project 3's resilience and the incentives that could facilitate a desirable change (see Section 4.3).

4.2. Quantitative Analysis (Scenarios A and B)

The difference between scenarios A and B is the re-mining stage. The quantitative analysis aims at determining whether mine waste re-mining is indeed beneficial with regards to the recovery of the mineral resource and the environmental performance. It is clear that a waste re-mining project would enhance the resource recovery at the mine site level. However, how would it affect the environmental footprint?

In scenario B, project 2 adds to the industrial activities on site and contributes to increasing the mine site's environmental footprint. However, and possibly the main argument in favour of scenario B is that project 2 is taking place on a site already heavily impacted by industrial activities rather than being implanted in another "green field". Thus project 2 is not responsible for the initial impact that turned the green field into a brown field. Furthermore, re-mining virtually prevents the opening of another mine.

Hence, minimising the mine site's overall environmental impact (E) is not the goal. The goal is to minimise the environmental impact embedded in metals, that is to say maximise the eco-efficiency (P/E). The eco-efficiencies of scenarios A and B may be compared:

$$\frac{P1 + P2}{E1 + E2} > \frac{P1}{E1 + E1'} \tag{1}$$

A mining project's eco-efficiency, and in particular its environmental footprint, may be calculated through the use of system-modelling tools such as Life Cycle Analysis (LCA) or Material Flow Analysis (MFA). A Material Flow Analysis at the mine site level would involve mapping of all material flows within the system, understand where metal losses occur and which processes are the most energy and material intensive. The Eurostat guidelines for Economy-Wide MFA or any similar methodology may be used [44]. MFAs usually do not connect material flows with environmental impacts. However, they can be coupled with LCAs to reach that goal.

In a Life Cycle Analysis the practitioner performs an inventory of the emissions of all processes occurring over the mine's life cycle. Processes are hence associated to their environmental impacts. LCA background data also allow accounting for indirect emissions occurring outside the mine site. However, while LCA presents a great potential to assist the mining industry in identifying areas of improvement, it may also suffer from methodological limitations as it is a standardized tool that has rarely been applied at the mine site level [14]. More research is needed to address these limitations and research efforts may benefit from practical case studies.

4.3. Qualitative Investigation (Scenarios B and C)

The qualitative analysis aims at comparing scenarios B and C and identifying incentives that can encourage a change towards scenario C. What distinguishes scenario C from scenario B is the planning stage. The preventive mine waste management system is to be planned in the early stages of the mining project. According to McLellan *et al.* [45] this is the only way to minimise mine waste. Indeed, in order to prevent and not only reprocess the waste once it has been generated, or mitigate its environmental impact, critical decisions are to be made upstream. For example, the cut-off grade determination will directly influence the amount and composition of waste rock.

The qualitative investigation targets the decision-making process and the factors that influence it. This can be done on three different scales that ensure the issue is dealt with in a comprehensive way.

4.3.1. Three Main Scales of Investigation

Laurence [18] argues for the need to focus on the mine site level in order to understand the factors and the practices that influence its resilience. Practices vary significantly and from one mine to another the resource may be managed sustainably or unsustainably. However, sustainable resource management is an issue that needs to be addressed over a wide range of scales [20]. A framework that considers the progress that the mineral industry is making towards sustainability requires a higher scale than the mine site level.

Hilson [46] distinguishes three main dimensions: technical, managerial and political, which illustrate well the different scales for such a framework (Figure 5). While the "technology and efficiency" scale represents what is implemented in practice at the operational level, it is mainly the result of the decisions taken at the management level. The management culture on the other hand is influenced by the political and regulatory context.

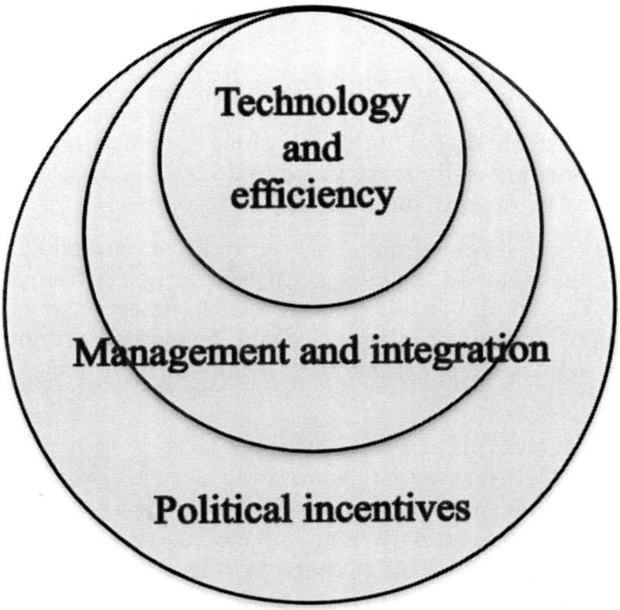

Figure 5. A three-scale qualitative framework.

Technology and Efficiency

Technological innovation may play a role in increased metal recovery rates (process efficiency) and eco-efficiency. For example, the Hall-Héroult and Bayer processes, the only available pathway to produce aluminium, are notoriously inefficient and energy consuming [47,48] despite aluminium being a relatively abundant metal. Research could potentially identify alternative technologies that would reduce wastage significantly. More generally, pre-concentration methods such as differential blasting may be added prior to the mineral processing plant and enhance metal recovery [49].

However, technological innovation is not sufficient and several authors have insisted on the need for innovation at a system level [50]. Moran and Kunz [20] emphasize that the improvement in several individual indicators does not automatically represent an improvement in the overall system. Efficiency can be improved not only by specific technological improvements but also by systemic innovation, which involves the planning and the management levels. The systemic approach also ensures a connection between the operational and managerial levels.

Management and Integration

Moran and Kunz [21] define four dimensions by which a mine site can be integrated and connected: by its value chain, its life cycle, by the spatial dimension, and by flows of information.

At the value chain level resource recovery may potentially be improved by changing the terms of the contract between the concentrator and the smelter. At the spatial level synergies with other businesses can be stimulated. For example a mine site could develop a partnership with a wastewater treatment company that would recover metal from mine wastewater [51].

At the life cycle level planning for recovery and anticipating a future use for the waste are necessary. Collaboration between mining disciplines (as well as with non-mining disciplines such as biochemistry, chemistry, hydrology, ecology) may allow innovative ideas to be integrated across the mine's life cycle [52]. This connects with the information dimension, where Moran and Kunz [20] emphasize the importance of policies, either governmental or corporate, as well as formal or informal networks within the company. Collaboration and exchanges of information between professionals need to be enhanced to find common ground. Interdisciplinary considerations should become dominant in all decisions [20]. The lack of data on mine waste also shows how monitoring mine waste would be helpful to achieve more sustainable mining operations. More generally Mitchell *et al.* [53] argue that companies can improve their performance only if they manage to gather and analyse the data on their operations.

The management team may make use of analytical industrial ecology tools such as Life Cycle Analysis or Material Flow Analysis in order to identify options for improvement and develop a strategy to enhance the local efficiency. LCA and MFA are system-modelling tools and thus enable innovation at a system level. LCA may be applied either to the life cycle of mining operations or to a metal product's value chain, relating to the first and second dimensions defined by Moran and Kunz [20]. MFA allows mapping the flows within spatial boundaries and may be used to understand better information flows (third and fourth dimensions).

This second scale is where the operations are integrated into the bigger picture and synergies with other actors are identified. Moran and Kunz [20] argue that integrating and connecting the different components in the mining system leads to resilience. This integration, aiming at multidisciplinary collaborations and system improvements, is critical to enable progress towards more sustainable operations. It is designed to "strengthen the inter-linkages between operations and their surrounding environment and community" [20] in order to satisfy all stakeholders. Finally, political incentives can influence managerial decisions into implementing sustainable practices and thus set the "rules of the game".

The Role of Policies

Policies can influence and provide the right incentives to change business practices. They are particularly important in the present context as resource recovery is not *a priori* the mining companies' concern. Mining companies have a commitment to shareholders to generate profits and, while efficiency measures and a longer life time might result in increasing benefits, nothing forces the operator to extract more than what it judges economically desirable. Mining companies have no obligation to make the most out of the resource they own and reprocessing waste is not part of their core business. Furthermore, the natural resource is not a purchased input and is therefore not included in productivity calculations [54].

Countries on the other hand are more concerned with the long-term accessibility of the national resources than corporations. Furthermore, local or provincial governments become responsible for the legacies of abandoned mine sites. An improved resilience at the mine level results in the mining company having sufficient time and the means to implement its planning for closure appropriately, which would remove the burden from the government. Dobbs *et al.* [55] point out various ways in which governments would gain in building stronger partnerships with the mineral industry and could result in improved resource extraction being a key goal of mining companies.

According to Bringezu [56], an improvement in process efficiency would result in reducing metal prices, which would consequently increase material consumption. For Bringezu, this is the reason why it is necessary to look at the overall performance of a country and not only at the mine site level. Resource-rich countries need to provide incentives that enhance resource recovery from a specific deposit.

Hilson [57] investigates the barriers and enablers available to apply cleaner production practices. The author identifies policies as of prime importance and so does Ehrenfeld [58]. Legislative efforts can however be counter-productive when they change too frequently, are inconsistent, or fail to take a holistic approach. Ehrenfeld [58] also underlines that regulations tend to narrow down the spectrum of options for businesses while they could instead encourage change by opening new opportunities. Despite that, when appropriately determined, political instruments have the potential to encourage technological and systemic innovation.

5. CONCLUSIONS

This paper questioned what sustainability and especially sustainable resource management means at the mine site level. The authors argue that a sustainable mine is, above all, a mine that:

- Recognises the scarcity of the non-renewable resource by optimising extraction and minimising the amounts of valuable materials in the waste stream;
- And is able to provide long-term social and economic stability to its stakeholders by prolonging and diversifying its operations. This means being resilient to economic changes.

These two aspects are interrelated since, for instance, a more careful extraction would require more time and workforce; and social and economic stability could potentially come from the diversification in activities and operations as a result of enhanced resource recovery.

A new perspective to mine waste management is then presented as a potential solution to the sustainable mining challenge. Such management should include a planning for future material recovery and consequently the handling of the waste as a potential future resource. This could include notably the collection of leachate from impermeable storage facilities. The technical aspects are not investigated; rather the paper presents an emphasis on how this new approach could be integrated across the mine's life cycle in order to reduce waste generation at earlier stages.

Sustainable management of resources and a prevention-oriented waste management system account for collateral inputs and outputs to the operations: water and energy use, infrastructure requirements, environmental burden. Enhanced metal recovery should not produce an unacceptable increase in these flows. This is why the authors propose the use of systemic and quantitative tools such as Material Flow Analysis to assess the benefits of increased metal recovery and identify a threshold of acceptability for the consequent increase in environmental impacts.

The system boundaries for such an analysis need to be determined carefully.

In addition to setting the ground for a quantitative analysis the authors finally propose a qualitative investigation of the non-technical factors that could influence such a change in waste management practices. Three main levels are defined to frame this investigation as factors can be technical, managerial or political. The next step will be to apply this framework on a set of case studies, and analyse the results to propose the necessary conditions, such as policy changes or incentives for innovation, that would be required to unable such a framework to be practically implemented.

ACKNOWLEDGMENTS

The authors would like to acknowledge the support of the Wealth from Waste Research Cluster, a collaborative program between the Australian CSIRO (Commonwealth Scientific Industrial Research Organisation); University of Technology, Sydney; The University of Queensland, Swinburne University of Technology, Monash University and Yale University.

AUTHOR CONTRIBUTIONS

This paper and the literature review it is based on have been written and developed by Éléonore Lèbre as part of her Ph.D. project. The advisory team is composed of Glen Corder (principal advisor) and Artem Golev (secondary advisor).

CONFLICTS OF INTEREST

The authors declare no conflict of interest.

REFERENCES

1. Dudka, S.; Adriano, D.C. Environmental impact of metal ore mining and processing: A review. *J. Environ. Qual.* **1997**,*26*, 590–602.

2. Hatayama, H.; Tahara, K.; Daigo, I. Worth of metal gleaning in mining and recycling for mineral conservation. *Miner. Eng.* **2014**, *76*, 58–64.

3. Norgate, T.E.; Jahanshahi, S.; Rankin, W.J. Assessing the environmental impact of metal production processes. *J. Clean. Prod.* **2007**, *15*, 838–848.

4. Giurco, D.; Prior, T.; Mudd, G.M.; Mason, L.; Behrisch, J. *Peak Minerals in Australia: A Review of Changing Impacts and Benefits*; Institute for Sustainable Futures: University of Technology, Sydney, Australia; Department of Civil Engineering: Monash University, Monash, Australia, 2010.

5. Mason, L.; Prior, T.; Mudd, G.M.; Giurco, D.; Littleboy, A.; Boyle, T.; Fyfe, J.; White, S. Availability, addiction and alternatives: Three criteria for assessing the impact of peak minerals on society. *J. Clean. Prod.* **2011**, *19*, 958–966.

6. Prior, T.; Giurco, D.; Mudd, G.M.; Mason, L.; Behrisch, J. Resource depletion, peak minerals and the implications for sustainable resource management. *Glob. Environ. Chang.* **2012**, *22*, 577–587.

7. Meadows, D.; Randers, J.; Meadows, D. *Limits to Growth—The 30 Year Update*; Earthscan: London, UK, 2005.

8. Norgate, T.E.; Jahanshahi, S. Low grade ores—Smelt, leach or concentrate? *Miner. Eng.* **2010**, *23*, 65–73.

9. Gordon, R.B.; Bertram, M.; Graedel, T.E. Metal stocks and sustainability. *Proc. Natl. Acad. Sci.* **2006**, *103*, 1209–1214.

10. European Commission. *Towards a Thematic Strategy on the Sustainable Use of Natural Resources*; COM(2003) 572 Final; European Commission: Brussels, Belgium, 2003.

11. Ayres, R.U. Metals recycling: Economic and environmental implications. *Resour. Conserv. Recycl.* **1997**, *21*, 145–173.

12. Norgate, T.E.; Lovel, R.R. *Water Use in Metal Production: A life Cycle Perspective*; Commonwealth Scientific and Industrial Research Organisation (CSIRO): Canberra, Australia, 2004.

13. Mudd, G.M. *The Sustainability of Mining in Australia: Key Production Trends and Their Environmental Implications for the Future*; Department of Civil Engineering, Monash University: Monash, Australia; Mineral Policy Institute: Girrawheen, Australia, 2009.

14. Norgate, T.E.; Haque, N. Energy and greenhouse gas impacts of mining and mineral processing operations. *J. Clean. Prod.* **2010**, *18*, 266–274.

15. Ayres, R.U.; Ayres, L.W.; Råde, I. *The Life Cycle of Copper, Its Co-Products and By-Products*; Mining, Minerals and Sustainable Development, International Institute for Environment and Development: London, UK, 2002; p. 210.

16. Mudd, G.M. Australia's mining legacies. *Arena* **2013**, *124*, 19–24.

17. Wills, B.A. *Wills' Mineral Processing Technology—An Introduction to the Practical Aspects of Ore Treatment and Mineral Recover*, 7th ed.; Butterworth-Heinemann: Oxford, UK, 2006.

18. Laurence, D. Establishing a sustainable mining operation: An overview. *J. Clean. Prod.* **2011**, *19*, 278–284.

19. Mining, Minerals and Sustainable Development (MMSD). *Breaking new ground: Mining, Minerals and Sustainable Development project*; Earthscan Publications Ltd: London, UK, 2002; p. 476.

20. Moran, C.J.; Kunz, N.C. Sustainability as it pertains to minerals and energy supply and demand: A new interpretative perspective for assessing progress. *J. Clean. Prod.* **2014**, 1–11.

21. Walker, B.; Holling, C.S.; Carpenter, S.R.; Kinzig, A. Resilience, adaptability and transformability in social-ecological systems. *Ecol. Soc.* **2004**, *9*. Article 5.

22. Lottermoser, B.G. *Mine Wastes: Characterization, Treatment And Environmental Impacts*; Springer: Berlin, Germany, 2010.

23. Wang, T.; Muller, D.B.; Graedel, T.E. Forging the anthropogenic iron cycle-Supporting information. *Environ. Sci. Technol.* **2007**, *41*, 5120–5129.

24. Worldwatch Institute. Global municipal Solid Waste Continues to Grow. Available online: http://www.worldwatch.org/global-municipal-solid-waste-continues-grow (accessed on 5 April 2015).

25. Hudson-Edwards, K.A.; Jamieson, H.E.; Lottermoser, B.G. Mine waste: Present, past and future. *Elements* **2011**, *7*, 375–380.

26. Bringezu, S. How Much Resource Extraction Would Be Safe? Approaches to Determine a Safe Operating Space of Mineral and Biomass Use. In Proceedings of the Industrial Ecology in the Asia-Pacific Century: Interdisciplinarity science for building

sustainable industrial systems and human settlements, Melbourne, Australia, 17–19 November 2014.

27. Global Reporting Initiative (GRI). *G4—Sustainability Reporting Guidelines. Part 1: Reporting Principles and Standard Disclosures*; Global Reporting Initiative (GRI): Amsterdam, The Netherlands, 2013; p. 94.

28. Brattebø, H.; Røine, K.; Opoku, H.; Ehrenfeld, J.R.; Kjelstrup, S.; Bohne, R.; Skonhoft, A.; Hermansen, J.; Karlsson, S.; Holmberg, J.; *et al. Textbook in Industrial Ecology. Introduction to Industrial Ecology—Theory, Methods and Applications*; Norwegian University of Science and Technology: Trondheim, Norway, 2007.

29. Simate, G.S.; Ndlovu, S. Acid mine drainage: Challenges and opportunities. *J. Environ. Chem. Eng.* **2014**, *2*, 1785–1803.

30. Unger, C.; Lechner, A.M.; Glenn, V.; Edraki, M.; Mulligan, D.R. Mapping and prioritising rehabilitation of abandoned mines in Australia. In Proceedings of the Life-of-Mine Conference, Brisbane, Australia, 10–12 July 2012.

31. International Commission On Large Dams (ICOLD). *Bulletin 121—Tailings Dams, Risk of Dangerous Occurrences. Lessons Learnt from Practical Experiences*; International Commission on Large Dams: Paris, France, 2001.

32. WISE. Chronology of major tailings dam failures. Available online: http://www. wise-uranium.org/mdaf.html(accessed on 8 April 2015).

33. Azam, S.; Li, Q. Tailings dam failures: A review of the last one hundred years. *Geotechnical News* **2010**, 50–53.

34. Fraser Institute. How are waste materials managed at mine sites? Available online: http://www.miningfacts.org/Environment/How-are-waste-materials-managed-at-mine-sites/ (accessed on 10 April 2015).

35. Reid, C.; Becaert, V.; Aubertin, M.; Rosenbaum, R.K.; Deschenes, L. Life cycle assessment of mine tailings management in Canada. *J. Clean. Prod.* **2009**, *17*, 471–479.

36. Fonseca, A.; McAllister, M.L.; Fitzpatrick, P. Measuring what? A comparative anatomy of five mining sustainability frameworks. *Miner. Eng.* **2013**, *46*, 180–186.

37. Dold, B. Sustainability in metal mining: From exploration, over processing to mine waste management. *Rev. Environ. Sci. Bio/Technol.* **2008**, *7*, 275–285.

38. Lottermoser, B.G. Recycling, reuse and rehabilitation of mine wastes. *Elements* **2011**, *7*, 405–410.

39. Folke, C.; Carpenter, S.; Elmqvist, T.; Gunderson, L.; Holling, C.S.; Walker, B. Resilience and sustainable development: Building adaptive capacity in a world of transformations. *AMBIO J. Hum. Environ.* **2002**, *31*, 437–440.

40. Boyle, P.; Wilson, G.; Jones, B. *Rasp Mine, Zinc-Lead-Silver Project—Waste Management Plan V2*; Broken Hill Operations PTY Ltd.: Broken Hill, Australia, 2012.

41. APS Group Scotland. *Guidance on applying the waste hierarchy*; The Scottish Government: Edinburgh, UK, 2013; p. 49.

42. Van Ewijk, S.; Stegemann, J. Limitations of the waste hierarchy for achieving absolute reductions in material throughput. *J. Clean. Prod.* **2014**.

43. Golev, A. Application of industrial ecology principles for enhanced resource efficiency in heavy industrial areas. Ph.D. Thesis, The University of Queensland, St Lucia, Australia, December 2012.

44. Eurostat. *Economy-Wide Material Flow Accounts (Ew-Mfa)—Compilation Guide*; European Commission: Brussels, Belgium, 2013.

45. McLellan, B.C.; Corder, G.D.; Giurco, D.; Green, S. Incorporating sustainable development in the design of mineral processing operations—Review and analysis of current approaches. *J. Clean. Prod.* **2009**, *17*, 1414–1425.

46. Hilson, G. Defining "cleaner production" and "pollution prevention" in the mining context. *Miner. Eng.* **2003**, *16*, 305–321.

47. Namboothiri, S.; Taylor, M.P.; Chen, J.J.J.; Hyland, M.M.; Cooksey, M. Aluminium production options with a focus on the use of a hydrogen anode: A review. *Asia-Pac. J. Chem. Eng.* **2007**, *2*, 442–447.

48. Ayres, R.U.; Ayres, L.W.; Masini, A. An application of exergy accounting to five basic metal industries. *Sustain. Met.s Manag.* **2006**, *19*, 141–194.

49. CRC ORE. Optimising resource extraction—Desktop study of differential blasting potential. Available online: http://www.crcore.org.au/site-studies-new/escondida (accessed on 20 October 2015).

50. Steen, J. Digging Deeper into the Productivity Crisis in the Global Mining Industry: Causes and Prospects for Improvement. In Proceedings of the SMI RHD Conference. Resource Extraction: Challenges and Opportunities, Sustainable Minerals Institute, University of Queensland, St Lucia, Australia, 21 November 2014.

51. BioteQ. Bioteq, environmental technologies. Key sectors. Available online: http://www.bioteq.ca/key-sectors/#mining(accessed on 20 October 2015).

52. Edraki, M.; Baumgartl, T.; Manlapig, E.; Bradshaw, D.; Franks, D.M.; Moran, C.J. Designing mine tailings for better environmental, social and economic outcomes—A review of alternatives approaches. *Clean. Prod.* **2014**, *84*, 411–420.

53. Mitchell, P.; Steen, J.; Moran, C.J.; Bradbrook, M.; Henderson, C.; MacAulay, S.; Higgins, L.; Kastelle, T.; Kunz, N.C.*Productivity in Mining: Now Comes the Hard Part. A Global Survey*; Sustainable Minerals Institute, University of Queensland: St Lucia, Australia; Imperial College London: London, UK, 2014.

54. Topp, V.; Soames, L.; Parham, D.; Bloch, H. *Productivity in the Mining Industry: Measurement and Interpretation*; Productivity Commission Government of the Commonwealth of Australia: Canberra, Australia, 2008.

55. Dobbs, R.; Oppenheim, J.; Kendall, A.; Thompson, F.; Bratt, M.; van der Marel, F. *Reverse the curse: Maximizing the Potential of Resource-Driven Economies*; McKinsey Global Institute: New York, NY, USA, 2013.

56. Bringezu, S.; Bleischwitz, R. *Sustainable Resource Management: Global Trends, Visions and Policies*; Greenleaf Publishing Limited: Sheffield, UK, 2009.

57. Hilson, G. Barriers to implementing cleaner technologies and cleaner production (CP) practices in the mining industry: A case study of the americas. *Miner. Eng.* **2000**, *13*, 699–717.

58. Ehrenfeld, J.R. Industrial ecology: A strategic framework for product policy and other sustainable practices. In Proceedings of the Green Goods: The Second International Conference and Workshop on Product Oriented Policy, Stockholm, Sweden, September 1994.

4

CHAPTER

Characterization of Black Sand Mining Activities and Their Environmental Impacts in the Philippines Using Remote Sensing

Estelle Chaussard [1,*] and Sara Kerosky [2]

[1]Department of Geology, State University of New York at Buffalo, Buffalo, NY 14260, USA

[2]Department of Political Science, University of California, San Diego, La Jolla, CA 92110, USA

ABSTRACT

Magnetite is a type of iron ore and a valuable commodity that occurs naturally in black sand beaches in the Philippines. However, black sand mining often takes place illegally and increases the likelihood and magnitude of geohazards, such as land subsidence, which augments the exposure of local communities to sea level rise and to typhoon-related threats. Detection of black sand mining activities traditionally relies on word of mouth, while measurement of their environmental effects requires on-the-ground geological surveys, which are precise, but costly and limited in scope. Here we show that systematic analysis of remote sensing data provides an objective, reliable, safe, and cost-effective way to monitor black sand mining activities and their impacts. First, we show that optical satellite data can be used to identify legal and illegal mining sites and characterize the direct effect of mining on the landscape. Second, we demonstrate that Interferometric Synthetic Aperture Radar (InSAR) can be used to evaluate the environmental impacts of black sand mining despite the small spatial extent of the activities. We detected a total of twenty black sand mining sites on Luzon Island and InSAR ALOS data reveal that out of the thirteen sites with coherence, nine experienced land subsidence at rates ranging from 1.5 to 5.7 cm/year during 2007–2011. The mean ground velocity map also highlights that the spatial extent of the subsiding areas is 10 to 100 times larger than the mining sites, likely associated with groundwater use or sediment redistribution. As a result of this subsidence, several coastal areas will be lowered to sea level elevation in a few decades and exposed to permanent flooding. This work demonstrates that remote sensing data are critical in monitoring the development of such activities and their environmental and societal impacts.

Keywords: black sand mining; magnetite; Philippines; optical remote sensing; InSAR; subsidence

1. INTRODUCTION

There are two types of iron ore most commonly used in the production of steel worldwide: hematite (Fe_2O_3, 69.9% Fe) and magnetite (Fe_3O_4, 72.4% Fe). Hematite deposits have been easily accessible and largely exploited until the last decade. Depletion of these deposits led to an increased demand for magnetite as the need for high-quality iron products continues [1,2]. In the Philippines, magnetite occurs naturally in black sand, which results from the weathering and erosion of metamorphic and igneous rocks. Black sand accumulates in streams and drainage systems and is carried and deposited onto beaches. As a result, much of the Philippine coast is composed of black sand beaches. In response to the demand for magnetite, black sand mining and processing activities have significantly increased in recent years and the extracted magnetite is largely exported to China's steel mills [2,3].

Black sand mining occurs both legally and illegally but monitoring presents significant technical challenges and social costs. Reports from local civil service organizations [4] and the recent indictment of a provincial governor on charges of graft over black sand mining [5] suggest these operations persist due to corruption. Thus, ground monitoring of illegal activities is not only hard to apply at large scales but also dangerous for the involved individuals. Here we test whether remote sensing data can be used to identify and measure the scope of mining activities in the Philippines.

Black sand mining disturbs marine and coastal ecosystems and increases erosion and associated geohazards. Coastal black sand dunes serve as habitat for small animals and plants whose loss implies a threat to their predators. These dunes are also natural barriers to salt water, without them the sea often floods inland at high tides. The mining activities increase coastal erosion not only by directly removing sand but also by disrupting the sediment budget, often depriving areas in the down-current direction of their sand input. As a result, coastal erosion often continues to affect the areas even decades after cessation of the mining activities. This removal of material and associated erosion also likely results in land subsidence, which makes local communities particularly vulnerable to floods, damage from seasonal typhoons, and sea level rise. Additionally, black sand operations are often associated with groundwater extraction, which results in lowering of the water levels and increases the land subsidence. Subsidence is expected to continue after cessation of the water extraction due to residual compaction [6]. Groundwater is used for processing (to wash sand, suppress dust, transport sand as a slurry, *etc.*) and for dewatering purposes for black sand to be extracted below the water table surface and for removing the water adhering to the sand grains [7].

We will use Interferometric Synthetic Aperture Radar (InSAR) to evaluate the vulnerability of local communities and the environment to black sand mining. InSAR has been used to successfully detect ground deformation associated with a number of geohazards, from earthquakes, to volcanic eruptions and land subsidence due to groundwater extraction [8,9,10,11,12,13,14,15]. Here, we test whether the mining activities result in any detectable ground deformation despite their small spatial extent and evaluate for the first time the environmental effects of black sand mining with a high spatial resolution. To protect coastal ecosystems and communities, black sand mining is illegal in the Philippines within 200 meters of the shore [16] but the oversight of mining activities is poor. Using the high spatial resolution of the InSAR data we will evaluate the spatial impact of the mining activities. We will also characterize the magnitude of the associated subsidence, which will enable estimation of the time until the affected coastal areas are lowered to sea level elevation and exposed to flooding.

In this paper we aim to (1) demonstrate the value of remote sensing for monitoring of black sand mining activities and their environmental impact; (2) contribute a dataset of known and suspected mining areas in the Philippines; and (3) identify the areas of particular environmental and social concern subjected to increased hazards.

2. METHODS

2.1. Identification of Mining Sites

The study of illegal activities presents a unique challenge since the behavior is intended to remain hidden. We use a multi-pronged approach to identify occurrences of black sand mining in the Philippines. First, we collected a list of known mining sites from various sources, including contacts on the ground, online news articles reporting either illegal mining activity or community protests related to illegal mining, and reports issued by the Philippine Mines and Geosciences Bureau (MGB) listing mining permits. Permits do not guarantee mining activity, but increase the likelihood by granting formal permissions to mining companies to work in the area. We also included potential sites based on iron ore or magnetite deposits reported in the United States Geological Survey (USGS) Mineral Resource Data System.

We then use optical satellite imagery to confirm the occurrence and precise locations of the mining sites, to characterize landscape changes over time, and to quantify the spatial extent of the mining activities. We rely on DigitalGlobe images provided by Google Earth and its historical imagery archive. These optical satellite data complement on-the-ground data by offering an objective form of measurement and enhanced spatial coverage of legal and illegal activities alike.

2.2. Evaluation of Environmental Effects

To identify the environmental impacts of black sand mining activities we rely on Interferometric Synthetic Aperture Radar (InSAR). InSAR allows us to characterize the extent and amplitude of land subsidence in areas with mining occurrence by enabling measurement of ground displacement in the radar line-of-sight (LOS) direction of a SAR satellite between different passes of the satellite over the same area [8]. The Advanced Land Observing Satellite-1 (ALOS) of the Japanese Space Exploration Agency (JAXA) acquired SAR data with global coverage between late 2006 and mid 2011 on a 46-day repeat orbit, imaging most of the world's continents up to 25 times [17]. We use ALOS data made available by the Alaska Satellite Facility (ASF) and process 14 ALOS frames on six tracks (Figure 1). Six SAR acquisitions are available on tracks 446, 447, and 448; nine acquisitions on track 449; and twenty-one acquisitions on tracks 445 and 450.

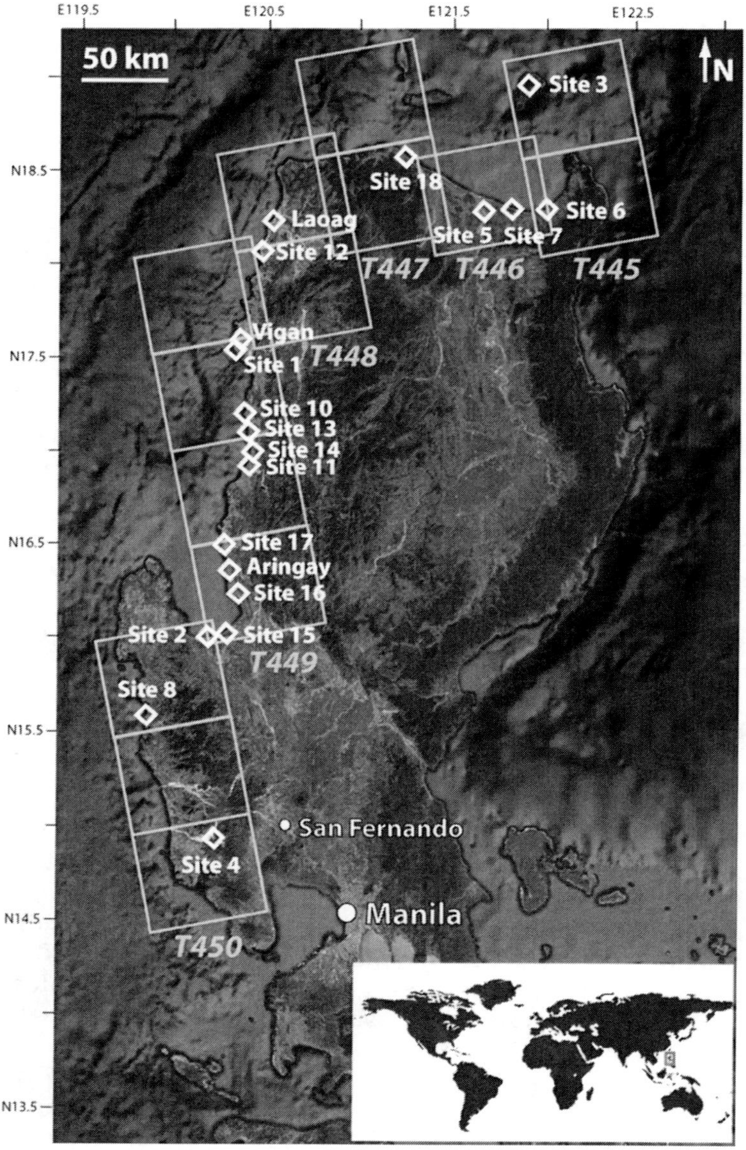

Figure 1. Northern island of the Philippines (Luzon Island) with the identified sites of interest (Table 1) marked by white diamonds, and the location of the ALOS PALSAR data marked by the grey squares. The track numbers are specified below the respective tracks. Larges cities are indicated for reference with white circles.

Table 1. Compilation of the sites of interest and their information. The first column shows the site identification, the second column its longitude and latitude, the third column whether the mining activities are legal (y) or illegal (n) or other type of activities, and the fourth column shows the source of information. The fifth column shows whether the InSAR mean velocity map has coherence in the area and the sixth column shows the detected LOS subsidence rates in cm/year. The last two columns show the name of the municipality with its population and the name of the province, respectively. The rows in grey correspond to sites out of the InSAR survey area or not associated with black sand mining activities.

Site	Lat/Long	Legal (y/n)	Source of Info.	Coher-Ence (y/n)	LOS Subs. cm/year	Municipality (Population)
1	17°32′37″N 120°21′25″E	n	Ground contacts	n		Caoayan (18,551)
2	16° 2′10″N 120°13′50″E	n	News articles	y	4.8	Lingayen (98,740)
3	18°56′53″N 121°55′12″E	n	News articles	n		n/a
4	14°58′54″N 120°16′50″E	n	News articles	y	1.3	San Marcelino (31,879)
5	18°16′27″N 121°40′30″E	n	News articles	n		Camalaniugan (23,404)
6	18°17′35″N 122° 0′53″E	n	News articles	n		Gonzaga (36,303)
7	18°17′27″N 121°49′46E	n	News + contacts	n		Buguey (28,455)
8	15°37′24″N 119°54′50″E	n	News articles	y	1.6	Candelaria (25,020)
9	9°31′51″N 123° 9′18″E	n	News articles	Out of scope		Tanjay City (79,098)
10	17°12′22″N 120°24′56″E	y	MGB permit	y	3.0	Candon CIty (57,884)
11	16°56′27″N 120°26′8″E	y	MGB permit	y	0	Tagudin (38,122)
12	18° 3′39″N 120°29′0″E	y	MGB permit	n		Paoay (23,956)
13	17° 6′46″N 120°26′8″E	y	MGB permit	y	4.3	Santa Lucia (24,981)
14	17° 1′4″N 120°27′3″E	y	MGB permit	y	0	Santa Cruz (37,911)
15	16° 3′7″N 120°19′57″E	y	MGB permit	y	4.3	Dagupan (163,676)
16	16°15′49″N 120°23′38″E	y	MGB permit	y	0	Santo Tomas (35,999)
17	16°31′6″N 120°18′46″E	y	MGB permit	y	0	Bauang (70,735)
18	18°33′57″N 121°14′51″E	y	MGB permit	n		Sanchez Mira (23,257)
Laoag	18°13′59″N 120°30′57″E	no mining	USGS rec. 10304314	y	0	Laoag City (104,904)
Vigan	17°35′59″N 120°22′57″E	no mining	USGS rec. 10134140	n		Vigan City (49,747)
Aringay	16°22′59″N 120°20′57″E	no mining fields	USGS rec. 10183093	y	3.9	Aringay (44,949)
InSAR 1	17°21′41″N 120°27′23″E	n		y	2.5	Santa Maria (28,597)
InSAR 2	16°27′10″N 120°34′33″E	other mining		y	4.5	La Trinidad (107,188)
InSAR 3	15°34′17″N 120° 5′37″E	n		y	1.8	Masinloc (44,342)
InSAR 4	15° 7′56″N 120°20′8″E	volcanic deposits		y	1.2	Botolan (54,434)
InSAR 5	14°54′15″N 120°34′48″E	fields		y	3.5	Lubao (150,843)
InSAR 6	14°41′36″N 120°33′37″E	n		y	2.6	Balanga (87,920)

First, ground displacement in the radar LOS direction is obtained from the phase difference of two SAR images acquired over the same area at different times (interferogram) [8]. We use the ROI_PAC software [18] to produce 40 to 250 interferograms on each track. We remove topographic contributions using the Shuttle Radar Topography Mission (SRTM) 1-arc second digital elevation model [19]. We co-register the interferograms of each frame to a master image, use the statistical-cost network-flow algorithm for phase unwrapping (SNAPHU; [20]), and reference all interferograms to the same pixel. We subtract a ramp from each interferogram to remove any long wavelength signal (larger than 10 km, associated with orbital errors and interseismic deformation due to the subduction zone) so the obtained velocity maps highlight only localized deformation.

To precisely track ground deformation between the first and the last SAR acquisition we use the SBAS time series technique [21]. The interferograms produced on each track are inverted to retrieve surface displacement through time relative to the first acquisition. We use only interferograms with small spatial and temporal baselines (*i.e.* position of the satellite close in space perpendicularly to the LOS at the different times (<1.6 km), and data acquired less than a year apart) to maintain high coherence (Figure 2). Since the PALSAR sensor onboard ALOS was L-band (wavelength of 23.6 cm), coherence is typically maintained longer than for X- and C-band sensors. Thus, we also add interferograms that span more than a year and cover the same season. The phase due to the digital elevation model (DEM) errors in SBAS time series is proportional to the perpendicular baseline history and, thus, it must be accounted for and corrected (Figure 2, right). We correct DEM errors in the in the time-domain following the method of *Fattahi and Amelung,* [22]. Lastly, we perform the final pixel selection based on a temporal coherence threshold of 0.7 to eliminate pixels affected by phase-unwrapping errors [23,24,25].

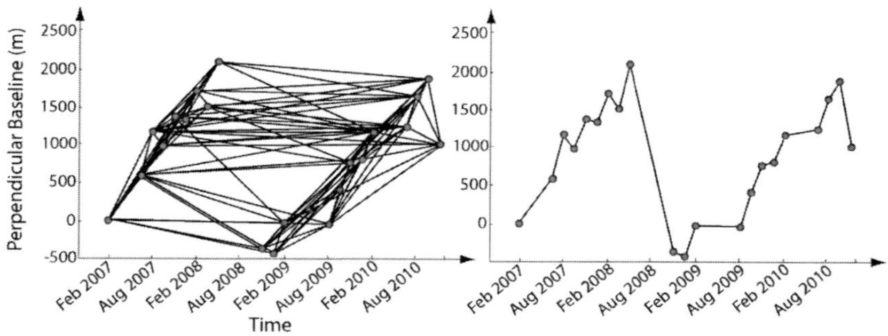

Figure 2. Example for Track 445 of a perpendicular baselines-temporal baselines plot of a triangulated network of interferometric pairs. The dots show the SAR acquisitions dates. The **left** plot shows the interferograms produced on this track (black lines) after applying the baseline thresholds. The **right** plot shows the perpendicular baseline history of the set of SAR acquisitions, which is proportional to the DEM errors.

The remaining signal contains noise contributions from atmospheric delay due to variation in water vapor content at different acquisition times. To evaluate the amplitude of the noise we follow the method of Chaussard *et al.* [12] and produce time series for pixels in non-deforming areas (metamorphic rocks). These time series provide a constraint on the amplitude of the atmospheric contamination at each epoch. Atmospheric noise can be as large as several centimeters in a single interferogram but is highly variable temporally and spatially, which implies that in non-deforming areas the trend through the time series has approximately a null slope. Thus, atmospheric noise corresponds to deviation from a longer trend, which represents deformation. Accordingly, we derive our estimates of ground subsidence using linear regressions through the time series, which are less affected by the atmospheric noise than single interferograms.

Chaussard *et al.* [13] showed that land subsidence due to groundwater pumping in Indonesia will result in flooding of urban areas within a few decades, this process being orders of magnitude larger than the rates of sea level rise. Here we calculate the approximate time for the mining sites areas associated with land subsidence to be at sea level elevation and, thus, permanently flooded and present this "time until flooding" so decision-makers can gain a sense of the time scales involved. This is the last step of our method, which is summarized in a flow chart in Figure 3. We evaluate the elevation based on the ASTER GDEM V2 Digital Elevation Model (DEM) and on the Global Multi-Resolution Topography (GMRT) [26]. These 30 m DEMs lead to uncertainties of ~2 m on the relative elevation estimates for the mining sites. The time until flooding is equal to $E/(S+R)$ with E the current ground elevation in cm, S the rate of subsidence in cm/year, and R is rate of sea level rise in cm/year. The two closest cities of Luzon Island that have current tide gauges monitoring (Legaspi and Davao) by the National Oceanic and Atmospheric Administration [27] show a mean sea level rise trend of 5.38 mm/year and 5.32 mm/year, respectively. These tide gauges are south of the region studied and can be affected by local subsidence e.g., [28]. Thus, we verify that these sea level rise rates are in agreement with spatial altimetry data and non-Boussinesq models.*Moon* and *Song* [29] showed that between 1993–2010 Luzon Island was affected by a sea level rise of ~5 mm/year in agreement with the tide gauges data. Worldwide maps of sea level change based on satellite altimetry [30] show even faster sea level rise rates in the western Pacific at up to 10 mm/year east of Luzon island due to La Niña conditions (enhanced trade winds and cooling in the eastern tropical Pacific, corresponding to the absence of prolonged El Niño events in 1997–1998 and 2011). We, thus, consider that the tide gauge data are a good representation of the regional sea level rise for long time periods and use an average linear sea level rise of 5.35 mm/year in the Philippines.

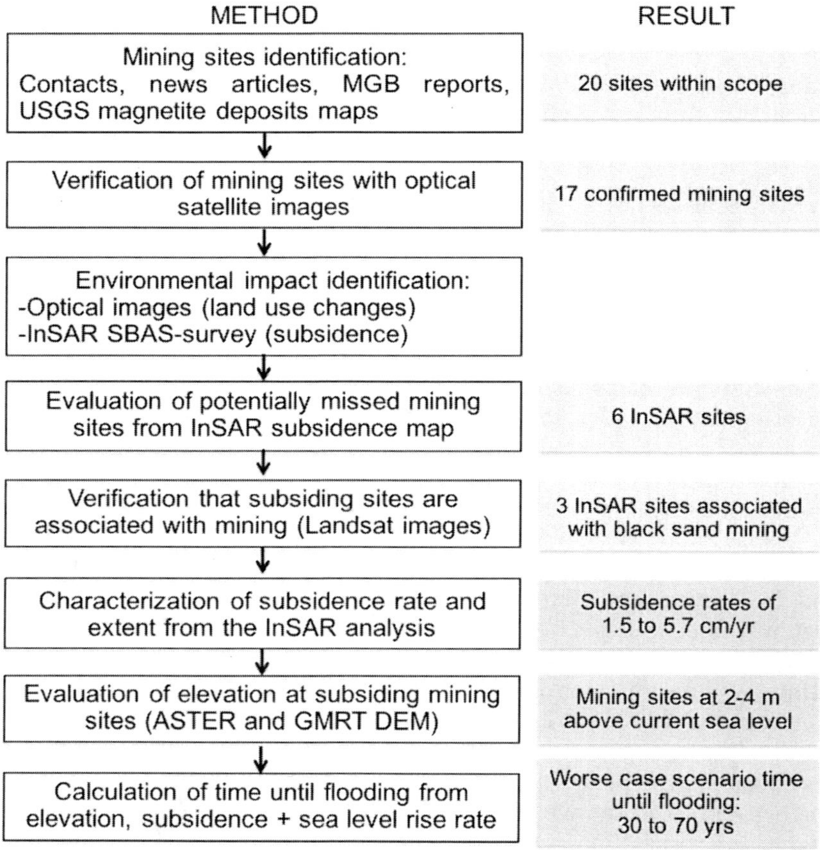

Figure 3. Flow chart summarizing the methods used and results obtained in our survey to characterize black sand mining activities and their environmental impacts in the Philippines. Optical images and InSAR time series data are combined and show that remote sensing data provides an objective, reliable, safe, and cost-effective way to monitor black sand mining activities. Optical satellite data lead to identification of legal and illegal mining sites and InSAR allows us to evaluate the environmental impacts of black sand mining in terms of land subsidence. The InSAR time series method is described in detail in the method section.

We consider that the land subsidence rates observed are linear in time and persist even after the mining activities are over. This approximation is in part justified by the fact that coastal erosion often continues to affect the areas even decades after cessation of the activities due to sediment redistribution and by the fact that the land subsidence due to mining-related groundwater extraction is expected to continue due to residual compaction even after cessation of pumping [6]. However, the linear extrapolation for time scales of half a century to a century may not be entirely appropriate as the mining sites are small and the aquifers involved are likely shallow and may consolidate leading to the subsidence to slow and stop. Thus, using the linear approximation may overestimate of the subsidence for long time scales and lead to lower end members of the values of time until flooding. We choose to give these lowest time values rounded to one standard deviation and present them as a "worst-case scenario" estimates representing what will happen if conditions are maintained as they are currently for decades. The development of new mining sites would result in new subsidence areas and could contribute in making these "worst-case scenario" estimates closer to being "average estimates" as sediment redistribution and aquifer compaction could become more wide-spread.

3. RESULTS

3.1. Identification of Mining Sites

We identified a total of twenty potential mining sites along the northern and northwestern coast of Luzon Island (Table 1). Precise coordinates for two mining sites were obtained from geo-tagged photos provided by contacts on the ground. Coordinates for six mining locations were identified from news articles referring to illegal black sand mining. Nine potential locations were identified from mining permits granted by the Mining and Geosciences Bureau in 2007, which were confirmed as mining sites with optical imagery. Locations of three coastal magnetite deposits were obtained from the USGS Mineral Resources Data System. However, for these magnetite deposits, the historical optical images did not reveal any clear mining activities for the 2003–2015 time span covered by the data available via Google Earth. Thus, out of the twenty potential sites, seventeen were confirmed as black sand mining sites with optical images (black labels in Table 1).

We use optical imagery to evaluate the amplitude of changes in landscape before and after mining is developed (Figure 4). We confirm that mining sites are very localized with perimeters on the order of a few hundreds of meters. Most of the sites are located on the coast, within the black sand dunes on the inland side of the beach (Figure 4). We also often notice the presence of water ponds nearby the extraction sites (Figure 4), which confirms the intensive use of water in the extraction process and the likely pumping of groundwater.

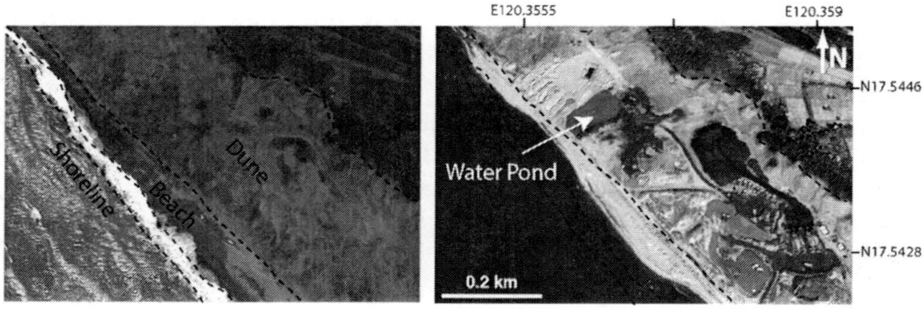

Figure 4. Example of optical images illustrating the coastal landscape before (**left**) and after (**right**) the development of black sand mining activities (site 1). The mining activities are located within the black sand dunes, on the inland side of the beach. The white arrow highlights a water pond developed near the mining site.

The main limiting factor for characterizing the development of new mining sites is the low temporal sampling of the optical images available through Google Earth. Nevertheless, our results show that optical imagery is useful for characterizing legal and illegal mining activities, their scope, and their extent. With the increasing accessibility of high-resolution satellite data, monitoring of black sand mining activities from optical images could become a routine practice. In this study, we relied on manual evaluation of changes via visual inspection of optical images, and limited our study to cases with precise coordinates from our list of potential sites. In the future, the development of automatic classification using tools available through the Google Earth Engine could lead to automated and systematic evaluation of mining sites as soon as the optical imagery becomes available.

3.2. Evaluation of Environmental Effects

The mean InSAR velocity map produced (Figure 5) shows the speed at which the ground is moving with a high spatial resolution and for the entire coastal area of the Luzon Island. We produce a Google Earth product with the velocity maps to illustrate the ground deformation associated with black sand mining activities (Supplementary Material), which enables browsing of the large spatial area covered and easy sharing of our results with partners on the ground.

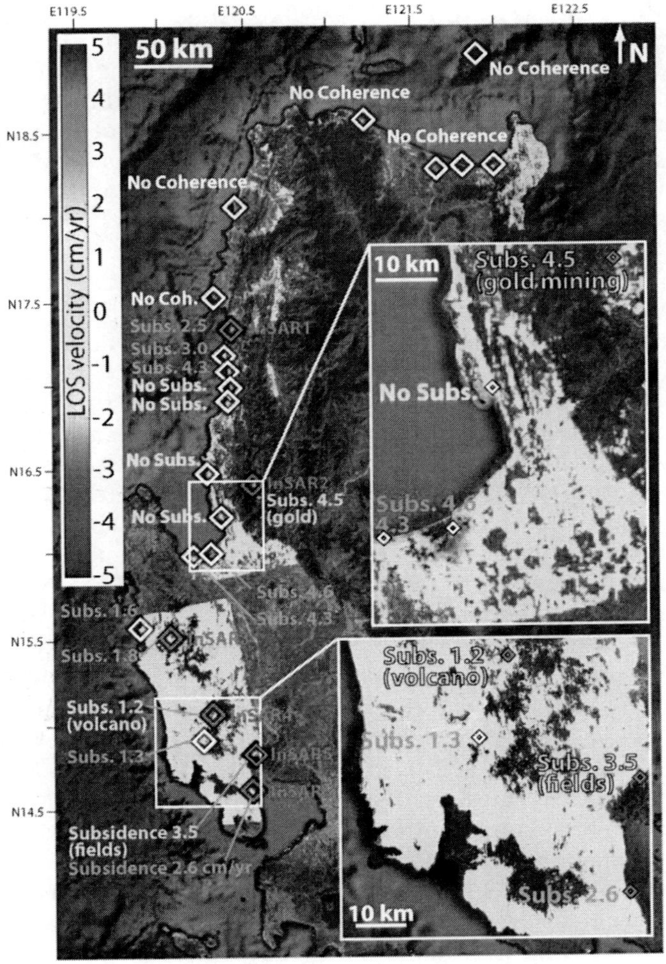

Figure 5. InSAR mean LOS ground velocity map. The background colors show the speed at which the ground is moving. Vertical motion corresponds to 1.2 times the LOS velocity. Blue colors show motion away from the satellite (subsidence) and red colors motion towards the satellite (uplift). White diamonds show the identified mining sites and red diamonds the subsiding sites detected in the velocity map but not originally identified (labeled InSAR sites). The insets show examples of subsiding areas (blue colors). White labels show sites with no coherence in the InSAR mean velocity map, cyan labels are sites with coherence but no detected subsidence, orange labels show subsiding mining sites with the subsidence rates in cm/year, and gray labels are areas with subsidence but not associated with black sand mining.

Decorrelation is the largest limitation of using InSAR for detection of subsidence associated with mining activities (Figure 5). Since the PALSAR sensor onboard ALOS was L-band (wavelength of 23.6 cm), it enables measurements through vegetation [12,15], but interferometry remains limited by temporal decorrelation of the radar signal; that is, changes in the characteristics of the ground between each satellite acquisition. When the ground characteristics change too much between satellites passes deformation cannot be evaluated. We found that seven out of the seventeen black sand mining sites surveyed with InSAR do not have sufficient coherence (Table 1, and Figure 5 sites marked as "no coherence"). The majority of the mining sites are located in compressible alluviums, which outline most of Luzon Island and in coastal areas or near river beds with relatively flat topography and similar types of vegetation. Thus, differences in geology and terrain are unlikely to be the main contribution to the observation of decorrelation at some mining sites but not others. Additionally, some inland sites located in mountainous areas preserve correlation (e.g., site 4 and InSAR site 3) while coastal sites to the north lose correlation. The number of SAR acquisitions available likely influences the coherence, as the more frequent the acquisitions are, the more likely ground characteristics are to be maintained between each satellite pass. The ALOS satellite was acquiring on a 46-day orbit but acquisitions were not systematically collected worldwide. As a result, the number of acquisitions per track varies. The two tracks with the most SAR acquisitions are the tracks 445 and 450 (twenty-one dates) and track 449 (nine dates). Three tracks have a lower amount of SAR acquisitions: track 446, 447, and 448 (six dates). We notice that these three tracks with the lowest amount of SAR dates have a high number of sites suffering from decorrelation (three on track 448, one on track 447, and two on track 446) while tracks with higher numbers of SAR acquisitions (449 and 450) have no sites suffering from decorrelation. This suggests that with more frequent SAR acquisitions (with for example the 14-day repeat of the L-band ALOS-2 satellite, and the planned 12-day repeat of the NiSAR mission) more sites would have coherence and could be monitored for subsidence. The exception being track 445, which has a large amount of SAR acquisitions but where two mining sites suffer from decorrelation. It is worth noting that the overall correlation for track 445 is more than for tracks 446, 447, and 448, the majority of the area being correlated. The decorrelation at site 3 (small island north of Luzon) can be explained by the significant topography and vegetation, which differentiate it from most of the other coastal sites. One explanation for the occurrence of decorrelation in areas with large amount of SAR acquisitions (such as site 6 on track 445) would be either very large deformation or recent mining development but, unfortunately, we do not have access to detailed mining data on the timing of the activities and their amplitude to investigate which parameters control the loss of coherence.

For each site with coherence we calculate a mean subsidence rate, which corresponds to the slope of the best-fitting linear regression through the time series to limit the effect of atmospheric noise (Figure 6). The time series correspond to averages over the pixels experiencing subsidence. Out of the ten mining sites with coherence, four show no clear sign of deformation (from north to south sites 14, 11, 17, and 16) (Figure 5 and Figure 6 first and second rows). Six out of the ten sites with coherence show subsidence at rates ranging from 1.3 cm/year to up to 4.6 cm/year in the Radar LOS (Table 1, Figure 5 andFigure 6). Since InSAR is more sensitive to vertical than horizontal movements we convert LOS (dLOS) into vertical displacement (dv) for every time series using the average ALOS incidence angle ($\theta = 34.3°$): dv = dLOS/cosθ. Vertical ground displacement is 21% more than LOS displacement, *i.e.*, 1 cm of LOS displacement corresponds to 1.2 cm of vertical displacement. The vertical subsidence associated with mining thus reaches up to 5.7 cm/year.

The InSAR velocity map also reveals subsidence in six additional sites (red diamonds on Figure 5) not originally identified. Out of these six sites, four were confirmed as mining sites with optical images (sites InSAR-1, 2, 3, and 5), but one was associated with mining activities other than black sand (gold extraction, site InSAR-2, located east of site 17) (Figure 5,Table 1). Subsidence in the radar LOS at the gold mining site reaches up to 4.5 cm/year (Figure 6; 5.4 cm/year vertically) and at the three additional black sand mining sites up to 2.6 cm/year (3.1 cm/year vertically) (Figure 6). These results show that InSAR is a good complement to other methods for identifying mining sites, but the distinction between black sand mining and other processes cannot be made based solely on the InSAR data (Figure 3). For example, based on optical images we confirmed that subsidence was also detected on volcanic deposits (InSAR site 4) and associated with agricultural activities (InSAR site 5 and Aringay).

The InSAR mean velocity map provides both a high spatial coverage (entire northwest coast of Luzon Island) and a high spatial resolution, which allows us to evaluate the spatial extent of the subsiding areas compared to the mining sites extent characterized from optical images (Figure 7). We observe that the subsidence patterns extend much beyond the mining sites with an approximate ~10- to 100-times larger spatial extent (mining sites mostly extend on less that 0.05 km^2 while subsiding areas cover up to 5–10 km^2, Figure 7). This larger extent of the subsiding

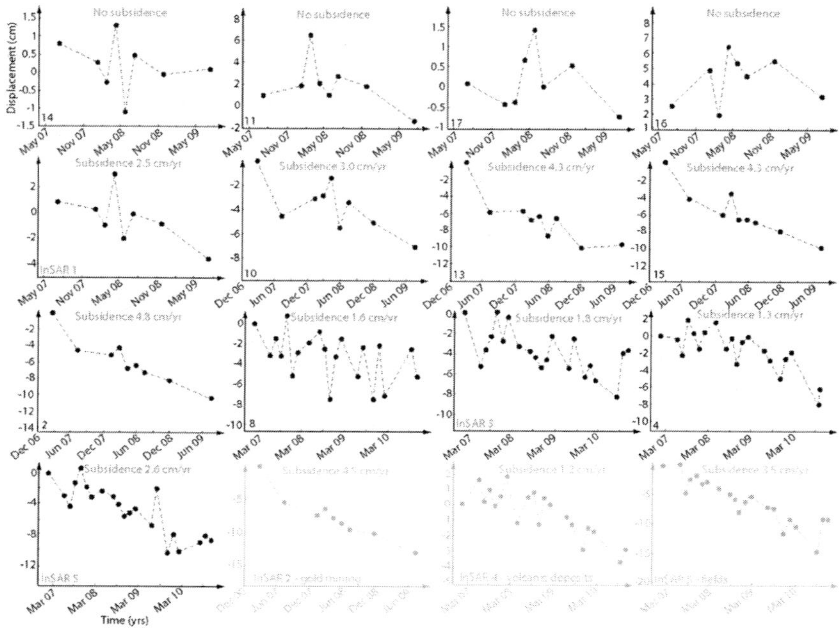

Figure 6. Ground displacement (in cm) as a function of time for the sites shown inFigure 5. Each dot marks a SAR acquisition time, the subsidence rates labeled correspond to the slope of the best fitting linear regression though each time series. The site names are labeled at the bottom left of each quadrant. The top four quadrants correspond to black sand mining sites with no clear deformation. The other sites are ordered from north to south. The quadrants labeled "InSAR" are sites displaying subsidence but which were not originally identified as black sand mining sites. The last three InSAR sites are not associated with black sand mining activities (gray).

areas likely reflects groundwater usage or sediment redistribution around the affected areas. We also notice that the spatial extent of the subsiding areas associated with the black sand mining sites is small compared to the extent of the land subsidence associated with other processes such as ground water extraction, e.g., [13,14,15]; oxidation of peatlands, e.g., [31]; natural sediment compaction e.g., [32]; slope motion e.g., [33]; sinkholes e.g., [34]; and hydrothermal or magmatic activity, e.g., [12,35]. Thus, the small spatial extent of the land subsidence associated with black sand mining activities can be used to detect black sand mining sites and isolate this process from others.

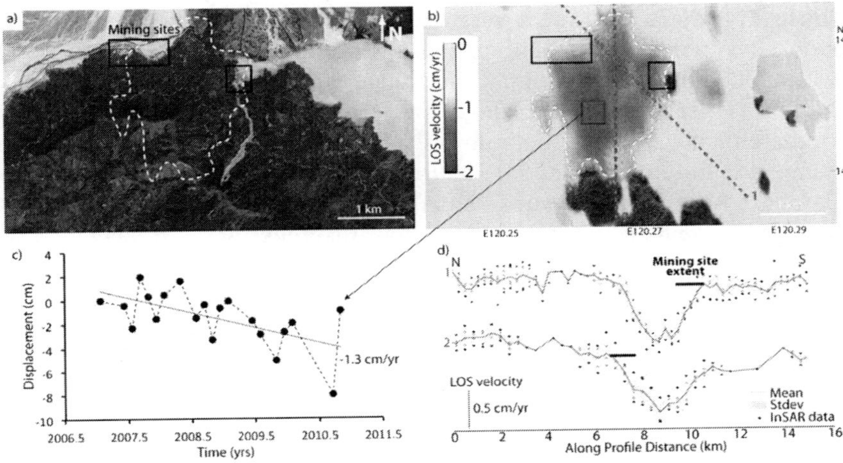

Figure 7. Example of an optical satellite image (**a**) showing the location and extent of the mining activities (black rectangles) and the corresponding InSAR mean velocity (**b**) for Site 4; (**c**) Mean times series for the maximum subsiding area and its best-fitting linear subsidence rate; (**d**) Example of profiles through the subsiding area showing the LOS velocity as a function of the distance along the profile (location shown with red dashed lines on (b). The top profile (1) is oriented NW-SE and the bottom profile (2) N-S. The black dots show the InSAR data, the gray area the standard deviation, and the red line the mean value. The horizontal black lines show the spatial extent of the mining site projected on the profile compared to the extent of the subsiding area.

We calculate, for the coastal sites with detected subsidence, the time until which the elevation will reach sea level and, thus, the area will be permanently flooded (Table 2). We follow the method described earlier and provide worse-case scenario estimates considering linear rates of land subsidence and of sea level rise (5.35 mm/year) to show what will happen if the conditions are maintained as they are currently for decades. We use the lowest bound of the current elevation provided by the DEM for these estimates and round them to one standard deviation. Sites with subsidence rates of 1.8 and 3 cm/year are projected to be underwater in ~50–70 years. Sites with subsidence rates of 4.3 and 4.6 cm/year are projected to be underwater in approximately 30–40 years. These results demonstrate the vulnerability of the coastal areas to black sand mining with only a few decades before coastal areas undergoing mining are permanently flooded.

Table 2. Projected time in years until a site reaches sea level and is exposed to permanent flooding. Column 1 is the site name; Column 2 the LOS subsidence rate; Column 3 the subsidence rate converted to vertical; Column 4 is the vertical subsidence taking into account the mean predicted sea level rise; Column 5 is the mean elevation above sea level in meters; Column 6 is the estimated time until the site is at sea level elevation based only on subsidence rate; Column 7 is the time until the site is at sea level elevation considering subsidence and sea level rise.

Site	Observed Mean LOS Subsidence (cm/Year)	Vertical Subsidence (cm/Year)	Vertical Subsidence + Sea Level Rise (0.5 cm/Year)	Mean Elevation above Sea Level (m)	Time to Sea Level Elevation (Years) without s.l. Rise	Time to Sea Level Elevation (Years) with Sea Level Rise
10	3	3.6	4.1	2–4	60	50
13	4.3	5.2	5.7	2–4	40	40
15	4.6	5.5	6.0	2–4	40	30
2	4.3	5.2	5.7	2–4	40	30
8	1.8	2.2	2.7	2–4	90	70

4. DISCUSSION

Combining optical images and InSAR analysis allows for an in-depth identification of legal and illegal mining activities and their environmental impacts. To facilitate rapid response to illegal activities these data should be made accessible quickly after acquisition and processing should be automated. Implementation of systematic identification of new mining sites based on optical images is in the process of being developed in the Google Earth Engine. Currently, no SAR satellite provides global coverage of the Philippines. The Sentinel-1A, ALOS-2, and future NiSAR satellites should eventually allow for global Earth coverage once the missions are fully operational. The L-band sensor onboard the ALOS-2 and future NiSAR satellites would be particularly adapted for continuous monitoring of mining activities and their impacts. Thus, rapid, systematic processing of the SAR data that are planned to be distributed quickly after acquisition could lead to remote monitoring of illegal mining activities worldwide.

Our analysis reveals subsidence at several cm/year collocated with mining activities. Processes such as tectonic subsidence, natural compaction, and artificial settlement do not account for more than a few mm/year of subsidence (2–5 mm/year [32,36]). Additionally, natural subsidence occurs uniformly over large areas with consistent sediment deposits. In the Philippines InSAR reveals subsidence rates of ten times greater magnitude and with relatively small spatial extent. Thus, these subsidence rates cannot be explained by natural processes and likely result from man-made activities [13]. Agricultural activities often contribute to rapid land subsidence [13,15], as confirmed at the InSAR site 5 and Aringay, labeled as fields and experiencing subsidence collocated with agricultural activities. However, with the exception of the subsiding the site 15, the mining sites are located close to the shore and not in the direct proximity of agricultural activities. Thus, groundwater extraction for agricultural activities are unlikely the cause of the observed subsidence. The colocation of the mining sites and the subsiding areas suggests a causal relationship with either the removal of material or associated effects, such as ground water pumping.

However, we observe subsidence associated with some mining sites but not all, which makes the causal relationship difficult to validate. Geology affects whether extraction results in land subsidence. For example, groundwater extraction results in compaction and subsidence only in compressible deposits [13,15]. In the Philippines, we notice that the majority of the mining sites (subsiding and not) are located in compressible alluviums, which outline most of Luzon Island (beige inFigure 8). Thus, the difference between subsiding and non-subsiding sites cannot be explained by differences in geology. Two of the black sand mining sites of this study are not located on compressible deposits and they experience subsidence (red squares in Figure 8). It is likely that some surficial compressible deposits exist at these sites but have not been precisely mapped.

In mining areas where land subsidence was not detected it is possible that the rate of mineral and water extraction was not great enough to trigger subsidence above our detection threshold (cm/year). Recent data collection in the field confirms that a significant increase in illegal mining activity occurred between 2010 and 2013 (Table 3), unfortunately outside the time covered by the ALOS data. Continuous monitoring with new satellite platforms would help to better constrain the relationship between mining and land subsidence.

Table 3. Number of villages per province (from north to south) with reported illegal mining activity for different time periods. Due to timing and information constraints, we are unable to report mining activity for La Union and Pangasinan.

Province (N to S)	Before 2007	2007–2010	2010–2013	2013–Present
Cagayan	1	1	27	18
Ilocos Norte	1	1	0	0
Ilocos Sur	6	17	13	4
Zambales				2

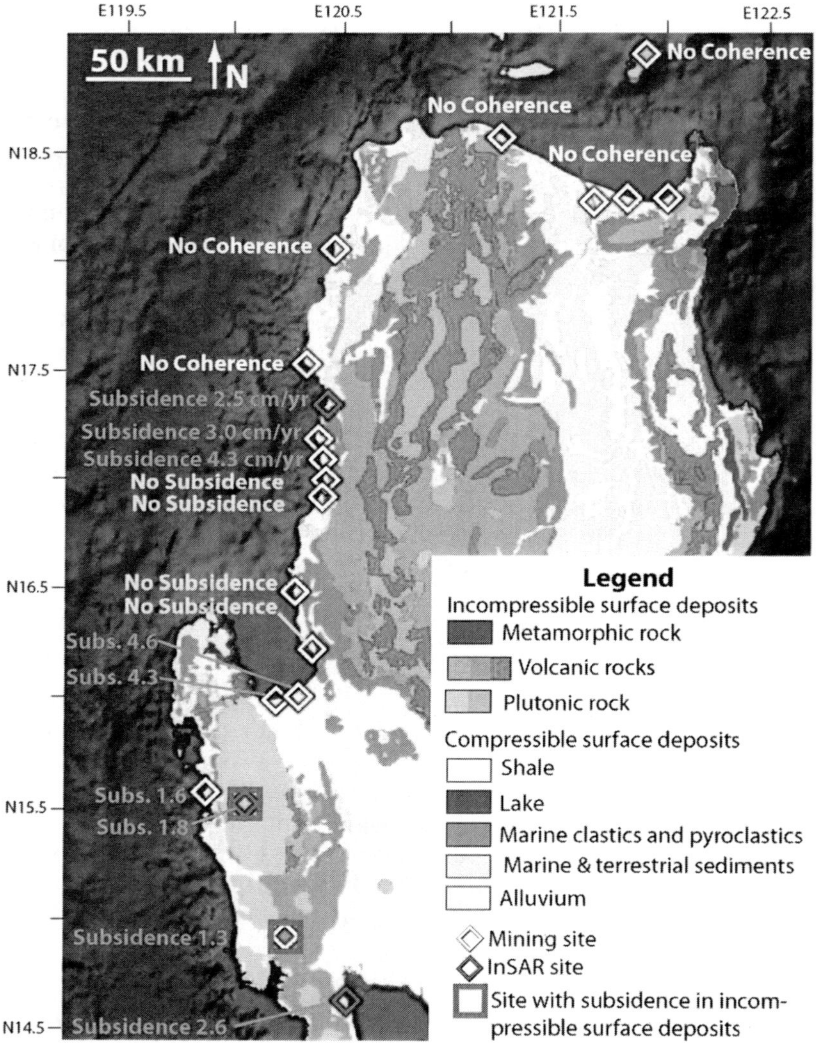

Figure 8. Map of the surficial geology of Luzon Island with the black sand mining sites identified with diamonds (geological data from the MGB of the Philippines). Labeled in white are mining sites with no coherence in the InSAR mean velocity. Labeled in cyan are the sites with coherence but no subsidence. Labeled in orange are the subsiding black sand mining sites with the rates indicated in cm/year. The two subsiding sites located in incompressible deposits are marked by red squares.

To better constrain the parameters that influence the occurrence of land subsidence additional data are needed. In particular, the amount of subsidence likely depends on the time since the development of the mining activities, the amount of mining, and the amount of water extraction. Unfortunately, such data were not available to us due to the limited control on the mining activities. Future collaborations with communities on the ground could help access such datasets and further constrain the parameters controlling the amount and extent of land subsidence associated with mining.

The observation that mining is not always associated with subsidence does not allow us to validate the direct causality between mining and subsidence. We are however able to identify communities exposed to very high risk of flooding by typhoons and sea level rise. Escalation of legal or illegal mining activities in these areas may further exacerbate this risk.

5. CONCLUSIONS

Remote sensing data are critical to monitor, control, and respond to black sand mining activities and their environmental and societal impacts. We show that optical images can be used to identify mining sites and respond to development of illegal activities. On Luzon Island we identified a total of twenty black sand mining sites, seventeen identified from contacts, news, and permits sources and confirmed with optical images and three identified in the InSAR mean velocity map and confirmed with optical images. We demonstrate that InSAR data can be used to characterize the land subsidence associated with the mining activities despite the small spatial extent of the mining sites. Seven of the twenty sites had no coherence in the InSAR data, thus no deformation could be measured, likely due to the limited number of acquisitions; four showed no ground subsidence above our detection threshold (cm/year); and nine experienced subsidence at rates ranging from 1.5 to 5.7 cm/year. The mean InSAR velocity map also highlights that the extent of the areas affected by subsidence is significantly larger than the mining sites themselves, likely associated with groundwater use or sediment redistribution around the affected areas.

Our results highlight the threat posed to coastal towns nearby black sand mining activities. Since most mining sites are at low elevation, the rapid subsidence results in high exposure to flooding and seasonal typhoons, and amplifies the effect of climate change–driven sea level rise. We show that several coastal areas will be at sea level elevation in a few decades due to the rapid subsidence. Since subsidence likely continues to affect the areas even decades after the cessation of mining activities due to the disruption of the sediment budget, characterization of the temporal evolution of land subsidence with longer SAR temporal coverage will be critical to mitigate environmental and societal effects of black sand mining activities.

ACKNOWLEDGMENTS

We thank the Center for Effective Global Action (CEGA) at UC Berkeley for support through the Spring 2014 Behavioral Sensing Fellowship. We also thank the Institute of International Studies, the Development Impact Lab, and the Behavioral Sensing Workgroup at UC Berkeley for supporting this project. E.C thanks the State University of New York at Buffalo Center for Computational Research for access to their computing resources for processing. The ALOS-PALSAR data are copyright of the Japanese Space Agency (JAXA) and the Japanese Ministry of Economy, Trade and Industry (METI) and was made available by the Alaska Satellite Facility (ASF). S.K. thanks Michael Davidson, Cesi Cruz, Ilocos Sur Collective Action for the Preservation of the Environment, Defend Ilocos Against Mining Plunder, and the Federation of Environmental Advocates of Cagayan for pertinent information and guidance in the field.

AUTHOR CONTRIBUTIONS

E.C. and S.K. conceived and designed the analysis. E.C. performed the InSAR analysis and interpretations and wrote the paper. S.K. performed the analysis of the mining sites locations and of the optical satellite images and contributed revisions and suggestions to the manuscript. E.C. and S.K. both contributed to the interpretations and discussion of the results.

CONFLICTS OF INTEREST

The authors declare no conflict of interest.

REFERENCES

1. Legodi, M.; Dewaal, D. The preparation of magnetite, goethite, hematite and maghemite of pigment quality from mill scale iron waste. *Dyes Pigments* **2007**, *74*, 161–168.

2. Domingo, E.G. The Philippine mining industry: Status and trends in mineral resources development. *J. Southeast Asian Earth Sci.* **1993**, *8*, 25–36.

3. Rappler Online Newspaper. Available online: http://www.rappler.com/business/special-report/whymining/whymining-latest-stories/74937-black-sand-mining-operations-under-scrutiny (accessed on 14 September 2015).

4. Rabe, A.; (The Parish of St. Nicholas of Tolentino, Sinait, the Philippines; The Ilocos Sur Collective Action for the Preservation of the Environment, Vigan, the Philippines). Personal communication, 2014.

5. Bernal, B. Pangasinan Gov to Face graft Cases over Black Sand Mining. Available online: http://www.rappler.com/nation/72659-black-sand-mining-espino-face-graft-charges (accessed on 30 December 2015).

6. Terzaghi, K. Principles of soil mechanics: IV. Settlement and consolidation of clay. *Erdbaummechanic* **1925**, *95*, 874–878.

7. Welhan, J.A. *Interactions near the Highway Pond Gravel Pit, Pocatello, Idaho*; Staff Report 01–3; Idaho Geological Survey: Moscow, Russia, 2001.

8. Massonnet, D.; Rossi, M.; Carmona, C.; Adragna, F.; Peltzer, G.; Feigl, K.L.; Rabaute, T. The displacement field of the landers earthquake mapped by radar interferometry. *Nature* **1993**, *364*, 138–142.

9. Tralli, D.; Blom, R.; Zlotnicki, V.; Donnellan, A.; Evans, D. Satellite remote sensing of earthquake, volcano, flood, landslide and coastal inundation hazards. *ISPRS J. Photogramm. Remote Sens.* **2005**, *59*, 185–198.

10. Lu, Z. InSAR imaging of volcanic deformation over cloud-prone areas—Aleutian Islands. *Photogramm Eng. Remote Sens.* **2007**, *73*, 245–257.

11. Amelung, F.; Galloway, D.L.; Bell, J.W.; Zebker, H.A.; Laczniak, R.J. Sensing the ups and downs of Las Vegas: InSAR reveals structural control of land subsidence and aquifer-system deformation. *Geology* **1999**, *27*, 483–486.

12. Chaussard, E.; Amelung, F.; Aoki, Y. Characterization of open and closed volcanic systems in Indonesia and Mexico using InSAR time series. *J. Geophys. Res.-Solid Earth* **2013**, *118*, 1–13.

13. Chaussard, E.; Amelung, F.; Abidin, H.; Hong, S.-H. Sinking cities in Indonesia: ALOS PALSAR detects rapid subsidence due to groundwater and gas extraction. *Remote Sens. Environ.* **2013**, *128*, 150–161.

14. Chaussard, E.; Burgmann, R.; Shirzaei, M.; Fielding, E.J.; Baker, B. Predictability of hydraulic head changes and characterization of aquifer-system and fault properties from InSAR-derived ground deformation. *J. Geophys. Res.-Solid Earth* **2014**, *119*, 6572–6590.

15. Chaussard, E.; Wdowinski, S.; Cabral-Cano, E.; Amelung, F. Land subsidence in central Mexico detected by ALOS InSAR time-series. *Remote Sens. Environ.* **2014**, *140*, 94–106.

16. *An Act Instituting a New System of Mineral Resources Exploration, Development, Utilization, and Conservation, Republic Act 7942*; The House of Representatives and the Senate of the Philippines in Congress: Manilla, the Philippines, 1995.

17. Chaussard, E.; Amelung, F. Characterization of geological hazards using a globally observing spaceborne SAR.*Photogramm. Eng. Remote Sens.* **2013**, *79*, 982–986.

18. Rosen, P.A.; Hensley, S.; Peltzer, G.; Simons, M. Updated repeat orbit interferometry package released. *Eos Trans. AGU***2004**, *85*, 47.

19. Farr, T.G.; Rosen, P.A.; Caro, E.; Crippen, R.; Duren, R.; Hensley, S.; Kobrick, M.; Paller, M.; Rodriguez, E.; Roth, L.; *et al.* The shuttle radar topography mission. *Rev. Geophys.* **2007**, *45*, RG2004.

20. Chen, C.W.; Zebker, H.A. Two-Dimensional Phase Unwrapping with use of statistical models for cost functions in nonlinear optimization. *J. Opt. Soc. Am. A* **2001**, *18*, 338–351.

21. Berardino, P.; Fornaro, G.; Lanari, R.; Sansosti, E. A new algorithm for surface deformation monitoring based on small baseline differential SAR interferograms. *IEEE Trans. Geosci. Remote Sens.* **2002**, *40*, 2375–2383.

22. Fattahi, H.; Amelung, F. DEM error correction in InSAR time series. *IEEE Trans. Geosci. Remote Sens.* **2013**, *51*, 4249–4259.

23. Pepe, A.; Lanari, R. On the Extension of the minimum cost flow algorithm for phase unwrapping of multitemporal differential SAR interferograms. *IEEE Trans. Geosci. Remote Sens.* **2006**, *44*, 2374–2383.

24. Casu, F.; Manzo, M.; Lanari, R. A Quantitative assessment of the SBAS algorithm performance for surface deformation retrieval from DInSAR data. *Remote Sens. Environ.* **2006**, *102*, 195–210.

25. Tizzani, P.; Berardino, P.; Casu, F.; Euillades, P.; Manzo, M.; Ricciardi, G.P.; Zeni, G.; Lanari, R. Surface deformation of Long Valley Caldera and Mono Basin, California, investigated with the SBAS-InSAR approach. *Remote Sens. Environ.***2007**, *108*, 277–289. [Google Scholar] [CrossRef]

26. Ryan, W.B.F.; Carbotte, S.M.; Coplan, J.O.; O'Hara, S.; Melkonian, A.; Arko, R.; Weissel, R.A.; Ferrini, V.; Goodwillie, A.; Nitsche, F.; *et al.* Global multi-resolution topography synthesis. *Geochem. Geophys. Geosyst.* **2009**, *10*.

27. National Oceanic and Atmospheric Administration Tide Gauges Monitoring. Available online: http://tidesandcurrents.noaa.gov/sltrends/sltrends.html (accessed on 14 September 2015).

28. Raucoules, D.; Le Cozannet, G.; Wöppelmann, G.; de Michele, M.; Gravelle, M.; Daag, A.; Marcos, M. High nonlinear urban ground motion in Manila (Philippines) from 1993 to 2010 observed by DInSAR: implications for sea-level measurement. *Remote Sens. Environ.* **2013**, *139*, 386–397.

29. Moon, J.-H.; Tony Song, Y. Sea level and heat content changes in the western north Pacific. *J. Geophys. Res. Oceans* **2013**,*118*, 2014–2022.

30. Cazenave, A.; Llovel, W. Contemporary sea level rise. *Annu. Rev. Mar. Sci.* **2015**, *2*, 145–173.

31. Dixon, T.H.; Amelung, F.; Ferretti, A.; Novali, F.; Rocca, F.; Dokka, R.; Sella, G.; Kim, S.-W.W.; Wdowinski, S.; Whitman, D. Space geodesy: subsidence and flooding in new Orleans. *Nature* **2006**, *441*, 587–588.

32. Teatini, P.; Tosi, L.; Strozzi, T. Quantitative evidence that compaction of Holocene sediments drives the present land

subsidence of the Po Delta, Italy. *J. Geophys. Res.* **2011**, *116*, B08407.

33. Lei, L.; Zhou, Y.; Li, J.; Burgmann, R. Monitoring slow moving landslides in the Berkeley hills with TerraSAR-X data. In Proceedings of the IEEE International Geoscience and Remote Sensing Symposium (IGARSS), Holonunu, HI, USA, 25–30 July 2010; pp. 230–232.

34. Nof, R.N.; Baer, G.; Ziv, A.; Raz, E.; Atzori, S.; Salvi, S. Sinkhole precursors along the dead sea, Israel, revealed by SAR interferometry. *Geology* **2013**, *41*, 1019–1022.

35. Pritchard, M.E.; Jay, J.A.; Aron, F.; Henderson, S.T.; Lara, L.E. Subsidence at southern Andes volcanoes induced by the 2010 Maule, Chile earthquake. *Nat. Geosci.* **2013**, *6*, 632–636.

36. Törnqvist, T.E.; Wallace, D.J.; Storms, J.E.A.; Wallinga, J.; van Dam, R.L.; Blaauw, M.; Derksen, M.S.; Klerks, C.J.W.; Meijneken, C.; Snijders, E.M.A. Mississippi delta subsidence primarily caused by compaction of Holocene strata. *Nat. Geosci.* **2008**, *1*, 173–176.

5

CHAPTER

Assessing Landform Alterations Induced by Mountaintop Mining

Aaron E. Maxwell[1], Michael P. Strager[2*]

[1]*Natural Resource Analysis Center, West Virginia University, Morgantown, USA*

[2]*Division of Resource Management, West Virginia University, Morgantown, USA*

ABSTRACT

A comprehensive impact analysis of mountaintop removal and valley fill (MTR/VF) mining requires an understanding of landform alterations since ecological impacts are so intricately linked. In this study we investigated mining in the Coal River Watershed, West Virginia, USA, using landform terrain analysis. Previous studies have relied on elevation differencing of preand postmining surfaces to assess absolute elevation and volumetric change. Our landscape analysis, utilizing light detection and ranging (LiDAR)- derived elevation data, indicated specific landform types and distributions that were significantly altered after MTR/VF mining and reclamation. The use of categorical landform data provides insights to assessing and understanding the extent of topographically altered mountaintops. Our study provides an opportunity to further examine the impact of MTR/VF on forest communities, terrestrial habitat, ecosystem health, and biodiversity.

Keywords: Mountaintop Mining; Landforms; Change Analysis

1. INTRODUCTION

In the Southern Coalfields of West Virginia, MTR/VF is currently the leading cause of land cover change [1-4]. Multiple watersheds in West Virginia have more than 10% of their surface area disturbed by surface mining [5], which results in the loss of forest and a conversion to barren land cover [6]. It has been estimated that all surface mining in Appalachia has resulted in a net loss of 420,000 ha of forest [3], and the interior character of the forest is threatened by the introduction of non-forest edges [7]. During the MTR/VF process, forests are cleared, top soil is removed, overburden material is blasted away to uncover coal seams, and overburden material is placed in adjacent valleys, filling stream segments and creating valley fills [5,8]. In Kentucky, it has been estimated that greater than 660 km of headwater streams were buried between 1985 and 1999 [9]. Later reclamation produces grasslands or shrub/scrub land cover; however, productivity is often limited due to poor soil conditions [6]. Attempts are often made to preserve soil; however, it becomes homogenized and soil horizons are not maintained [10].

In addition to the land use and land cover (LULC) change, the landscape and terrain are recontoured with modified watershed ridges, altered vegetation conditions, and modified soil character [8]. References [1,11] estimated that surface mining was responsible for displacing more material in the Southern Coalfields of West Virginia than river systems and natural geomorphic processes. Furthermore, MTR/VF methods have resulted in more moved material and faster landscape alterations as compared to traditional surface mining methods, such as auger, contour, and highwall mining [12].

Elevation differencing of Digital Elevation Models (DEMs) representing different temporal conditions has been explored to describe topographic change. For example, [13] utilized NASA Shuttle Radar Topographic Mission (SRTM), US Geologic Survey National Elevation Dataset (NED), and Terrain Resource Information Management Program (TRIM) data to describe and map alpine glacier changes in southeast Alaska and northwest British Columbia. Reference [14] utilized LiDAR-derived DEMs to calculate dune volume changes over a 1-year period at sites along the Cape Hatteras National Seashore.

Landscape and geomorphic change resulting from MTR/VF disturbance were specifically analyzed in Perry County, Kentucky, USA [15] utilizing NED (pre-mining) and SRTM (post-mining) data. The study highlighted the complexity of such an analysis when timestamps of the NED data are variable since they were created relative to best available data and 1:24000 scale topographic quadrangle maps. In a similar study, the West Virginia

Department of Environmental Protection (WVDEP) utilized interferometric synthetic aperture radar (IFSAR) and elevation raster data produced from digital line graph (DLG) hypsography to map valley fill extents throughout nine counties in southern West Virginia. Because the radar data did not adequately penetrate the tree canopy, it was necessary to remove forested areas from the analysis. It was found that a complete inventory of the fills required additional visual classification [16].

To the best of our knowledge no previous work has quantified the post-mining landscape in terms of changes in terrain characteristics from pre-mining conditions using landscape-scale categorical terrain data. An understanding of the terrain alteration using landform data is appropriate to assess the impact of the topographically altered mountaintops.

This paper expands upon earlier differencing and topographic change work in the MTR/VF region by implementing a methodology relying on a categorical representation of the landscape as landforms. This data differentiates the landscape into the following classes: cliff, steep slope, slope crest, upper slope, flat summit, sideslope, cove, dry flat, moist flat, wet flat, and slope bottom. This method was adopted after [17] because such features adequately represent the Southern Coalfields at the landscape scale.

Attempts have been made to link landform data to habitats using predictive modeling. For example, [17] attempted to link the ecological community types developed by the Nature Conservancy to landforms, elevation, and lithology. These categories represent landforms of ecological significance at the landscape scale. Our goal was to investigate how the distribution of these categories was impacted by surface mining. Our results may help to reestablish a post-mining landscape that can benefit terrestrial habitat, ecosystem health, and biodiversity since these have been shown to be linked to landform clasess [17].

2. METHODOLOGY

2.1. Study Area

The Coal River Watershed is a 230,755 ha Hydrologic Unit Code (HUC) 8 headwater watershed completely within West Virginia, USA. The Coal River Watershed exists within the Appalachian Plateau physiogeographic province, a dissected, westward-tilted plateau dominated by Pennsylvanian bedrock. Pennsylvanian stratigraphy is characterized as cyclic sequences of sandstone, shale, clay, coal, and limestone [18]. The terrain is dissected by a dendritic stream network and shows fine texture with moderate to strong local relief. In comparison to the northern Appalachian Plateau, the Southern Coalfields is generally more rugged due to resistant strata [19]. The terrain analysis using landforms suggests that this terrain is naturally dominated by steep slopes.

According to LULC estimates derived from aerial photography, 8.8% of this watershed was disturbed by active surface mining or mine reclamation in 2009. It should be noted that this estimate does not take into account historical mining areas which have since been reforested.**Figure 1** shows the watershed location within West Virginia. This watershed was selected as a case study because it is heavily impacted by mining, there has been continued mining and reclamation between 2003 and 2010, and because LiDAR data were available for the extent.

2.2. Digital Elevation Data Utilized

The LiDAR data, representing recent topographic conditions, were collected between April 9th, 2010 and April 18th, 2010 at a flight height of 1524 meters above ground level (AGL), a pulse frequency of 70 kHz, a scan frequency of 35 Hz, and a scan angle (full field of view) of 36°. The swaths were flown with a 30% overlap, at an average speed of 250 km/hr, and with an average width of 979 m. An Optech ALTM 3100 C sensor was used to collect the data. The scan and flight specifications were selected to support a 0.7 m contour interval and a 1 m nominal ground post spacing. Ground points were filtered from the all return data to produce ground point files in LAS 1.2 format.

An interpolated raster surface was created from point data using ArcMap 10, and average point spacing of 0.01 m. Inverse distance weighting (IDW) was used to interpolate a raster surface since sample points were in an evenly distributed pattern. This process resulted in a 1 m resolution, floating point elevation dataset.

2.3. Assessing Landform Changes between 2003 and 2010

Landform changes were assessed in filled and cut or excavated areas resulting from mining and reclamation between 2003 and 2010. The 2010 LiDAR-derived DEM was compared to a 2003 photogrammetrically-derived DEM representing 2003 conditions. The 2003 DEM data were provided by the West Virginia Statewide Addressing and Mapping Board (SAMB) who contracted BAE Systems ADR to create a stereo photogrammetric-derived digital terrain model (DTM) from statewide, spring, leaf-off, 1:4800 scale aerial photography. The DEM was created in compliance with National

Dataset standards (1/9th arc second) and produced from mass points and breaklines. This dataset supports a vertical accuracy of +/− 3.048 m, root mean square error (RMSE) and has a cell size of 3 m [20].

Figure 1. Study Area.

In order to match the resolution of the SAMB DEM, we resampled the LiDAR-derived raster to 3 m resolution using bilinear interpolation and snapped the grid to match the extent of the SAMB DEM. This process assured that the raster grids were completely aligned. This process resulted in two 3 m DEMs representing 2003 and 2010 conditions that could be compared. A vertical data transformation was unnecessary since both DEMs had a vertical reference of NAVD88 orthometric.

In order to detect systematic error between the 2003 and 2010 DEM data, a total of 281 points were collected in the field throughout the Coal River Watershed on flat, paved surfaces using Pacific Crest realtime kinematic (RTK) survey equipment. The elevation measurements from the 2003 and 2010 elevation raster data were obtained at these locations. The measurements from each DEM were then compared. The mean difference was −0.4 m with a maximum difference of 0.8 m and a minimum difference of −2.0 m. Based on this analysis we did not correct for systematic difference between DEMs.

Once DEMs were obtained, processed, and prepared, they were subtracted using Raster Calculator within the Spatial Analyst Extension of ArcMap [21]. This produced a grid of elevation differences throughout the watershed. Negative values indicated potential cuts or excavations while positive values indicated potential fills. However, difference could simply have resulted from errors in the DEMs or in methodology. As a result, once this elevation difference model was calculated, it was necessary to determine a tolerance or threshold that would constitute true change and not simply error or noise between the digital elevation datasets. The photogrammetric DEM had an error tolerance (RMSE) of +/− 3.048 m while the LiDAR DEM had and error tolerance of only 15 cm (0.15 m). An equation suggested by [22] was used to estimate this threshold. The RMSE of each dataset was squared, the results were summed, and the square root of the sum was then taken. The result was then multiplied by three to determine a cut off representing values that were greater than three standard deviations from the mean. This method assumes a Gaussian distribution. Elevation change measurements outside of this range were considered true elevation differences and not a result of error or noise. Reference [15] used the same methodology to derive an error tolerance for the analysis conducted in Perry County, Kentucky, USA.

According to this method, it is not certain whether differences less than +/− 9.2 m represented true topographic change. As a result, the elevation difference grid was reclassified so that values between −10 m and +10 m were considered no change, error between the DEMs, or noise. Pixels with values less than −10 m were considered potential cuts while pixels with values greater than 10 m were considered potential fills. This process produced a cut and fill mask.

A LULC dataset was created for the region from 2009 National Agriculture Imagery Program (NAIP) orthophotography. Optimally, imagery collected at the time of the LiDAR collection would have been utilized; however, such data were not available. The imagery was classified using an object-based feature extraction methodology, augmented with GIS decision rules and manual digitizing. Forest, grass, and barren land cover were extracted from the raw imagery.

Combining these data with mine permit boundaries from the West Virginia Department of Environmental Protection (WVDEP) made it possible to delineate barren and grassland land cover in mine permits. Grasslands in permits represent potential reclamation while barren areas represent areas of active mining or areas that have not yet been reclaimed. The extents of valley fill faces and slurry impoundments were digitized. According to an error assessment based on manual aerial photograph interpretation at 100 randomly selected points, the LULC dataset had an overall accuracy of 94% (K_{Hat} of 93%).

It was possible to further remove erroneous pixels as potential cut or fills by utilizing additional data, such as the high resolution land cover. For example, potential cut or fill pixels could be removed if they were outside of disturbance resulting from surface mining. The potential cut or fill pixels that existed within areas of mine disturbance (barren, grasslands, or valley fill faces) were considered. Slurry impoundments were excluded because of errors associated with water level. To complete this analysis, it was necessary to convert the raster data to polygons that represented contiguous areas of cut or fill. No smoothing or simplification was applied.

Small areas of topographic change could also be removed as error or noise. Thresholds were determined from the hand digitized valley fill faces. The average size of these faces was 73,257 m^2(7.3 ha) with a standard deviation of 67,013 m^2 (6.7 ha). Based from this distribution, any contiguous area of cut or fill that was smaller than 60,000 m^2 (6 ha) was considered noise and removed. Although this threshold was somewhat arbitrary, it was selected since this study focused on detecting large expanses of excavation and fills.

The result of this analysis was a vector layer of potential cut or fill extents in which topographic change had occurred between 2003 and 2010. The extents were used to define areas of potential terrain change for analysis and comparison of the distributions of landforms. Although there was error associated with this methodology, it was adequate for this analysis since we were not interested in conducting an assessment of absolute elevation change or volumetric change but in defining areas in which to compare categorical terrain data and their impact on forest communities.

2.4. Creation of Landforms

The DEM data were used to classify the terrain into the following landforms: cliff, steep slope, slope crest, upper slope, flat summit, sideslope, cove, dry flat, moist flat, wet flat, and slope bottom. The methodology described by [17] provides a means to utilize DEM data to classify the landscape into different units of ecological significance. DEM derivatives including slope in degrees, a hydrologically filled DEM, flow direction, and flow accumulation were created using ArcMap 10. The slope and flow accumulation grids were utilized to calculate a moisture index. The moisture index is a relative measure of the moisture of a specific cell, and it assumes that the moisture level is a function of how much water flows into the cell, predicted by flow accumulation, and how fast the water can flow out, described by slope [20-25].

Landscape position was calculated following [26] to divide the landscape into the following categories: ridge, wide ridge, slope/flat, slope/cove. The approach is based on a local neighborhood analysis of elevation values in which a cell is compared to the mean elevation of all values within 3 by 3 windows. Landscape position was combined with slope and moisture index data to derive the final landform classes. This procedure was conducted for both the 2003 and the 2010 DEMs. The comparisons of the derived landforms included:

- Areas filled between 2003 and 2010 (mining related);
- Areas cut or excavated between 2003 and 2010 (mining related);
- Areas of mine reclamation in 2009, areas of mining disturbance or reclamation in 2009, and areas in WVDEP mine permits but not disturbed (forested in 2009).

This enabled the comparison of post-mining landform categories with pre-mining conditions.

3. RESULTS

Figure 2 shows the distribution of landforms in a MTR/VF site representing 2010 conditions. Comparing the mine site with the surrounding landscape, the distribution of landforms is greatly altered by mining even after reclamation has occurred. **Figure 3** shows a different MTR/VF site and offers a comparison of landforms within areas cut or filled between 2003 and 2010, and this visualization also shows a redistribution of landforms.

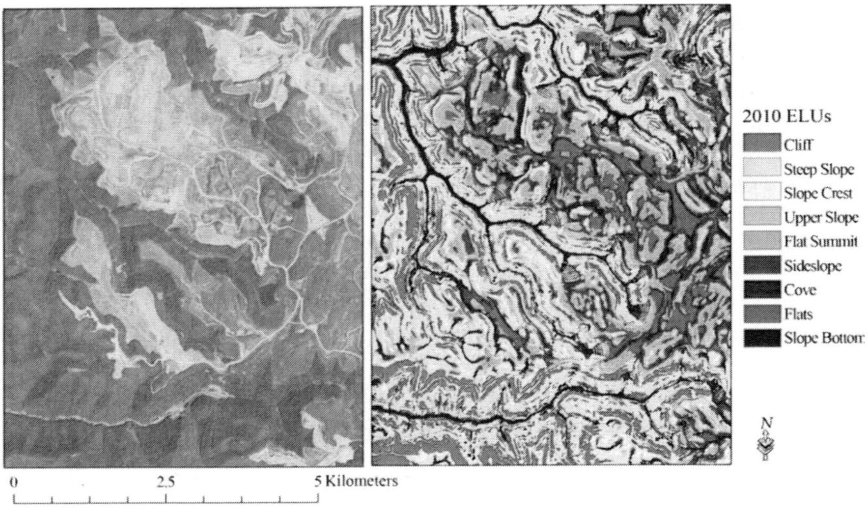

Figure 2. Example of landform distribution after landscape alteration and MTR/VF. The imagery is orthophotography representing growing season 2011 conditions.

Landforms were compared for areas of cut or fill between 2003 and 2009 within mine disturbance. Such areas represent potential areas of true topographic change. The results are described in Tables 1 and 2 and Figures 4 and 5. Chi-square tests were performed on the categorical data for this observational study, and the results suggest a statistically different distribution of landforms after volumetric change ($\alpha = 0.001$).

It was possible that differences in the DEM production methodology could have resulted in differences in the resulting landform classification. This was observed by visual inspection of the results and was a source of error in the data represented in Figures 4 and 5. As a result, differences in classification and distribution of landforms could result from true terrain change or could be a product of the differences in DEM production methodology, photogrammetric or LiDAR-derived. In order to resolve such ambiguity, terrain classification differences were assessed for different surface mining LULC categories (2009) using the 2010 ELU data only. The results are described in **Table 3**and **Figure 6** in which landform classes were compared for areas in mine permits but still forested or not disturbed,

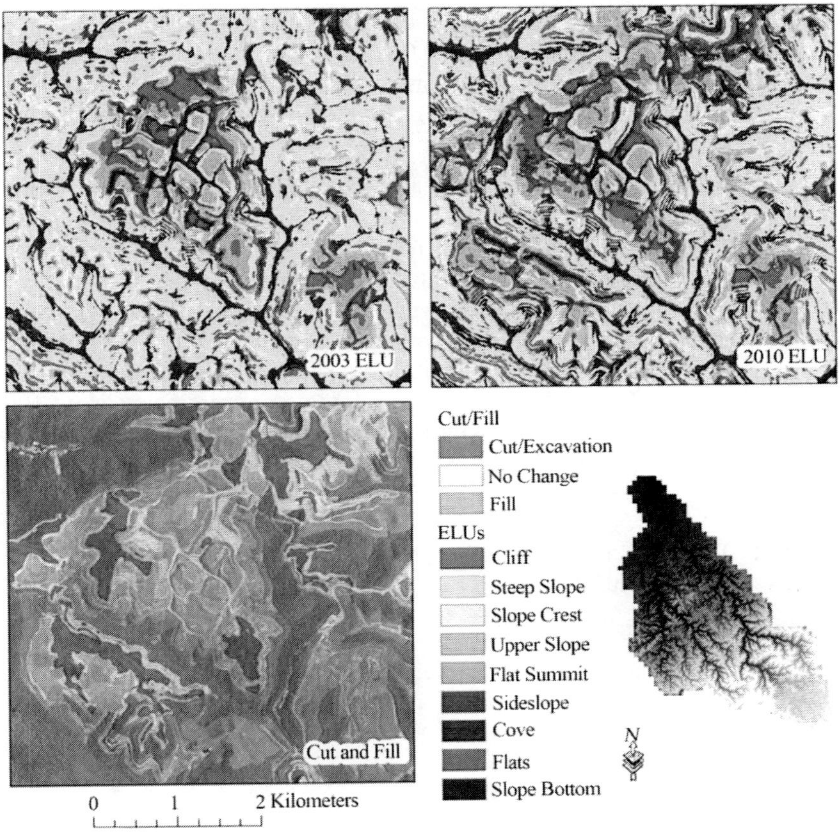

Figure 3. Comparison of landforms at mine site. The imagery is orthophotography representing growing season 2011 conditions.

areas of mine reclamation, and areas of mine reclamation or mining disturbance. The assumption was that areas permitted for surface mining should generally have similar topography. As a result, undisturbed areas within the mine permits representing pre-mining terrain were compared to mined or reclaimed areas representing post-mining terrain. This analysis showed a marked change in landforms from steep slopes to more flat topography, such as moist flats, upper slopes, and flat summits. We argue this, this is a more valid method to assess topographic change induced by surface mining and excavation because this technique was not impacted by differences in DEM methodology.

A chi-square test was performed to test for a statistical difference in the distributions of landforms associated with **Table 3** and **Figure 6**. The test suggest a statistically significantly different categorical distribution of landforms preand post-mining (α = 0.001). The landform classes were then reclassified to only slopes and flats, and a second chi-square test suggests a statistically significant increase in flats and decrease in slopes postmining (α = 0.001).

Two sample Student's t-tests were used to determine if preand post-mining average slope conditions were statistically different. A comparison of slopes for all pixels in areas that were not mined in 2003 but were predicted as reclaimed relative to summer 2009 conditions showed a statistically significant different average slope at the 99.9% confidence level. However, as noted above, difference could have been a result of different methodol ogy utilized to produce the DEM data. As a result, we also compared all pixels that were forested in permits, relative to growing season 2009 conditions, and areas that were either reclaimed or mined disturbed or reclaimed.

Table 1. Distribution of landforms for areas within mines and filled between 2003 and 2010.

	2003		2010	
ELU	Ha	Percent	Ha	Percent
Cliff	257.34	9.20%	90.97	3.25%
Steep Slope	1128.87	40.34%	444.05	15.87%
Slope Crest	14.16	0.51%	84.20	3.01%
Upper Slope	101.72272	3.63%	413.88	14.79%
Flat Summit	66.14	2.36%	379.32	13.55%
Side Slope	291.75	10.43%	568.09	20.30%
Cove	602.42	21.53%	202.18	7.22%
Dry Flat	3.00	0.11%	76.29	2.73%
Moist Flat	106.85	3.82%	427.81	15.29%
Wet Flat	82.40	2.94%	34.46	1.23%
Slope Bottom	143.80	5.14%	77.15	2.76%
TOTAL	2798.45	100%	2798.40	100%

Table 2. Distribution of landforms for areas within mines and cut or excavated between 2003 and 2010.

	2003		2010	
ELU	Ha	Percent	Ha	Percent
Cliff	261.11	10.59%	214.42	8.69%
Steep Slope	1025.26	41.57%	314.84	12.76%
Slope Crest	564.54	22.89%	134.42	5.45%
Upper Slope	280.71	11.38%	419.97	17.03%
Flat Summit	169.72	6.88%	446.11	18.09%
Side Slope	96.98	3.93%	364.3	14.76%
Cove	41.38	1.68%	165.93	6.73%
Dry Flat	1.04	0.04%	56.22	2.28%
Moist Flat	12.07	0.49%	212.46	8.61%
Wet Flat	11.01	0.45%	20.89	0.85%
Slope Bottom	2.76	0.11%	117.29	4.76%
TOTAL	2466.58	100%	2466.58	100%

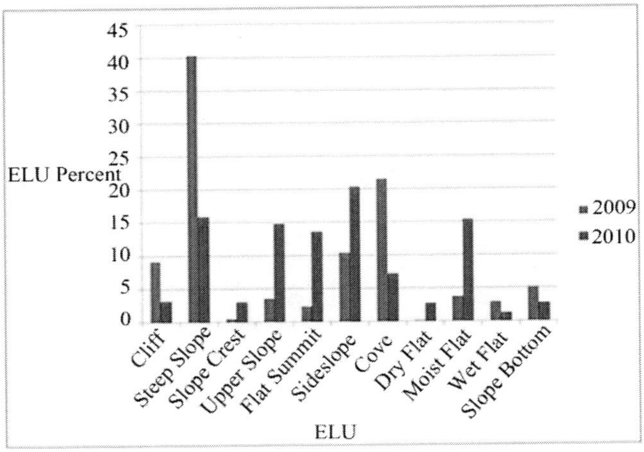

Figure 4. Comparison of landforms for areas within mines and filled between 2003 and 2010.

Table 3. Distribution of landforms within different land cover categories (2009) relative to 2010 DEM.

ELU	Forest in Permit Ha	Reclaimed Ha	Mine Ha
Cliff	3306.74	596.99	1165.89
Steep Slope	11328.50	2777.78	4274.67
Slope Crest	1939.41	909.98	1443.73
Upper Slope	1638.62	2042.73	3413.90
Flat Summit	464.11	1277.28	2697.18
Side Slope	1587.01	2262.82	3723.33
Cove	2493.24	1388.31	2268.33
Dry Flat	75.05	152.48	406.80
Moist Flat	105.30	887.71	2015.08
Wet Flat	11.87	76.44	174.32
Slope Bottom	418.58	488.21	1040.06

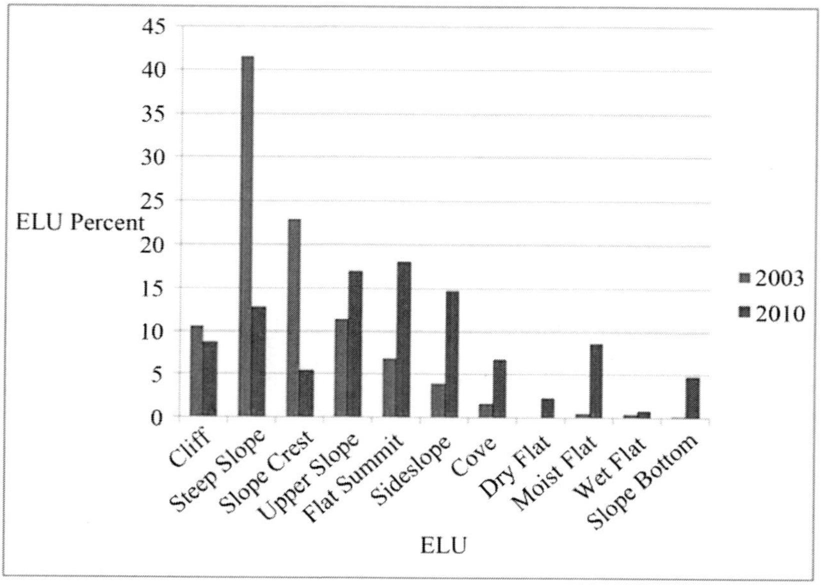

Figure 5. Comparison between landforms for areas within mines and cut or excavated between 2003 and 2010.

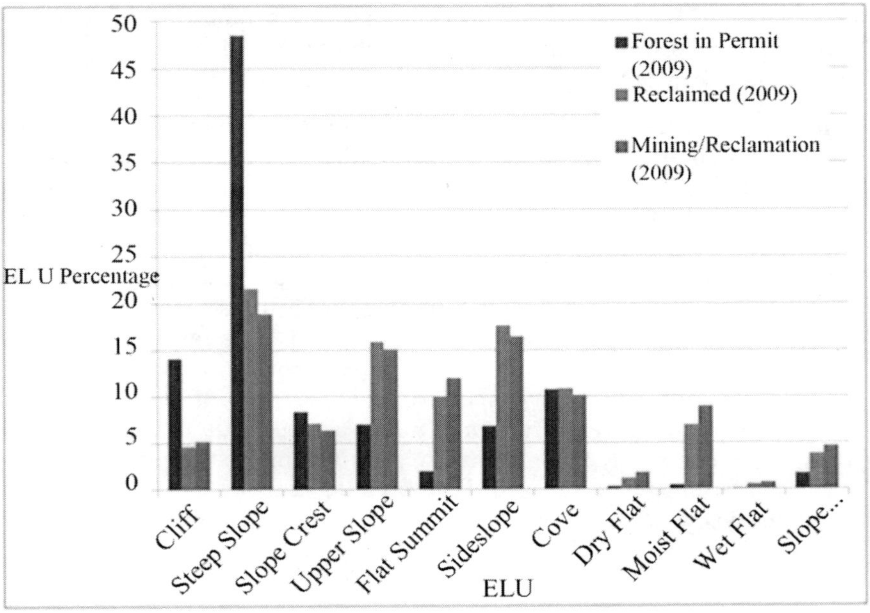

Figure 6. Comparison of landforms within different land cover categories 2009 relative to 2010 DEM.

4. DISCUSSION

The Southern Coalfields of West Virginia are characterized by steep slopes and narrow valleys [19]; however, this description often does not accurately describe the post-mining terrain and geomorphology. This study helped to indicate the specific landform classes that are altered in our representative study area.

The geomorphic complexity and steep slopes that characterize this region contribute to the biological richness of the Appalachian Plateau and the mixed mesophytic forest biogeographic region [27]. The Southern Coalfields of West Virginia exists within one of the most biodiverse regions within the temperate zone, and more than 2000 vascular plants exist on the landscape [28,29]. This region is globally recognized as significant for biodiversity conservation [30]. The variety of forest types is in part a result of the elevation changes and steep topography of this landscape [27].

There has been a shift in focus from rare or endangered species management to ecosystem and landscapescale management, which requires that a diversity of ecological processes be considered [31]. The methodology of this study provides an opportunity to understand landscape and terrain alterations and the ability now to examine the critical links of landforms to terrestrial habitat, ecosystem health, and biodiversity.

The question arises as to how this terrain alteration will impact biodiversity and ecosystem health. If forests are reestablished on these topographically altered mountaintops, will pre-mining forest communities and ecosystems be reestablished? Although traditional reclamation often results in grasslands or scrub/shrub lands on the reclaimed and topographically altered mountaintops [6,32] has suggested a forest reclamation approach that includes the creation of soils to support a forest community, loose grading to avoid soil compaction and increased bulk density, use of less competitive ground cover to allow for forest growth, planting of a wider variety of trees and native species, and use of proper tree planting techniques to aid in reforestation of mine scarred lands. This technique is being more commonly implemented to reclaim mine disturbed areas [32]. If such methodology is successful at creating stable and mature forest communities, will these forests represent pre-mining conditions or serve the same ecological function if the topography has been greatly altered? This question is yet to be fully explored. Assuming that topographic factors, such as slope and landscape position, are correlated with forest community type as described by [33], we suggest that it is reasonable to assume that topographic changes will result in forest community changes, and forest community variety and distribution are related to biodiversity as suggested by [27]. Terrain alteration should be considered along with the impact of introducing non-forest edge and disrupting the interior nature of the forest since MTR/VF has been shown to have the potential to alter community faunal composition [34].

5. SUMMARY

This study identified specific landform classes that were impacted due to cut/excavation or filled material at surface mined sites in a West Virginia, USA watershed. The landform analysis methodology provided a means to quantify and spatially assess terrain alterations using categorical landscape data derived from DEMs, a potential first step for assessing the impact of MTR/VF on terrestrial habitat and ecosystems at the landscape scale. Research is needed to understand the relationship between terrain and forest communities in the Appalachian region at both the local and regional scales. A more complete understanding of these relationships may aid in an understanding of the impact of MTR/VF on forest communities, terrestrial habitat, ecosystem health, and biodiversity.

6. ACKNOWLEDGEMENTS

This study was sponsored by the Appalachian Research Initiative for Environmental Science (ARIES). ARIES is an industrial affiliates program at Virginia Tech, Blacksburg, VA, USA, supported by members that include companies in the energy sector. The research under ARIES is conducted by independent researchers in accordance with the policies on scientific integrity of their institutions. The views, opinions and recommendations expressed herein are solely those of the authors and do not imply any endorsement by ARIES employees, other ARIES-affiliated researchers or industrial members. Information about ARIES can be found at http://www.energy.vt.edu/ARIES.

We would also like to acknowledge funding support provided by US Environmental Protection Agency and the West Virginia Experiment Station. This work was completed with the assistance of Adam Riley, Elise Austin, and Charles Yuill who are Remote Sensing Analyst, GIS Analyst and Landscape Analyst respectively at West Virginia University.

Lastly, we thank the anonymous reviewers who helped to improve the manuscript.

REFERENCES

1. Hooke, R.L. (1994) On the efficacy of humans as geomorphic agents. GSA Today, 4, 224-225.
2. Saylor, K.L. (2008) Land cover trends project: Central Appalachians. US Department of the Interior, Washington. http://landcovertrends.usgs.gov/east/eco69Report.html
3. Drummond, M.A. and Loveland T.R. (2010) Land-use pressure and a transition to forest-cover loss in the eastern United States. BioScience, 60, 286-298.
4. Townsend, P.A., Helmers, D.P., Kingdon, C.C., McNeil, B.E., de Beurs, K.M. and Eshleman, K.N. (2009) Changes in the extent of surface mining and reclamation in the Central Appalachians detected using a 1976-2006. Landsat time series. Remote Sensing of Environment, 113, 62- 72.
5. Palmer, M.A., Bernhardt, E.S., Schlesinger, W.H., Eshleman, K.N., Fourfoula-Georgiou, E., Hendryx, M.S., Lemly, A.D., Likens, G.E.,

Loucks, O.L., Power, M.E., White, P.S. and Wilcock, P.R. (2010) Mountaintop mining consequences. Science, 327, 148-149.

6. Simmons, J.A., Currie, W.S., Eshleman, K.N., Kuers, K., Monteleone, S., Negley, T.L., Pohlad, B.R. and Thomas, C.L. (2008) Forest to reclaimed mine land use change leads to altered ecosystem structure and function. Ecological Applications, 18, 104-118.

7. Wickham, J.D., Ritters, K.H., Wade, T.G., Coan M. and Homer, C. (2007) The effect of Appalachian mountaintop mining on interior forest. Landscape Ecology, 22, 179- 187.

8. Bernhardt, E.S. and Palmer, M.A. (2011) The environmental costs of mountaintop mining valley fill operations for aquatic ecosystems of the Central Appalachians. Annals of the New York Academy of Sciences, 1223, 39-57.

9. USEPA (2005) Mountaintop mining/valley fills in Appalachia. Final programmatic environmental impact statement. Region 3, US Environmental Protection Agency, Philadelphia. http://www.epa.gov/region3/mtntop/index.htm

10. Fox, J.F. and Campbell, J.E. (2009) Terrestrial carbon disturbance from mountaintop mining increases lifecycle emissions for clean coal. Environmental Science and Technology, 44, 2144-2149.

11. Hooke, R.L. (1999) Spatial distribution of human geomorphic activity in the Unites States: Comparison with rivers. Earth Surface Processes and Landforms, 24, 687- 692.

12. Fritz, K.M., Fulton, S., Johnson, B.R., Barton, C.D., Jack, J.D., Word, D.A. and Burke, R.A. (2010) Structural and functional characterisitcs of natural and constructed channels draining a reclaimed mountaintop removal and valley fill coal mine. Journal of the North American Benthological Society, 29, 673-689.

13. Larson, C.F., Motyka, R.J., Arendt, A.A., Echelmeyer, K.A. and Geissler, P.E. (2007) Glacier changes in southeast Alaska and northwest British Columbia and contribution to sea level rise. Journal of Geophysical Research, 112, F01007.

14. Woolard, J.W. and Colby, J.D. (2002) Spatial characterization, resolution, and volumetric change of coastal dunes using airborne LiDAR: Cape Hatteras, North Carolina. Geomorphology, 48, 269-287.

15. Gesch, D.B. (2005) Analysis of multi-temporal geospatial data sets to assess the landscape effects of surface mining. National Meeting

of the American Society of Mining and Reclamation, ASMR, Lexington, 415-432.

16. Shank, M. (2004) Development of a mining fill inventtory from multi-date elevation data. Advanced Integration of Geospatial Technologies in Mining and Reclamation, Surface Mining Reclamation and Enforcement, Technical Innovation and Professional Services, Denver, 1-16.

17. Anderson, M.G., Merrill, M.D. and Biasi, F.B. (1998) Connecticut River Watershed analysis: Ecological communities and neo-tropical migratory birds. Eastern Conservation Science. The Nature Conservancy, 36.

18. WVGES (West Virginia Geologic and Economic Survey) (2005) Physiographic provinces of West Virginia.

19. Strausbaugh, P.D. and Core, E.L. (1977) Flora of West Virginia. Seneca Books Inc. Morgantown, 1079.

20. Grayson, R.B., Moore, I.D. and McMahon, T.A. (1992) Physically based hydrologic modeling: 1. A terrain-based model for investigative purposes. Water Resources Research, 28, 2639-2658.

21. Environmental Systems Research Institute (ESRI) (2010) ArcGIS ArcMap Version 10.0. Redlands.

22. Moore, D.M., Lees, B.G. and Davey, S.M. (1991) A new method for predicting vegetation distributions using decision tree analysis in a geographic information system. Environmental Management, 15, 59-71.

23. Boer, M., Del Barrio, G. and Puigdefabregas, J. (1996) Mapping soil depth classes in dry Mediterranean areas using terrain attributes derived from digital elevation models. Geoderma, 72, 99-118.

24. O'Loughlin, E.M. (1986) Prediction of surface saturation zones in natural catchments by topographic analysis. Water Resources Research, 22, 794-804.

25. Parker, A.J. (1982) The topographic relative moisture index: An approach to soil-moisture assessment in mountain terrain, Physical Geography, 3, 160-168.

26. Fels, J. and Zobel, R. (1995) Landscape position and classified landtype mapping for the statewide DRASTIC mapping project. North Carolina State University, Raleigh, 1-8.

27. Hinkle, C.R., McComb, W.C., Safely, J.M. Jr. and Schmalzer, P.A. (1993) Mixed mesophytic forests. In: Martin, W.H., Boyce, S.C. and Echternacth, A.C., Eds., Wiley, New York, 203-253.

28. Master, L.L., Flack, S.R. and Stein, B.A. (1998) Rivers of life: Critical watersheds for protecting freshwater biodiversity. Then Nature Conservancy, Arlington.

29. Stein, B.A., Kutner, L.S., and Adams, J.S. (2000) Precious heritage: The status of biodiversity in the united states. Oxford University Press, New York.

30. Ricketts, T.H., Dinerstein, E., Olson, D.M., Loucks C.J., Eichbaum, W., DellaSala, D., Kavanagh, K., Hedao, P., Hurley, P.T., Carney, K.M., Abell, R. and Walters, S. (1999) Terrestrial ecoregions of North America: A conservation perspective. Island Press, Washington.

31. Poiani, K.A., Richter, B.D., Anderson, M.G. and Richter, H.E. (2000) Biodiversity conservation at multiple scales: Functional sites, landscapes, and networks. American Institute of Biological Sciences, 50, 133-146. [

32. Zipper, C.A., Burger, J.A., Skousen, J.G., Angle, P.N., Barton, C.D., Davis, V. and Franklin, J.A. (2011) Resorting forests and associated goods and services on Appalachian coal surface mines. Environmental Management, 47, 751-765.

33. Whittaker, R.H. (1956) Vegetation of the great smoky mountains. Ecological Monographs, 26, 1-80.

34. USEPA (US Environmental Protection Agency) (2003) Draft programmatic environmental impact statement on mountaintop mining/valley fills in Appalachia.

6

CHAPTER

Distribution of Artisanal and Small-Scale Gold Mining in the Tapajós River Basin (Brazilian Amazon) over the Past 40 Years and Relationship with Water Siltation

Felipe de Lucia Lobo [1,2,*], Maycira Costa [1], Evlyn Márcia Leão de Moraes Novo [2] and Kevin Telmer [1,3]

[1]*Spectral Lab, Department of Geography, University of Victoria, 3800 Finnerty Road, Victoria, BC V8P 5C2, Canada*

[2]*Remote Sensing Division, National Institute for Space Research (INPE), Av. dos Astronautas 1758, São José dos Campos, SP 12227-010, Brazil*

[3]*Artisanal Gold Council, 101-732 Cormorant St., Victoria, BC V8W 4A5, Canada*

ABSTRACT

An innovative remote sensing approach that combines land-use change and water quality information is proposed in order to investigate if Artisanal and Small-scale Gold Mining (ASGM) area extension is associated with water siltation in the Tapajós River Basin (Brazil), containing the largest small-scale gold mining district in the world. Taking advantage of a 40-year period of the multi-satellite imagery archive, the objective of this paper is to build a normalized time-series in order to evaluate the influence of temporal mining expansion on the water siltation data (TSS, Total Suspended Solids concentration) derived from previous research. The methodological approach was set to deliver a full characterization of the ASGM expansion from its initial stages in the early 1970s to the present. First, based on IRS/LISSIII images acquired in 2012, the historical Landsat image database (1973–2001) was corrected for radiometric and atmospheric effects using dark vegetation as reference to create a normalized time-series. Next, a complete update of the mining areas distribution in 2012 derived from the TerraClass Project (an official land-use classification for the Brazilian Amazon) was conducted having IRS/LISSIII as the base map with the support of auxiliary data and vector editing. Once the ASGM in 2012 was quantified (261.7 km^2) and validated with photos, a reverse classification of ASGM in 2001 (171.7 km^2), 1993 (166.3 km^2), 1984 (47.5 km^2), and 1973 (15.4 km^2) with the use of Landsat archives was applied. This procedure relies on the assumption that ASGM changes in the land cover are severe and remain detectable from satellite sensors for decades. The mining expansion area over time was then combined with the (TSS) data retrieved from the same atmospherically corrected satellite imagery based on the literature. In terms of gold mining expansion and water siltation effects, four main periods of ASGM activities were identified in the study area: (i) 1958–1977, first occurrence of mining activities and low water impacts; (ii) 1978–1993, introduction of low-budget mechanization associated with very high gold prices resulting in large mining area expansion and high water siltation levels; (iii) 1994–2003, general recession of ASGM activities and exhaustion of easy-access gold deposits, resulting in decreased TSS; (iv) 2004 to present, intensification of ASGM encouraged by high gold prices, resulting in an increase of TSS.

Keywords: land use change; gold mining; water siltation; time-series analysis; Landsat; IRS/LISSIII

1. INTRODUCTION

Artisanal and Small-scale Gold Mining (ASGM) refers to the mining practiced using rudimentary technology by individuals, groups, or communities. This practice is primarily found in developing countries of South America, Asia, and Africa [1,2,3]. In the Amazon, substantial ASGM activities started in the 1950s at a few sites, called *garimpos*. Currently, hundreds of thousands of people are directly involved in ASGM in the Amazon Basin due to the relatively high rise in the gold price (USD 1100/oz) in recent years [2,4,5]. In Brazil alone, ASGM gold production is responsible for 30 tonnes of gold per year, of which about 26% is produced in the Tapajós River Basin in the Amazon by approximately 50,000 miners (or*garimpeiros*) who are distributed in more than 300 mining sites [4,5,6,7].

Despite its contribution to the local economy, ASGM causes several environmental impacts, such as mercury contamination, water siltation, and landscape degradation [2,8]. The ASGM local practice results in high water siltation due to the discharge of sediment from the margins of the rivers, where exploitation of alluvial deposits usually takes place [4]. As the miners exploit a particular area of the alluvial plain, new pits are opened, and the mine tailings (water + sediment) are usually discharged back into the river drainage, or to adjacent small ponds/lakes. The discharge of sediment into the water results in severe impacts on the water quality, such as decreasing light availability for primary production [9], and changing benthic [10] and fish communities [11].

The associated ASGM changes in the landscape (roads, aerial strips) generally include deforestation of the area in order to access gold deposits and settlements. Since several of these settlements are informally operating inside remote protected areas and indigenous lands in the Amazon [5,12], delineation of historical mining areas using satellite images is the only possible tool that allows for better understanding of changes in river water quality associated with gold mining practices in the region. The combined set of information, land use change and water quality change associated with ASGM, is of considerable interest to the land management community for the purposes of evaluation regarding social and environmental implications of gold mining activities in the region. At the same time, it provides information for the proposition of public policies which aim for the recovery of degraded areas and sustainable use of water and mineral resources [5,8,12].

Therefore, with the benefit of the available 40-year period of the Landsat archive, the objective of this paper is to build a normalized imagery time-series in order to extract land use change for four tributaries in the Tapajós River Basin (Jamanxim, Crepori, Tocantinzinho and Novo river basins) and evaluate its influence on the water siltation observed in previous research [12]. To accomplish this, the quality and the time span of the satellite images are of key importance. Studies that aim to detect land use change based on a variation of the target's reflectance requires radiometric and atmospheric correction with further normalization or intercalibration of the satellite images [13,14]. Moreover, detection of environmental changes caused by anthropogenic activities requires data that covers at least the entire period before the initial activities (pristine condition) and up to its present state. In the case of the Tapajós region (the geographic focus of this research), the first gold rush started in the late 1970s after the first Landsat MSS images were acquired (early 1970s). In this context, all historical Landsat archive images (1973–2001) were inter-calibrated with IRS/LISSIII images acquired in 2012, for which atmospheric corrections were previously validated with in situ radiometric measurements [12]. The inter-calibration process used dark vegetation as reference to create a normalized time-series to assure that temporal changes are only due to anthropogenic activities and not to atmospheric or radiometric effects. The corrected Landsat satellite imagery from 1973 (pristine conditions), 1984, 1993, 2001, and 2012 (IRS/LISSIII images) was used as a base map to identify and quantify the ASGM areas. The spatial and temporal effects of ASGM on water siltation (derived from satellite images, [12]) were then conducted based on the land-cover dynamic, general mining techniques, and socio-economic framework throughout the decades.

2. STUDY AREA

The lower section of the Tapajós River Basin covers about 130,370 km^2 and is located at the center of the Brazilian Amazon (Figure 1a). The first occurrence of mining activities started in 1958 and were based on manual efforts, without any mechanization [15]. In 1978, mechanization, such as dredges, was introduced, mostly in the Tapajós and larger tributaries. In the 1980s, exploitation of colluvial ore deposits began with water-jet systems [8] resulting in a boom of the mining population, which reached approximately 150,000 in 1989 [8], with gold production reaching approximately 25 tonnes a year. During this decade, more than 400 airstrips and over 600 mining sites (*garimpos*) were reported [8]. The gold rush in Tapajós compelled the Brazilian government to create the Gold Mining Reserve in 1983, along with other policies, to support local miners and to reduce socio-environmental impacts, with no satisfactory results [5]. In the late 1990s, ASGM decreased drastically, mainly discouraged by the low price of gold and the exhaustion of easy-access alluvial gold deposits. Consequently, gold production reduced gradually to approximately 10 tonnes per year in 2000.

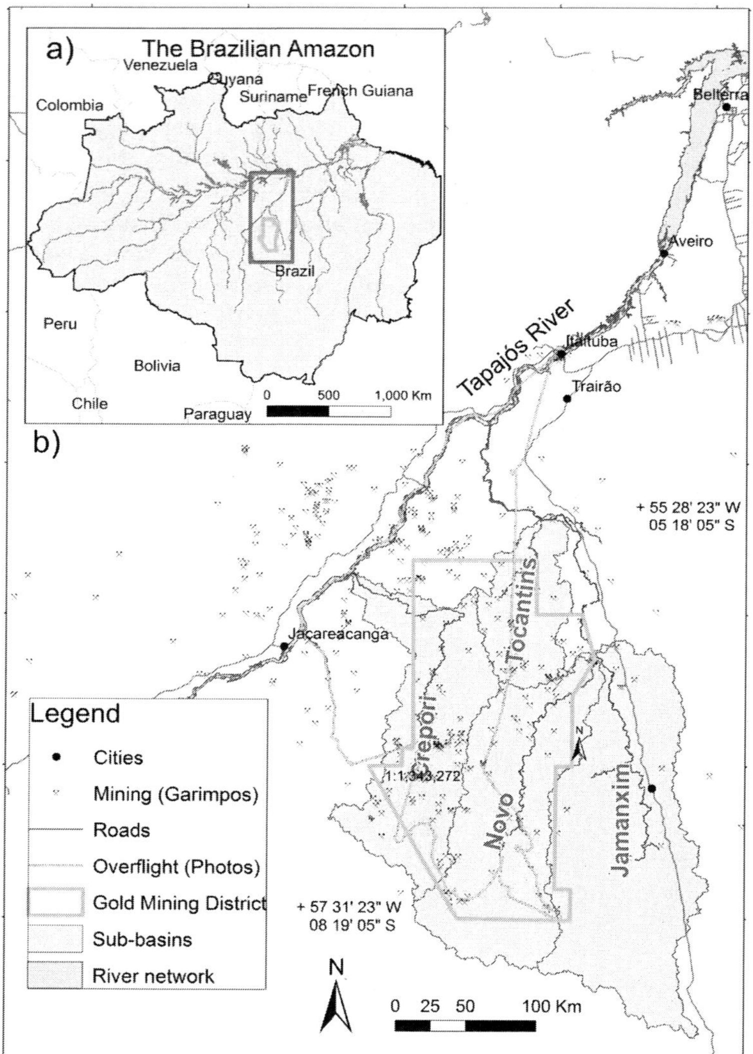

Figure 1. (**a**) Study area located in the Brazilian Amazon with indication of the Gold Mining District created in 1983 [6] by the Brazilian Federal Government. The four sub-basins of the Tapajós River Basin subject to this research are within: Jamanxim, Novo, Tocantins and Crepori sub-basins are depicted in (**b**), along with mining sites, roads, river network, and overflight path for aerial photos used in the land-cover validation process.

This region has been subject to high investments on agri-business and mining, resulting in several social conflicts for the access to natural resources and protected lands. Currently, more than 300 small-scale mines, with the participation of more than 50,000 miners, produce gold within three main sub-basins: the Novo, the Crepori, and the Tocantinzinho (abbreviated to Tocantins) [4,16]. For comparison purposes, another sub-basin (Jamanxim), characterized by low mining activity and high deforestation rates, was also included in the study area (Figure 1b).

3. MATERIAL AND METHODS

The methods used in this research involved six steps (Figure 2): (i) The building of the database including satellite images (1973–2012), aerial photos and auxiliary data (river network, soil type, etc.); (ii) Radiometric and atmospheric correction of satellite imagery having dark vegetation was used as a reference to create a normalized time-series for land use and water siltation detection; (iii) Mapping the gold mining area of the four sub-basins in 2012 based on the IRS/LISSIII images and validated with aerial photos; (iv) Classifying the historical Landsat images (2001, 1993, 1984 and 1973) and generating historical maps of mining area distribution using the 2012 map as a starting point; (v) Evaluating the influence of the historical mining area and sediment sources on satellite-derived historical water siltation (TSS, Total Suspended Solids concentration) from [16]; and (vi) Lastly, intersecting historical maps with the river network and soil type to assess possible sources of the sediment discharged into the rivers due to ASGM.

3.1. Database

The database organized for this research includes (Table 1): satellite images from 1973 to 2012 for identification of mining areas; aerial photos taken to validate classification of ASGM in 2012; ancillary data such as roads, river network and mining sites to support ASGM identification; an official land-cover classification in the Brazilian Amazon provided by the TerraClass project; and historical TSS estimates for the main tributaries of Tapajós River from [12] (see Section 3.2).

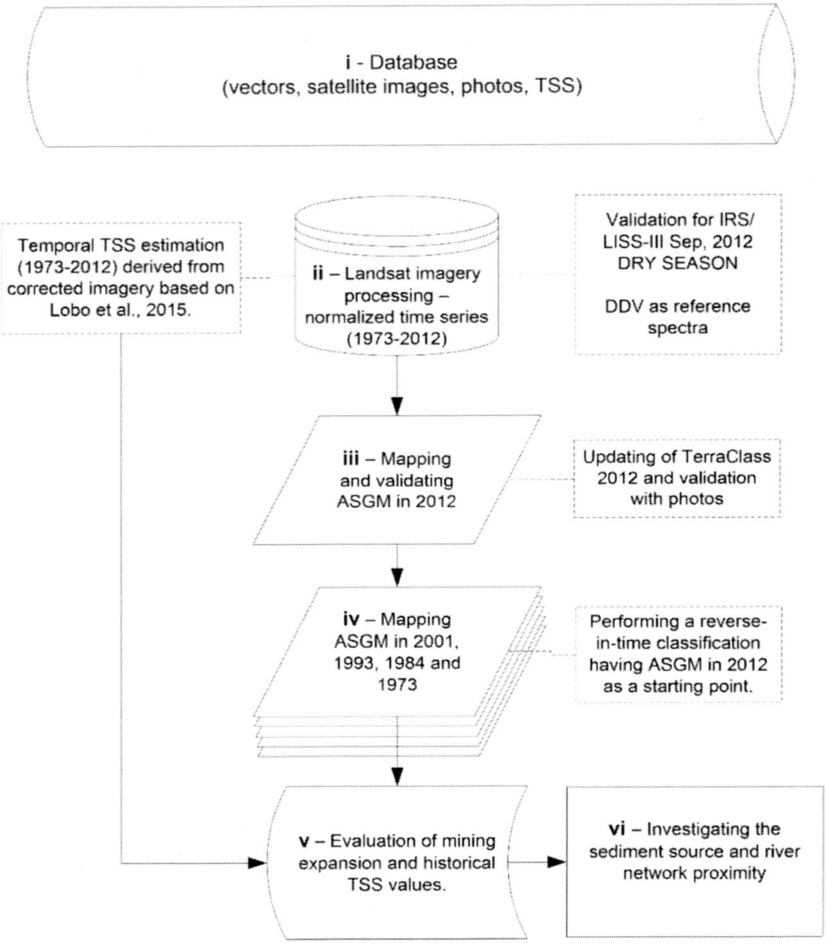

Figure 2. Flowchart of the methodology applied in this paper. (**i**) Database including satellite images, auxiliary vectors, photos and TSS from [12]; (**ii**) Landsat image processing to create a normalized time series from 1973 to 2012 from land use and water siltation extraction; (**iii**) Map ASGM in 2012 and validated with photos; (**iv**) Conducted a reverse in time classification of ASGM in 2001, 1993, 1984 and 1973; (**v**) Combined temporal analysis between TSS and ASGM expansion; (**vi**) Investigation of the sediment source resulted of ASGM activities.

Table 1. Description of the geographical database used including type of data, date of acquisition, general description, scale and source of data.

Type of Data	Date	Description	Scale/Resolution	Source
Satellite Images (5 tiles per date, raster files)	August 1973	Landsat 1 (MSS)	80 m	DGI/INPE [17]
	August 1984 September 1993 August 2001 September 2012	Landsat 5 (TM) Landsat 5 (TM) Landsat 5 (TM) IRS/LISSIII	30 m 30 m 30 m 24 m	
Aerial photos	May 2012	Overflight performed in May 2012. Total of 40 photos and 70 mining polygons identified.		Shared by ICMBio
Ancillary data (vector files)	2013 2010 2010 2006 2012	River network Water Bodies Roads Soil Mining sites and gold mining district	1:50,000 1:250,000 1:250,000 1:250,000 1:250,000	This study [18] [18] [19] [20]
Land use classification (vector files)	2012	TerraClass Project classified land-cover use for the Brazilian Amazon.	1:50,000	Available at CRA/INPE [21]
Historical TSS concentration (raster files)	1973–2012	Estimation of TSS concentration for Tapajós River Basin retrieved from satellite images.	24–80 m	[12]

In order to have a land use classification for every decade since the 1970s using Landsat archive, five satellite image mosaics were constructed. The first set of images was acquired by the Landsat/MSS (Multispectral Scanner) in 1973. In the decades to follow, cloud-free images acquired by the Landsat/TM (Thematic Mapper) were available for the years 1984, 1993, and 2001. In 2012, a group of images acquired by the IRS/LISSIII (Indian Remote Sensing/Linear Imaging Self Scanning Sensor) was used as an alternative due to unavailability of Landsat images. The pixel size of utilized images ranged from 80 m (Landsat/MSS) to 24 m (IRS/LISSIII) (Table 1). For each mosaic, at least five tiles were downloaded from INPE/DGI and corrected for geographic, radiometric, atmospheric deviations and normalized according to methodology proposed by [19] (see Section 3.2).

An overflight performed in May 2012 by ICMBio agents (Brazilian National Agency of Protected Areas) involved collecting about 40 photographs for use in identification of mining activities within protected lands (see overflight path inFigure 1). These photos were shared with us (the authors), and a total of 70 polygons were used in this research to validate land-cover classification in 2012 based on IRS/LISSIII imagery. The visual identification of the polygons in the photos considered the spatial distortion caused by the oblique view of the camera used in comparison to the satellite images acquired with a perpendicular view (nadir).

To support ASGM and deforestation identification, several auxiliary data (vector files) were used such as roads, water bodies, gold mining district areas, mining sites and soil types. Except for the river network generated in this study based on SRTM images [22], all data derives from official national institutes in Brazil (Table 1).

As an attempt to classify and quantify the different land use in the whole Brazilian Amazon, the national government created the TerraClass project led by National Institute for Space Research (INPE). The land use classification system follows the one recommended by FAO/UNESCO (Land Cover Classification System) and includes mining, pasture, agriculture areas, and secondary forest [23]. The classification is based on satellite imagery (usually Landsat archives) used for deforestation detection by PRODES program (Deforestation Monitoring Program in Brazil) and applies raster-based algorithms, such as Linear Spectral Mixing Models to map secondary forest [24], Normalized Difference Vegetation Index (NDVI) to map agriculture areas, and supervised classification Bhattacharya to map pasture areas. After mapping secondary forest, agriculture and pasture areas, the remaining deforested areas were classified into urban and mining regions based on visual interpretation of the satellite images used as background [21]. As the TerraClass classification is freely available at CRA/INPE [21] and represents one of the most important information sources for land use analysis in the Brazilian Amazon, it was used as a starting point to identify and quantify the mining and other land uses in 2012.

3.2. Imagery Time-Series for Monitoring Land Use and Water Quality Changes

To address the atmospheric correction and imagery normalization issues, the proposed method started with a set of images IRS/LISSIII acquired in September 2012, which were atmospherically corrected and validated based on in situ surface reflectance, ρsurf, as described by [12]. The atmospheric correction was performed using a physical-based algorithm in FLAASH-ENVI [25] and considered ranges of water vapor values and aerosol optical thickness (AOT) from the closest AERONET (AErosol RObotic NETwork) station [26], near Belterra city (3°06′S, 55°03′W). After validation, the 2012 IRS/LISSIII images were used as reference for the inter-calibration of the historical Landsat images (MSS and TM). For Landsat-TM images acquired from 1984 to 2001, the atmospheric correction was performed with water vapor and aerosol optical thickness

(AOT), optimized to match the DDV (Dense Dark Vegetation) spectra of those from previously corrected IRS/LISSIII.

Following the atmospheric correction of the imagery database, the DDV spectra from the historical images were extracted and compared with the reference images (IRS/LISSIII). The historical images in which ρsurf(red) did not fall within an acceptable range (up to 50% difference at red band, which corresponds to a difference of up to 1.5% in ρsurf(red)) and were subjected to another iteration of atmospheric correction by optimizing the water vapor and visibility parameters (Figure 3a). The acceptable difference in threshold was defined for the red band since the red band of all the sensors used in this analysis present similar ranges (center at ~650 nm). Furthermore, the same image data set was processed for the water siltation analysis and the red band presented the highest correlation to increasing TSS [12].

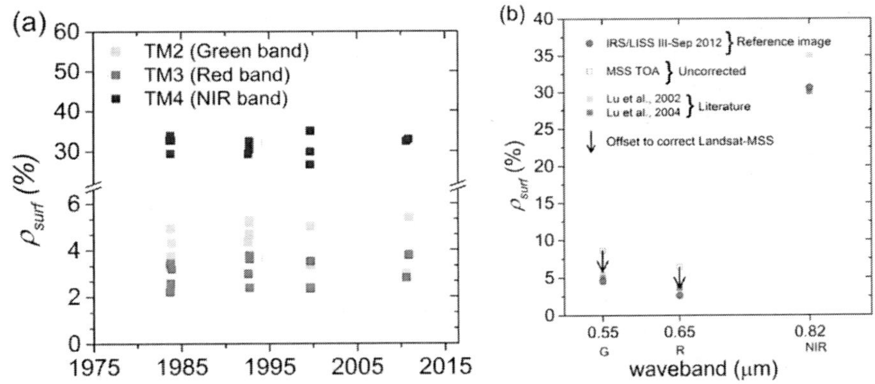

Figure 3. (**a**) DDV (Dark Dense Vegetation) spectra for dry seasons (August–October) extracted from historical Landsat-5 TM images having IRS/LISSIII spectra shown in (**b**) as reference; (**b**) DDV extracted from dry season (IRS/LISSIII) used as reference for optimizing atmospheric parameters on Landsat historical data. Examples of uncorrected Landsat 1-MSS (ρTOA), and dense forest from the Brazilian Amazon found in literature using Landsat TM data [27,28] are also shown. Adapted from [12].

Landsat 1-MSS specifications were not available for the physical method employed in the atmospheric correction of the other images. As such, the Landsat 1-MSS images (1973) image sets were radiometrically corrected to reflectance at the Top-of-Atmosphere (TOA), ρTOA. The atmospheric effects were corrected by applying offsets only at the green and red bands (Figure 3b), since the MSS infra-red waveband (750 nm) is not comparable to those from Landsat-TM and IRS/LISSIII centered at 820 nm. Once all the images were corrected and inter-calibrated, the analysis of land use (Section 3.2 and Section 3.3) and water siltation (Section 3.2) temporal detection was performed.

Historical Total Suspended Solids (TSS) Estimated from Normalized Imagery Time-Series.

The historical variation of TSS along the Tapajós River and main tributaries were extracted from [12] for the same period used in this paper (1973–2012), after an extensive sampling of water bodies impacted and non-impacted by gold mining tailings.

The relationship between TSS and satellite data (surface reflectance at red band, ρsurf(red)) was established after two field campaigns performed in April 2011 (high water level) and September 2012 (low water level) in the Tapajós River and main tributaries for TSS and surface reflectance measurements. Two sets of satellite images were acquired and calibrated during the field campaigns, using the in situ data (water surface reflectance at red band) as ground truth. As TSS increases, the ρsurf(red) increases according to the power function depicted in the Figure 4.

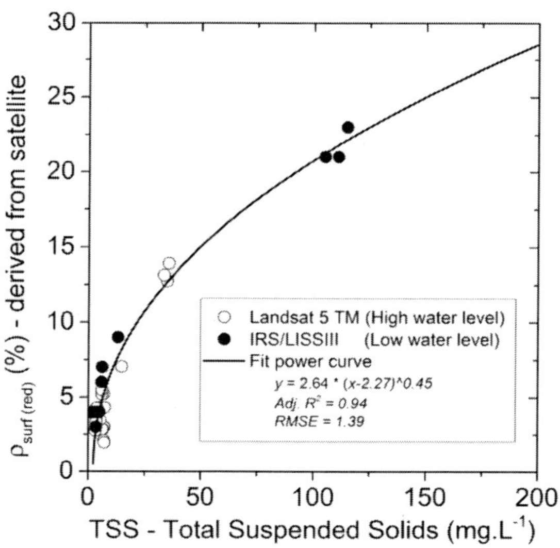

Figure 4. Non-linear fit between TSS (n = 39, Adj. R^2 = 0.94, RMSE = 1.39) and reflectance (red band) derived from satellite sensors (Landsat TM data for high water level, April 2011; and IRS/LISSIII data for low water level season, September 2012). Adapted from [12].

The historical water siltation estimation corresponds to the low water level period, as that is the time of year when mining activities are most intense [12]. The Landsat MSS (1973), TM (1984–2001) and IRS/LISSIII (2012) images were corrected for atmospheric and sun-glint effects, and ρsurf(red) at red band was used to retrieve TSS concentrations ($mg \cdot L^{-1}$) based on this empirical equation [12].

3.3. Mapping Mining Areas in 2012

Prior to having the 'ground truth' information for validation purpose of mining areas in 2012, we first analyzed the aerial photos to identify the polygons of ASGM and other land uses. A total of 70 polygons were identified in the photos, varying in size and shapes, and were used for validation of ASGM and other uses in 2012 (Figure 5). The scheme of land-cover classification used throughout this paper was defined as: (1) ASGM, which includes all land use related to active or inactive mining operations, such as mining areas themselves, airstrips, degraded areas, mining-related ponds, mining settlements; and (2) deforested areas that are not associated with mining, which includes pasture and other agriculture-related areas. Along with users' expertise in visual identification of mining areas in the IRS/LISSIII image (2012), the mapping process used auxiliary information such as *garimpos* distribution, roads, and other secondary information for a more accurate identification. All the visual interpretations of land use between ASGM and others were based on RBG composition (R/G/B: Red/NIR/Green bands) where ASGM followed an irregular pattern, bright signal (white), and mostly along the drainage network (Figure 5bi). On the other hand, other uses such as pastures present a regular shape; usually, between the drainage network, and not as bright as ASGM areas because of the presence of grass and other vegetation (Figure 5bii).

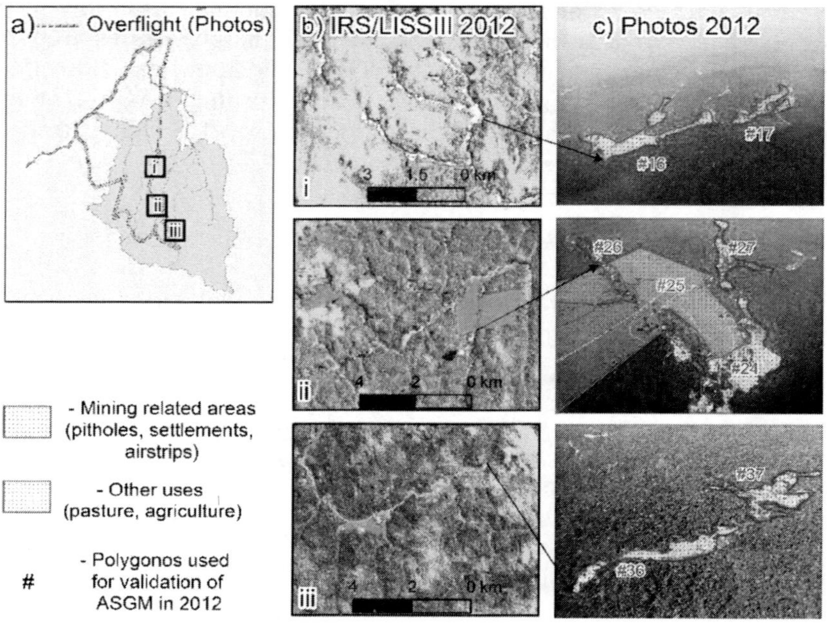

Figure 5. (**a**) An overflight was taken in the study area to identify ASGM and validate classification of land use based on IRS/LISSIII images (R/G/B: Red/NIR/Green bands) (**b**). Examples of eight polygons extracted from three different areas (**i–iii**) includes mining-related areas (**in red**) and other land uses that are not associated with mining (**in purple**). The visual identification of the polygons in the photos taken in 2012; (**c**) considered the spatial distortion caused by the oblique view of the camera used.

To identify the small-scale gold mining in the whole study area in 2012 using satellite images, the initial step that was considered was the usage of the TerraClass maps (named as *TerraClass-2012*), which provides land use classification for the Brazilian Amazon in 2012 (Table 1). We considered the use of *TerraClass-2012* as a starting point in identifying and mapping ASGM distribution in 2012 as it is an official classification that is freely available, and provides land use validated with field works. As the TerraClass project covers the whole Brazilian Amazon on a large scale, it lacks in detecting very small mining areas. To overcome that and provide an up-to-date identification of ASGM in 2012 (named as *MapASGM-2012*), *TerraClass-2012* was set as a base classification to guide the manual vector editing having the satellite image IRS/LISSIII acquired in 2012 as background. Then, validation of both *MapASGM-*

2012 and the *TerraClass-2012* maps took place with the 70 polygons of different sizes corresponding to mining and deforested areas identified in the aerial photos.

3.4. Mapping Mining Areas in 2001, 1993, 1984, and 1973

The procedure to generate the time series of mining area expansion followed a 'reverse-in-time' approach, having started with the use of the validated *MapASGM-2012* as a reference for the identification of the mining areas in the 2001 Landsat images. Repetition of the same procedure occurred for the other dates: 1993, 1984 and 1973. This procedure relies on the assumption that ASGM changes in the land cover are severe and remain detectable from satellite sensors for decades (Figure 6). In most cases, once there was an ASGM mining site in the past, it is likely to have remained a mining site to the present day.

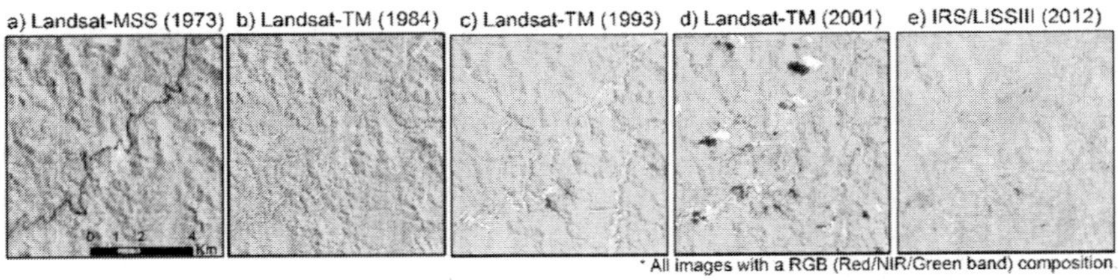

Figure 6. Example of how severe ASGM land changes remains detectable by satellite images. It also illustrates ASGM expansion over the decades from 1973 (**a**); 1984 (**b**); 1993 (**c**); 2001 (**d**); to 2012 (**e**). RGB composition of Landsat and IRS/LISSIII images.

The up-to-date validated *MapASGM-2012* can be used as a guide for mapping ASGM in the previous decades when no information for validation purposes is available. For example, to distinguish ASGM and deforested areas in 2001 (Figure 7a), an unsupervised classification (IsoData) using red and near-infrared bands was conducted to detect all land-cover use that is not forest into one class (named as *land use mask 2001*) (Figure 7b). In order to distinguish mining and deforested areas properly, this *land use mask 2001* was intersected with the *MapASGM-2012* (Figure 7c) followed by a precise manual vector editing at 1:25,000 scale to incorporate land use areas that were not included in the intersection with 2012, such as the case of regrown secondary forest, resulting in the *MapASGM-2001* (Figure 7d). The manual vector editing, based on visual interpretation and ancillary data, is also essential in assigning the correct classification in the case of the areas, which had been converted, to other land uses. Once the *MapASGM-2001* is completed, the same process (steps Figure 7a–d) was repeated for the remaining years (1993, 1984, and 1973) for the sub-basins of study: Crepori, Novo, Tocantins, and Jamanxim (Figure 1).

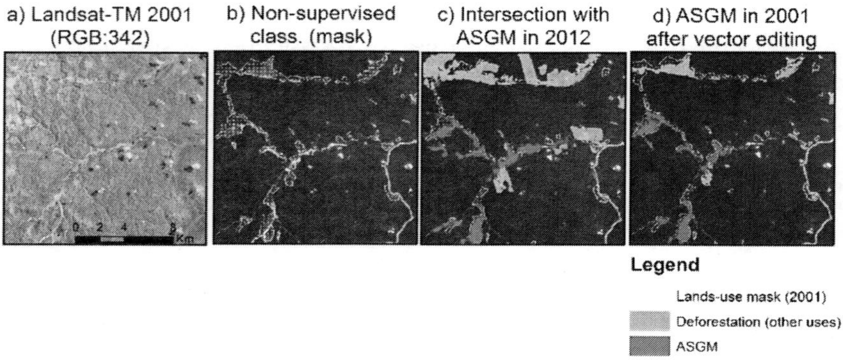

Figure 7. Mapping mining areas (and deforested areas that are not mining) using Landsat historical images (2001, 1993, 1984, and 1973). Satellite image database in 2001 (R/G/B: Red/NIR/Green band) (**a**) was submitted to an unsupervised classification where all land use classes, except forest, are included into one class, creating a *land use mask 2001* (in **yellow**) (**b**). This mask was then intersected with the *MapASGM-2012* (**c**) followed by a vector editing in order to include all mining (**red** polygons) and deforested areas (**pink** polygons) in 2001 resulting in the *MapASGM-2001* (**d**). The same procedure was applied to the imagery Landsat-TM database in 1993, 1984, and Landsat-MSS data in 1973.

3.5. Relationship between Historical Mining Areas and Water Siltation

Since neither historical nor present data on sediment transport of mining tailings into the river network is available, the historical mapped mining areas were spatially associated with corresponding satellite-based water siltation (total suspended solids-TSS) from [12] for the tributaries, Jamanxim, Crepori, Novo, and the Tocantins. Further investigation of spatial and temporal effects of ASGM on water siltation were conducted based on the land use dynamic (generated in this paper), mining techniques [29], gold production [6], and socio-economic framework [30,31] throughout the decades.

3.6. Investigating the Source of the Sediment Derived from ASGM

To assess the source of the sediment carried from mining areas to the drainage, the temporal gold mining areas were overlayed with thematic maps such as soil type and connectivity with the river network. The map relating to soil type, from the Brazilian national classification [24] in a 1:250,000 scale, was clustered according to the international classification [32] in order to be comparable to other research around the world. Buffering-zones (100, 250, 500 and 1000 m) along the river network (generated in this study, scale 1:50,000) were created to define proximity of mining area to the river network. According to the recent New Forest Code (National Law n°12.651/2012), a buffer distance of at least 100 m from rivers with a width between 50 and 200 m is permanently protected by law and no kind of land use is allowed because it directly affects water quality. Similarly, buffering-zones were created along roads [18] in order to investigate their relevance to water siltation.

4. RESULTS

4.1. Validating Gold Mining Area in 2012

The polygons identified in the photos were used to validate the land use cover classification in 2012. According to the validation matrix (Table 2), the *TerraClass-2012* underestimated the mining area. *TerraClass-2012* failed to classify mining in 21 out of 53 polygons that are identified as ASGM in the photos, with an overall accuracy of 64% (Table 2). The validation of the *MapAGSM-2012* classification showed that the overall accuracy increased to 93%, since manual editing included mining areas of different sizes from very small (5 ha) to exceedingly large (1000 ha), and updated mining areas in 2012. Omission errors total 6% (3/53), whereas commission errors are less than 4% (Table 2).

Figure 8 depicts one example of omission error and another example of misclassification of the *TerraClass-2012* in comparison to the updated *MapASGM-2012*.

Table 2. Matrix of validation of *TerraClass-2012* and *MapASGM-2012* land use classification having aerial photos as ground-truth. Overall accuracy is 64% for*TerraClass-2012* and 93% for *MapASGM-2012*.

		Classification of *TerraClass-2012*			
		Mining	**Others Uses**	**Not Classified**	**Total**
photos	Mining	32	13	8	53
	Others uses	3	14	0	17
	Total	35	27	8	70
		Classification of *MapASGM-2012*			
		Mining	**Other Uses**	**Not Classified**	**Total**
photos	Mining	50	3	0	53
	Other uses	2	15	0	17
	Total	52	18	0	70

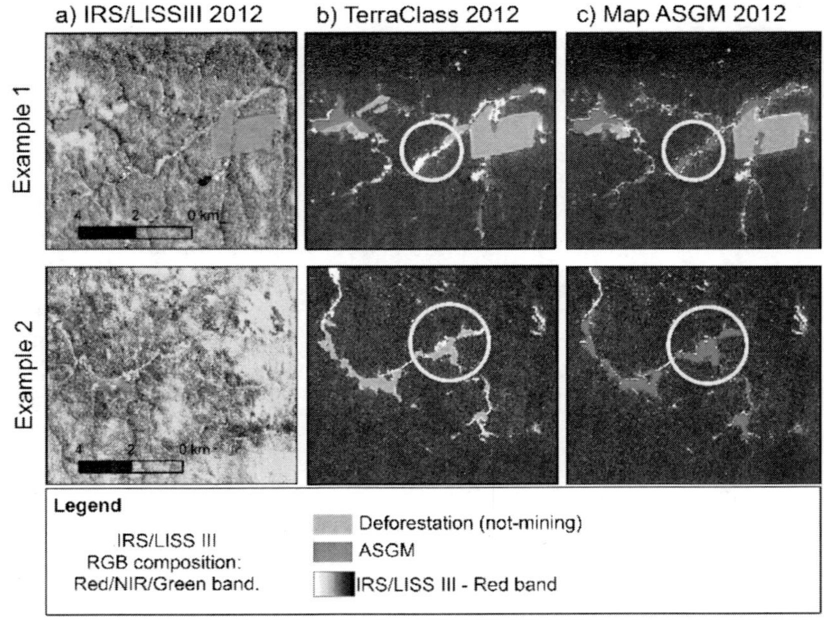

Figure 8. (**a**) The IRS/LISSIII was used as a background image to support ASGM identification in 2012. Examples of performance of *TerraClass-2012* (**b**) and updated*MapASGM-2012* (**c**), are given in two image subsets to illustrate the omission (Example 1) and misclassification errors (Example 2) observed for *TerraClass-2012* in comparison to updated *MapASGM-2012* (yellow circles).

4.2. Gold Mining Area from 1973 to 2012

The satellite-derived mining area in the four sub-basins of study during the past 40 years—1973, 1984, 1993, 2001, and 2012—are presented in Figure 9 and Table 3. In 1973, dominant forested conditions were observed with a mining area of 15.4 km^2 (0.03% of the total area of study), mainly limited to the Crepori (6.3 km^2) and Novo (7.0 km^2) sub-basins. Eleven years later, the mining area increased to 47.5 km^2 (0.09% of the total area), and again the Crepori and the Novo sub-basins showed the largest mining areas with 22.8 and 16.1 km^2, respectively. In 1993, the mining area tripled to 166.3 km^2 (0.30% of the total area), with the Crepori sub-basin alone showing ASGM up to 77.0 km^2, followed by the Novo (69.2 km^2), the Tocantins (11.9 km^2), and the Jamanxim (8.2 km^2). Contrary to the trend in previous decades, the next decade (to 2001) saw the total mining area increase only slightly, to 171.7 km^2 (0.31% of the total area). The stagnation period during the late 1990s and early 2000s rapidly changed within the following decade, resulting in an increase in the mining area to 261.7 km^2 (0.48% of the total area) in 2012. It is interesting to point out that the Jamanxim sub-basin, while presenting the lowest mining area, had the greatest deforestation rates throughout the five-decade period (20% of the total area in 2012), mostly along the BR-163 (South-North) highway.

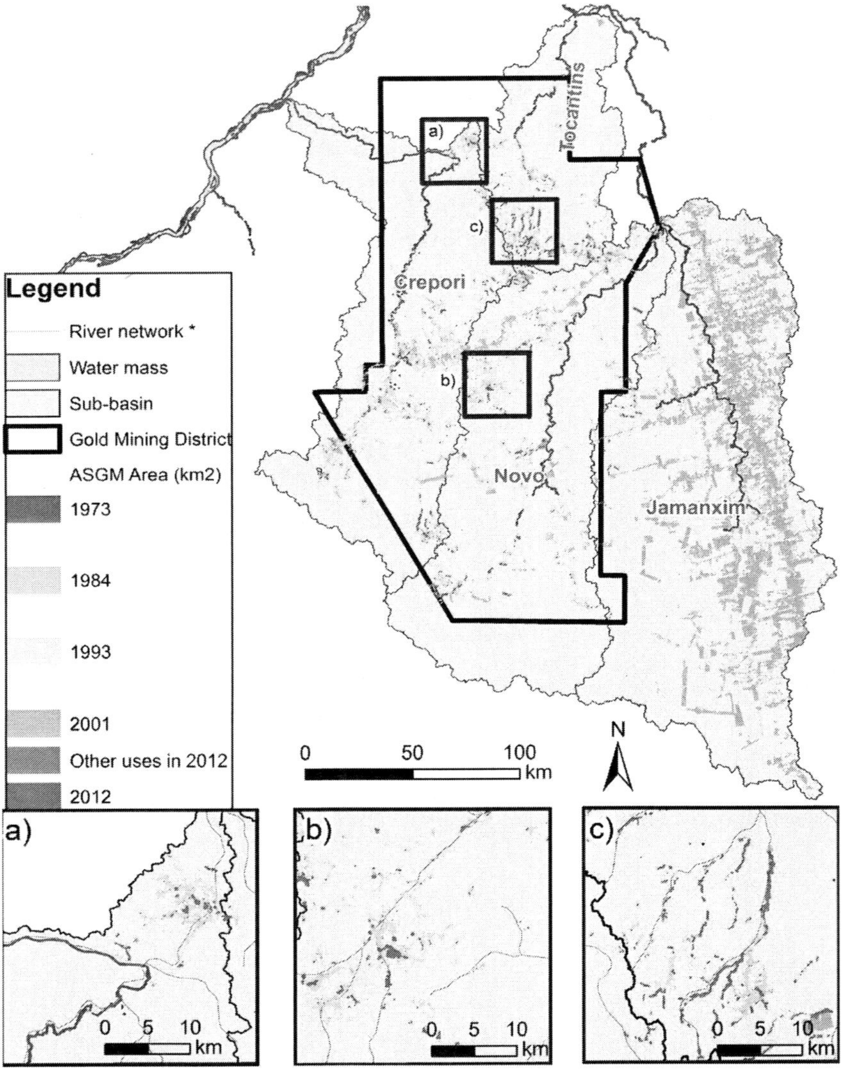

Figure 9. Map of the ASGM expansion (1973, 1984, 1993, 2001, and 2012) in the area of study. Note that the expansion occured during different periods at different regions: for Cuiú-cuiú area (**a**) encountered the highest expansion occured until 1984; for the Surubim area (**b**), most of the expansion happened between 1984 and 1993; and for the Tocantinzinho area (**c**) the expansion is more prominent between 1993 and 2012. Other land uses in 2012 are also plotted in dark gray. * River network only displayed on (**a–c**).

Table 3. Tabulation of historical ASGM and other uses not including mining for the four sub-basins of study: Crepori, Novo, Tocantins and Jamanxim.

		Mining Area (km^2)			
	Jamanxim	**Crepori**	**Novo**	**Tocantins**	**Total**
1973	0.1	6.3	7.0	2.0	15.4
1984	2.2	22.8	16.1	6.3	47.5
1993	8.2	77.0	69.2	11.9	166.3
2001	7.3	77.9	63.1	23.3	171.7
2012	10.9	115.5	96.2	39.1	261.7
		Other uses (km^2)			
	Jamanxim	**Crepori**	**Novo**	**Tocantins**	**Total**
1973	3.8	0.5	0.5	0.0	4.8
1984	185.1	11.7	10.2	2.9	209.9
1993	579.2	52.5	62.8	14.6	709.1
2001	1338.2	69.2	107.3	38.8	1553.5
2012	4212.0	192.2	365.7	54.9	4824.8

4.3. Historical Gold Mining Area and Water Siltation

The historical gold mining area maps, along with historical TSS concentration from [12] were combined in order to investigate the temporal relationship between mining area expansion and water siltation in the four tributaries: Jamanxim, Crepori, Novo, and the Tocantins sub-basins. Figure 10 shows the percentage of ASGM increase between 1973 and 2012 calculated as $(A_1 - A_0)/A_0$, where A_0 is the ASGM area at the initial date and A_1 is the ASGM area in the following decade, plotted along with historical TSS concentration.

Although other uses prevail and mining activity does occur in very small areas in the Jamanxim River sub-basin, the impact on water siltation is relatively minor compared to the other basins and TSS remains under 5 $mg \cdot L^{-1}$ throughout the period studied (Figure 10d).

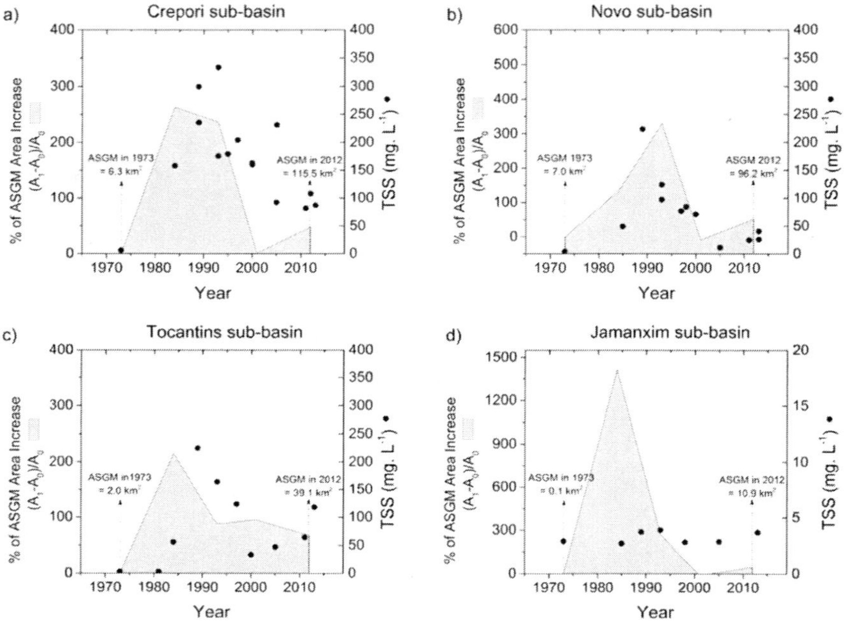

Figure 10. Percentage of ASGM area increase from 1973 to 2012 plotted along with historical TSS derived from [12] for the four sub-basins studied (**a–d**). Note different TSS scale for the Jamanxim River Basin.

In Crepori and Novo rivers, the ASGM started with a few mining locations in 1973, with no resultant perceived impact on water turbidity; TSS concentrations were <5.0 mg·L^{-1}, similar to upstream Jamanxim which can be considered relatively pristine in regards to ASGM. However, in 1984, mining areas increased by about two to three times for the Novo and Crepori rivers, respectively, and within that period, high TSS concentrations were observed corresponding to >50 mg·L^{-1} for Crepori, and >15 mg·L^{-1} for Novo (Figure 10a,b).

In the early 1990s, the Crepori and the Novo rivers exhibited a similar trend; TSS concentration was above 100 mg·L^{-1}, while mining areas increased from around 20.0 km^2 to 75.0 km^2 in both sub-basins. In the Tocantins sub-basin, the increase in mining area was less abrupt than in the Crepori and Novo sub-basins until the late 1980s. However, in 1989 the TSS concentration increased to concentrations as high as 220.0 mg·L^{-1}, indicating a significant increase in mining activities in the Tocantins sub-basin during this period (see Figure 11 for TSS maps).

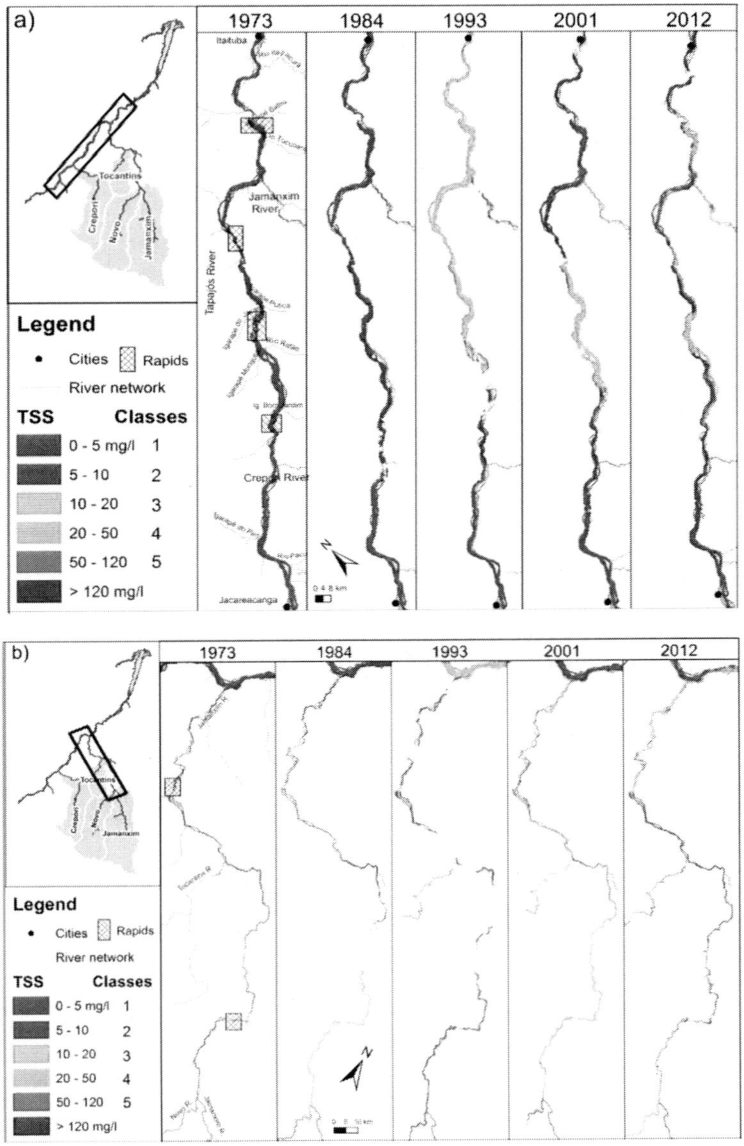

Figure 11. Spatio-temporal distribution of TSS along the Tapajós (**a**) and Jamanxim rivers (**b**). Sections of the river with rapids that often cause water mixing are also indicated. TSS maps based on five mosaics of satellite images: 1973 (Landsat-MSS), 1984, 1993, 2001 (Landsat-TM), and 2012 (IRS/LISSIII). All images are from the dry season (August to October); see [12] for more information.

The period between 1993 and 2001 shows scant expansion of the mining area in the sub-basins as seen in Figure 10. For all impacted sub-basins, a tendency of TSS reduction was observed during this period of ASGM recession. Even with reduction of mining expansion, the TSS values do not return to initial values in 1973 (TSS < 10 mg·L^{-1}) indicating that gold mining kept active, or even that inactive degraded areas (mining tailings) could still contribute to the high TSS values observed.

However, the recent gold rush, occurring over the past 10 years (represented by the percentage of gold mining area increase from 2001 to 2012), has influenced TSS concentrations. The Novo and the Tocantins rivers showed increasing TSS concentrations (mg·L^{-1}) in this period. This tendency was not observed for the Crepori, but this tributary still presented high TSS concentrations (around 100 mg·L^{-1}) throughout the past decade (Figure 10 and Figure 11).

Figure 11 shows the spatial and temporal distribution of TSS along the Tapajós River (Figure 11a) and the Jamanxim River (Figure 11b) with indication of its main tributaries, and illustrates the TSS variation observed in Figure 10.

4.4. Characterization and Distribution of Gold Mining Area: Proximity to Roads and Drainage Network, and Soil Types

The quantification of the mining area along the river network buffer zones showed that, in 2012, 40% of total mining areas were within a 500 m distance from the river network (Figure 12). Moreover, within the first 100 m, ~9% of the total mining area was mapped in 2012. Together with the increment of the total mining area over the years, the percentage along the river network also increased from 1973 to 2012. As an example, within the first 500 m from the river network, we found 12% of ASGM in 1973, and 31% in 1984. In the years 1993 and 2001, the percentage was similar to that of 2012 (40%), which indicates that the mining area increment in 2012 followed a pattern of gold exploitation similar to that of the previous two decades. Mapped mining areas are not predominantly located along the road network. For instance, approximately 5% of the total mining area in 2012 was mapped within a buffer of 1000 m along the roads, mostly along the 'Trans-Garimpeira' road (East-West). On the other hand, the distribution of other land uses are highly associated with the road distribution as evident in the Jamanxim basin with deforestation areas along the BR-163 (North-South) [31].

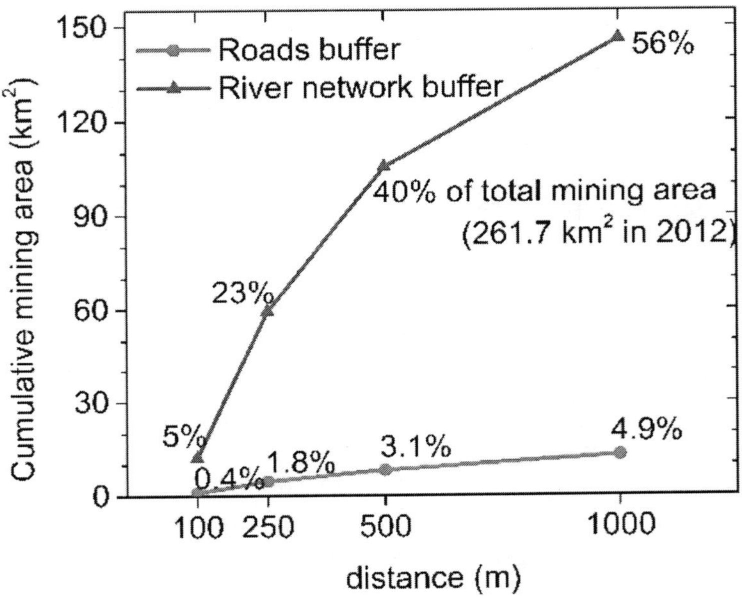

Figure 12. Cumulative mining and deforestation areas for 2012 along river and road buffer zones for 100, 250, 500, and 1000 m.

The intersection of the most recent gold mining map over the soil type map [32] indicates that mining predominantly occurs in ferrasol soils (Table 4). Overall, ferrasols are characterized with high clay content (up to 70% of total weight) and covers more than 90% of the study area. Another heavily mined area occurs in gleysol soils related to alluvial systems. Gleysols are generally also clay-rich, but with higher water content as they are subject to flooding several months a year.

Table 4. Tabulated mining area in 2012 over different soil types according the international classification [32].

Soil Type (World Reference Base, 1998)	Mining Area in 2012 (km^2)
Acrisol	1.2
Arenosol	2.6
Gleysol	53.7
Ferrasol	180.5

5. DISCUSSION

5.1. Satellite Imagery Time-Series for Estimation of Land Use and Water Siltation Change

The proposed inter-calibration of historical imagery, using DDV as a reference target, demonstrated that forest spectra from historical Landsat-5 TM (1984–2011), behaved as an invariant spectral feature over decades (Figure 3). Consequently, the ρsurf(λ) variation observed for mining areas and water bodies were attributed not to environmental conditions but only to the land use change and to variation in optical constituents in the water, respectively [12]. These procedures provided the necessary normalized database for a complete investigation on the impacts of ASGM expansion on the water siltation extracted from the same imagery database in the Tapajós River Basin.

The ASGM area distribution in 2012 was identified and quantified based on IRS/LISSIII images. We demonstrated that the TerraClass classification failed to classify small mining areas, and misclassified other polygons to other uses when validated with photos. To address this issue, a complete update of the *Terraclass-2012* occurred based on the brightness (spectra), shape, and location criteria for identification of ASGM (Figure 5) in comparison to other uses, such as pasture, resulting in the *MapASGM-2012*. Based on the fact that ASGM remains detectable for several decades by satellite images (Figure 6), the proposed methodology provided reliable classification to perform reverse-in-time identification and quantification of ASGM (2001, 1993, 1984, and 1973) imagery dataset having the MapASGM-2012 as a starting point (Figure 7). Although this method requires user training for proper identification of ASGM, it can be conducted with only a few steps and provides a reliable quantification and distribution of ASGM in the Amazon.

The imagery-derived TSS from the Tapajós River Basin [12] suggests that its variability is mainly affected by the concentration of inorganic suspended solids and is also increased due to mining activities. We observed that the possible increase in TSS caused by deforestation is not at the same magnitude as the increase in TSS caused by mining activity. For example, the Jamanxim River, which is characterized by highly deforested areas but with low mining activity, presented low TSS values. Generally, the natural clear water condition of the Tapajós River Basin [33] facilitates the detection of sediment plume caused by gold mining as opposed to naturally white water rivers which also receives mining tailing, such as the Madre de Dios River in Peru.

The whole methodological approach of creating a normalized (inter-calibrated) time series, followed by a simultaneously extraction of land use and water siltation change, provided very satisfactory information for a further discussion on the ASGM impacts on the water quality changes over the decades.

5.2. Spatio-Temporal Analysiof ASGM and Water Siltation (1973–2012)

The spatio-temporal extent of ASGM and its relationship with river water siltation is discussed in a chronological order considering the history of gold exploitation and the physical processes of sediment transportation (mining methods, and soil type) that affect water siltation of the Tapajós River system. Combining the historical context of mining activities from [5,8] and data from this research, four main periods of ASGM in this region were identified (Figure 13):

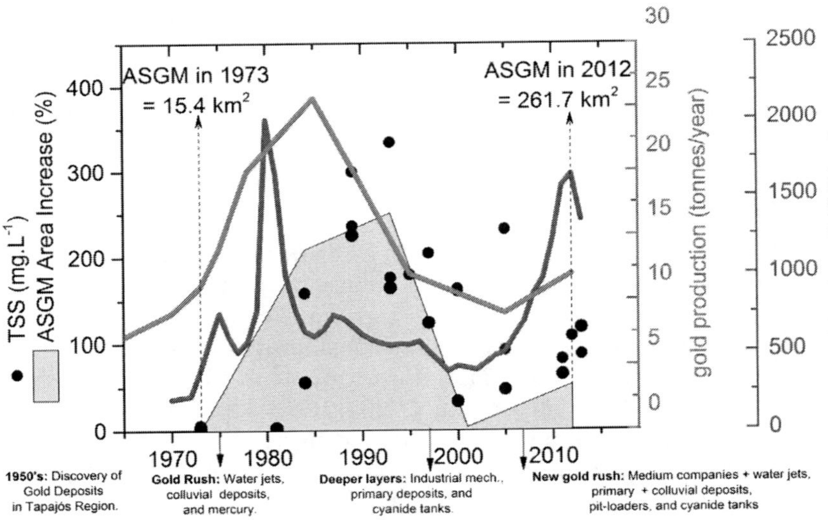

Figure 13. Plot of the percentage ASGM for the whole study area, along with the TSS concentration at the Crepori and Tocantins rivers, gold production in tonnes/year [4] in the Tapajós Area, and gold price (USD/oz) adjusted for inflation from 1970 to 2013 [34]. Chronological events of mining techniques, type of gold deposits exploited and type of miners are indicated by the arrows. Adapted from [12].

Period from the 50s to mid-70s—As an alternative to the decline of latex production, the first ASGMs were established around 1958 along the Tropas and the Crepori rivers [29]. Following the National Integration Program [35,36], the next decade experienced the construction of several highways crossing the Brazilian Amazon, such as the Cuiabá-Santarém Highway (BR-163) [31]. These investments encouraged a significant migration from the northeast region of Brazil to the Tapajós River basin to start new mining pits. As a result, in the early 1970s the estimated area of ASGM was approximately 15.4 km^2 (Table 3), with a gold production of 5 tonnes/year, approximately [5,7,8].

Until 1977, ASGM exploited mostly the alluvial deposits of secondary gold ore by manual and rudimentary methods, such as those involving the use of pans and picks [29]. The low capacity of ore processing during this period explains the observed low water siltation impact observed. TSS concentrations lower than 5.0 $mg \cdot L^{-1}$ were estimated for the Crepori River (Figure 10a), similar to the expected concentrations for 'clear water' according to the Amazonian water classification scheme [37].

Satellite-based estimated TSS concentrations in 1973 data for the remaining tributaries were contaminated by radiance adjacency effect [12] due to MSS image pixel size (80 m). Despite potential uncertainty of the 1973 satellite-based estimates for narrow rivers, the estimated TSS concentrations were similar to those of Crepori and upstream Tapajós River, thus corroborating that the establishment of ASGM (15.4 km^2) and employed techniques from 1958 to 1973 resulted in low water siltation impact.

Period from late 70s to early 90s—As a result of the introduction of a new low-budget technology based on the use of water-jets and rafts, and the concurrence of factors such as high gold prices and government support [4,5], the following decade had considerable spatial expansion of all mining locations. Our data shows that in 1984, ASGM area corresponded to 47.5 km^2 and in 1993 (166.3 km^2) maps showed a spatial expansion of all mining locations compared with the 1973 map (15.4 km^2) (Figure 9 and Table 3). This period corresponds to the peak of gold mining production (on average, 20 tonnes a year), when approximately 100,000 *garimpeiros* were reported in the studied region [4,5]. It is important to note that most of this ASGM expansion occurred along the river network. In 1993, for example, 55% of the mining areas were located within 1000 m of the river network (Figure 12) and it is worth noting that the soil type in the alluvial is mostly gleysol (Table 4), with high amounts of clay (>70%) and inorganic fine particles that remain suspended in the water for many kilometers [38]. Other studies mapping ASGM in the Amazon also indicate that most of the mining areas are distributed along the river network with high potential for direct impact on the aquatic system [38,39,40].

In the sub-basins where the ASGM area increased, the TSS increased more than one order of magnitude. For example, in the Crepori River, TSS increased from 6.5 $mg{\cdot}L^{-1}$ in 1973 to over 200.0 $mg{\cdot}L^{-1}$ between 1989 and 1993 (Figure 10). The effects of ASGM were also observed for the Novo and the Tocantins rivers, where TSS increased to levels above 100 $mg{\cdot}L^{-1}$ during the same period (Figure 10 and Figure 11).

The impacts of increased water siltation on these aquatic environments are numerous. Increased water siltation has a strong impact and direct consequences for how light is attenuated in the water and consequently the depth of the euphotic zone. For instance, Lobo et al. [41] have defined that in the Tapajós River basin, the euphotic zone can decrease from 6.0 m to 1.0 m in non-impacted and ASGM impacted rivers, respectively. These changes can potentially change the phytoplankton community [41], and primary production [42].

It is worth noting that, in the Jamanxin sub-basin, high rates of deforestation rates were observed, increasing from 3.8 km^2in 1973 to 579.2 km^2 in 1993, but impact on the TSS was not identified, since concentrations were lower than 5.0 $mg{\cdot}L^{-1}$ for all years (Figure 10). Although a few studies show that other uses such as pasture and agriculture fields can have an impact on the water quality [43,44,45], considering the dataset of the case of Tapajós River Basin, water siltation caused by other uses can be assumed to be negligible when compared to water siltation caused by ASGM.

Period from mid-90s to mid-2000s—the high spatial extension of ASGM areas and associated increase in TSS concentrations in the rivers observed in the previous period changed to a more stagnant condition in the following decade (Figure 10). From 1990 to 2000, gold production in this region decreased gradually from approximately 20 to 8 tonnes per year due to a reduction in the international gold price in the principal stock markets around the world [8]. In addition, the exhaustion of easy-access alluvial and colluvial gold deposits forced the miners to start exploiting primary deposits of gold, which require high mechanization levels and an increase in the costs of production [8]. These factors contributed to the observed minimal change in ASGM area and the decrease in suspended sediment in the riverine waters. Consequently, TSS was reduced by half or more in comparison with the previous decade in all impacted rivers (Figure 10 and Figure 11). Estimated TSS concentrations were still much higher than the initial period. This is likely a result of (although reduced) still active ASGM [46], as well as a result of mining tailings derived from abandoned mining areas (degraded areas). For the Jamanxim River, neither the sparse ASGM nor the very high deforestation rate from 1993 to 2001 showed any significant effects on TSS concentrations.

Period from mid-2000s to present—A new gold mining rush characterizes this period, when an increase of 50% in ASGM was observed from 2001 (171.7 km^2) to 2012 (261.7 km^2). Encouraged by the new rise in the price of gold, the Tapajós region has attracted significant investments by medium/large national and foreign companies to extract primary gold in rock deposits with the use of pit-loaders, heavy mills and trucks, characterizing a new period of 'industrial' mining [29]. Presently, about 35 negotiations have been conducted (joint ventures) involving 19 mining companies [4]. This type of mining operation requires higher mechanization levels with the use of several load pitchers (approximately 50, according to the local environment protection agency, ICMBio) and cyanide tanks.

Within this period, the economic scenario has been very favorable, not only for companies exploiting primary deposits, but also for ASGM to take place over unexploited alluvial and colluvial deposits. Therefore, the total mining area increase from 2001 to 2012 is possibly a result of a combination of the increase in primary ore (industrial), and an increase in mining operations using water-jets systems on hillsides. The increase in ASGM area, associated with the new mining techniques, resulted in a TSS increase from about 30.0 to above 100.0 $mg \cdot L^{-1}$ in the Tocantins River, for example (Figure 10). The change in the Novo River was smaller, but high TSS values were also observed (~50.0 $mg \cdot L^{-1}$). In the Crepori River, the large extent of the mining area has had a direct impact on the TSS concentrations throughout the decades, however, the TSS concentrations in 2012 (~75.0 $mg \cdot L^{-1}$) were not as high as in 1993 (>300.0 $mg \cdot L^{-1}$) (Figure 10 and Figure 11). This can be attributed to a few factors: (i) reduced numbers of miners (*garimpeiros*), and thus fewer active mining sites (*garimpos*); (ii) presence of medium-sized mining companies that have more control over their tailings, thus avoiding high sediment/tailing release into the waters; and (iii) reduced numbers of rafts along the tributaries in 2012, which are currently prohibited [4,5].

5.3. Management Recommendations and Future Research

One of the most important roles that temporal series of Landsat and IRS/LISSIII archive database plays is in the recovery of information concerning the temporal and spatial anthropogenic activities to support research on the socio-environmental impacts. With regards to ASGM, the consequences of water siltation to the aquatic environment, more specifically, to underwater light field and phytoplankton production, have been evaluated by [41]. In short, strong light attenuation reduces euphotic zone from 6 m to, approximately, 1 m in non-impacted and impacted water bodies, which can potentially change the phytoplankton community and decrease the photic zone and, thus, primary production. As a consequence of light attenuation, water siltation can change the fish community as suggested by [42]. The information produced in this paper indicates which rivers are subject to higher ASGM intensities that require more attention and investments for water and landscape recovery. Moreover, knowing the spatial distribution of sediment plume and the location of ASGM allows IBAMA and ICMBio (National Environmental Agencies) to monitor and control illegal/informal ASGM in protected areas, for example. Providing quantitative TSS values is strong evidence for improving monitoring/controlling efforts (making them more effective), and indicative of a need to change techniques as suggested by [2], that reduces the mining tailings.

The introduction of high amounts of sediment can create an additional problem of erosion and deposition considering the projected five dams in the Tapajós River Basin. The creation of an artificial lentic (reservoir) environment combined with mining tailings can cause an impaired reservoir storage capacity [47] but can potentially create an optimum condition for algae bloom considering low water flow and increased nutrient input from mining tailings. This is without mentioning the potential mercury contamination that is carried out, adsorbed by the fine particles [38,48,49,50].

Similar analytical frameworks incorporating historical satellite imagery, geophysical characteristics, and optical data can be applied to other mined areas of the world, for example, in the French-Guiana, Suriname, Roraima and Mato Grosso in Brazil. Future research on modelling the sediment transport in pristine and mined sub-basins will be performed in order to investigate the contribution of mining tailing to total sediment transportation and support local and regional policies for a balanced environment.

6. CONCLUSIONS

Despite the severe consequences for the aquatic system, the quantification of mining-related areas in the Amazon and its consequential changes to water quality using satellite images is rarely performed. This is either due to a lack of water quality data (for example, TSS and Chlorophyll-a); or because of limitation of satellite images specifications such as spatial resolution (small rivers require spatial resolutions of about 5 m to be properly studied) [2,9,15].

This research paper presents an innovative remote sensing approach that combines land-use change and water quality information in order to investigate whether ASGM area extension is associated with water siltation. With the benefit of a 40-year period of the multi-satellite imagery archive, the use of DDV as a reference target to inter-calibrate the historical images was satisfactorily applied and presented reliable outputs to perform a multi-temporal analysis on land use and water siltation changes. In the case of the Tapajós River, the tributaries and the main channel are wide enough to be detected by medium-resolution satellite images such as Landsat (30 m) and IRS/LISSIII (24 m), and historical TSS were retrieved by [12].

The identification and quantification of ASGM in four sub-basins started with an official land use classification (TerraClass 2012) as a starting point. This classification, however, failed in identifying small ASGM polygons and a high misclassification rate was observed when validated with photos. Alternatively, Terraclass 2012 classification was updated with the support of auxiliary data and precise vector editing and used as baseline for a reverse in time classification of ASGM in 2001, 1993, 1984, and 1973. This procedure relies on the assumption that ASGM changes in the land cover are severe and remain detectable from satellite sensors for decades as demonstrated in this research.

In terms of the ASGM expansion and water siltation impacts, we observed that from 1973 to 1993, as mining areas increased, so did TSS concentrations for all sub-basins with significant ASGM. However, for the following decades, this association also depends on several other factors regarding ASGM activities, such as the applied mining technique (pans, drafts, water-jets, industrial methods), type of exploited gold deposits (alluvial, colluvial, quartz veins deposits), and intensity of gold mining, represented by the number of miners and gold production. The combination of these factors allows for the identification of four main chronological periods of change in the water quality of the studied rivers. Direct impacts of ASGM on the water quality is more evident during the first two decades (up to 1993) when the natural system was more susceptible. In the following decades, even with the increment of ASGM, the impact on the water is less evident because most of the new ASGM exploitation occurs further from the river network, and in which direct impacts are less perceptive. The period from 2004 to the present day is characterized by a major intensification of ASGM encouraged by high gold prices, with the presence of mining companies (industrial mining) exploiting primary deposits. The impact of this recent gold rush on water quality is evident, mainly in the Tocantins and Novo tributaries, which raises awareness of the need for implementation of better mining techniques for water quality control and strict law enforcement in the present and near future.

ACKNOWLEDGMENTS

This research was developed in a partnership among the University of Victoria (UVic), the Brazilian Institute for Space Research (INPE), and the Artisanal Gold Council. The authors would like to acknowledge logistical support by ICMBio (Brazilian Agency of Protected Areas). We also acknowledge financial support from the National Sciences and Engineering Research Council of Canada (NSERC) to Maycira Costa and NSERC Discovery Grant to Kevin Telmer. The São Paulo Research Foundation (FAPESP) grants to Evlyn Novo for field activities (Process n. 2011/23594-8 and 2008/07537-1). The CNPq (National Counsel of Technological and Scientific Development) awarded the lead author with a PhD scholarship (CNPq- Science without Borders n. 237930/2012-9), followed by a Post-Doctoral position (CNPq-PDJ n. 150835/2015-9) at INPE/Brazil.

AUTHOR CONTRIBUTIONS

This research is part of *Felipe de Lucia Lobo*'s PhD Thesis defended in July 2015. He participated in every step of this article from planning, data collection, analysis and writing the manuscript as the first author. *Maycira Costa* was the main supervisor of this research and had an active participation in planning, data analysis, and text editing. *Evlyn Márcia Leão de Moraes Novo* was the co-supervisor of this research and was fundamental in defining the research approach and data analysis. *Kevin Telmer* is the Executive Director of the Artisanal Gold Council and provided key information for data analysis and discussion of the results.

CONFLICTS OF INTERESTS

The authors declare no conflict of interest, and no copyright issues with the data sources and documents cited.

ABBREVIATIONS

AERONET	Aerosol robotic network
AOT	Aerosol Optical Thickness
ASGM	Artisanal and Small-scale Gold Mining
DDV	Dense dark Vegetation
DN	Digital Number
FLAASH	Fast Line-of-sight Atmospheric Analysis of Hypercubes
IRS/LISSIII	Indian Remote Sensing/Linear Imaging Self Scanning Sensor
NIR	Near-Infrared
RGB	Red, Green, Blue
TOA	Top-of-Atmosphere
TSS	Total Suspended Solids

REFERENCES

1. Veiga, M.M. *Mercury in Artisanal Gold Mining in Latin America: Facts, Fantasies and Solutions*; UNIDO: Vienna, Austria, 1997.
2. Sousa, R.N.; Veiga, M.M. Using performance indicators to evaluate an environmental education program in artisanal gold mining communities in the Brazilian amazon. *Ambio* **2009**, *38*, 40–46.
3. Telmer, K.; Stapper, D. *Evaluating and Monitoring Small Scale Gold Mining and Mercury Use: Building a Knowledge-Base with Satellite Imagery and Field Work*; United Nations Industrial Development Organization: Victoria, BC, Canada, 2007.

4. Silva, A.R. *A Indústria Mineral no Pará*; Alberto Rogério Benedito da Silva: Belém, Pará, Brazil, 2012.

5. Fernandes, F.R.C.; Alamino, R.D.C.J.; Araújo, E.R. *Recursos Minerais E Comunidade: Impactos Humanos, Socioambientais E Econômicos*; CETEM/MCTI: Rio de Janeiro, Brazil, 2014.

6. CPRM. Província Mineral do Tapajós: Geologia, Metalogenia E Mapa Previsional Para Ouro Em SIG. Available online:http://www.cprm.gov.br (accessed on 15 April 2013).

7. Araújo Neto, H. *Perfil do Ouro*; Ministério de Minas e Energia: Brasília, Brazil, 2009.

8. Rodrigues, R.M.; Mascarenhas, A.F.S.; Ichihara, A.H.; Souza, T.M.C. *Estudo Dos Impactos Ambientais Decorrentes do Extrativismo Mineral E Poluição Mercurial no TAPAJÓS—Pré-Diagnóstico*; CETEM/CNPq: Rio de Janeiro, Brazil, 1994.

9. Roland, F.; Esteves, F.D. Effects of bauxite tailing on PAR attenuation in an amazonian crystalline water lake.*Hydrobiologia* **1998**, *377*, 1–7.

10. Tudesque, L.; Grenouillet, G.; Gevrey, M.; Khazraie, K.; Brosse, S. Influence of small-scale gold mining on french guiana streams: Are diatom assemblages valid disturbance sensors? *Ecol. Indic.* **2012**, *14*, 100–106.

11. Mol, J.H.; Ouboter, P.E. Downstream effects of erosion from small-scale gold mining on the instream habitat and fish community of a small neotropical rainforest stream. *Conserv. Biol.* **2004**, *18*, 201–214.

12. Lobo, F.L.; Costa, M.P.F.; Novo, E.M. Time-series analysis of landsat-MSS/TM/OLI images over Amazonian waters impacted by gold mining activities. *Remote Sens. Environ.* **2015**, *157*, 170–184.

13. Moran, M.S.; Bryant, R.; Thome, K.; Ni, W.; Nouvellon, Y.; Gonzalez-Dugo, M.P.; Qi, J.; Clarke, T.R. A refined empirical line approach for reflectance factor retrieval from landsat-5 TM and landsat-7 ETM+. *Remote Sens. Environ.***2001**, *78*, 71–82.

14. Hadjimitsis, D.G.; Clayton, C. Assessment of temporal variations of water quality in inland water bodies using atmospheric corrected satellite remotely sensed image data. *Environ. Monit. Assess.* **2009**, *159*, 281–292.

15. Nery, M.A.C.; Silva, E.A.D. *Balanço Mineral Brasileiro: Ouro*; DNPM (Departamento Mineral de Produção Mineral): Brasília, Brazil, 2002.

16. ICMBIO. *Plano de Manejo-Floresta Nacional do Crepori-State of Pará*; Resumo Executivo: Curitiba, Brazil, 2010.

17. DGI/INPE. Divisão de Geração de Imagens. Available online: www.dgi.inpe.br (accessed on 1 May 2013).

18. MMA. Geographical Data: I3geo. Available online: http://mapas.mma.gov.br/i3geo/datadownload.htm (acc essed on 2 March 2013).

19. SiBCS. *Sistema Brasileiro de Classificação de Solos*; Embrapa Solos: Rio de Janeiro, Brazil, 2006.

20. CPRM. Geobank. 2012. Available online: http://geobank.cprm.gov.br (accessed on 15 March 2013).

21. Coutinho, A.C.; Almeida, C.; Venturieri, A.; Esquerdo, J.C.D.M.; Silva, M. *Uso e Cobertura da Terra nas Áreas Desflorestadas da Amazônia Legal: Terraclass 2008*; EMBRAPA: Brasília, Brazil, 2013.

22. INPE. Topodata: Banco de Dados Geomorfométricos do Brasil. Available online: http://www.dsr.inpe.br/topodata(accessed on 13 March 2013).

23. Di Gregorio, A.; Jansen, L.J.M. Land Cover Classification System (LCCS). *Classification Concepts and User Manual for Software Version 1.0*; FAO/UNEP: Rome, Italy, 2000.

24. Shimabukuro, Y.E.; Smith, J.A. The least-squares mixing models to generate fraction images derived from remote sensing multispectral data. *IEEE Trans. Geosci. Remote Sens.* **1991**, *29*, 16–20.

25. Chander, G.; Markham, B.L.; Helder, D.L. Summary of current radiometric calibration coefficients for landsat MSS, TM, ETM+, and EO-1 ALI sensors. *Remote Sens. Environ.* **2009**, *113*, 893–903.

26. NASA/GSFC. Aerosol Robotic Network (Aeronet). Available online: http://aeronet.gsfc.nasa.gov (accessed on 13 March 2013).

27. Lu, D.; Mausel, P.; Brondizio, E.; Moran, E. Assessment of atmospheric correction methods for landsat TM data applicable

to amazon basin lba research. *Int. J. Remote Sens.* **2002**, *23*, 2651–2671.

28. Lu, D.; Mausel, P.; Brondízio, E.; Moran, E. Relationships between forest stand parameters and landsat TM spectral responses in the Brazilian Amazon basin. *For. Ecol. Manag.* **2004**, *198*, 149–167.

29. Bezerra, O.; Veríssimo, A.; Uhl, C. *Impactos da Garimpagem de Ouro na Amazônia Oriental*; Imazon: Belém/PA, Portuguesa, 1998.

30. Villas Bôas, R.C.; Beinhoff, C.; Silva, A.R. *Mercury in Tapajos Basin*; CNPq/CYTED: Rio de Janeiro, Brazil, 2001.

31. Fearnside, P.M. Brazil's Cuiabá-Santarém (BR-163) Highway: The environmental cost of paving a soybean corridor through the Amazon. *Environ. Manag.* **2007**, *39*, 601–614.

32. WRB/FAO. *World Reference Base for Soil Resources*, 1st ed.; International Soil Reference and Information Centre, ISRIC: Rome, Italy, 1998.

33. Junk, W.J.; Piedade, M.T.F.; Schöngart, J.; Cohn-Haft, M.; Adeney, J.M.; Wittmann, F. A classification of major naturally-occurring amazonian lowland wetlands. *Wetlands* **2011**, *31*, 623–640.

34. DNPM. Sumário Mineral, 32. Available online: http://www.dnpm.gov.br (accessed on 5 April 2013).

35. Browder, J.O. Public policy and deforestation in the Brazilian Amazon. In *Public Policy and the Misuse of Forest Resources*; Gillis, R.R.A.M., Ed.; Cambridge University Press: Cambridge, UK, 1988; pp. 247–298.

36. Fearnside, P.M. Environmental destruction in the Brazilian Amazon. In *The Future of Amazonia: Destruction or Sustainable Development*; The Future of Amazonia: Macmillan, London, UK, 1990; pp. 179–225.

37. Junk, W. General aspects of floodplain ecology with special reference to amazonian floodplains. In *The Central Amazon Floodplain*; Junk, W., Ed.; Springer: Berlin/Heidelberg, Germany, 1997; Volume 126, pp. 3–20.

38. Telmer, K.; Costa, M.; Simões Angélica, R.; Araujo, E.S.; Maurice, Y. The source and fate of sediment and mercury in the Tapajós river, Pará, Brazilian Amazon: Ground- and space-based evidence. *J. Environ. Manag.* **2006**, *81*, 101–113.

39. Almeida, R.; Shimabukuro, Y.E. Digital processing of a landsat-tm time series for mapping and monitoring degraded areas caused by independent gold miners, Roraima State, Brazilian Amazon. *Remote Sens. Environ.* **2002**, *79*, 42–50. [

40. Finer, M.; Novoa, S. Gold Mining Deforestation Continues to Expand in la Pampa (Madre de Dios, Peru). Available online: http://maaproject.org/2015/03/gold-mining-deforestation-expand-peruvian-amazon/ (accessed on 26 January 2016).

41. Lobo, F.; Costa, M.P.F.; Novo, E.M.; Telmer, K.H. Effects of artisanal small-scale gold mining tailings on the underwater light field in the Tapajós River basin. *Limnol. Oceanogr.* **2016**. in preparation.

42. Alho, C.R.; Reis, R.; Aquino, P.U. Amazonian freshwater habitats experiencing environmental and socioeconomic threats affecting subsistence fisheries. *Ambio* **2015**, *44*, 412–425.

43. Biggs, T.W.; Dunne, T.; Martinelli, L.A. Natural controls and human impacts on stream nutrient concentrations in a deforested region of the Brazilian amazon basin. *Biogeochemistry* **2004**, *68*, 227–257.

44. Deegan, L.A.; Neill, C.; Haupert, C.L.; Ballester, M.V.R.; Krusche, A.V.; Victoria, R.L.; Thomas, S.M.; Moor, E. Amazon deforestation alters small stream structure, nitrogen biogeochemistry and connectivity to larger rivers. *Biogeochemistry***2010**, *105*, 53–74.

45. Farella, N.; Lucotte, M.; Louchouarn, P.; Roulet, M. Deforestation modifying terrestrial organic transport in the rio tapajós, Brazilian amazon. *Org. Geochem.* **2001**, *32*, 1443–1458.

46. Telmer, K.H. The Effect of Changing Gold Prices on Artisanal Mining. Available online: http://artisanalgold.blogspot.ca (accessed on 2 April 2015).

47. Shalash, S. Effects of sedimentation on the storage capacity of the high aswan dam reservoir. *Hydrobiologia* **1982**, *91*, 623–639.

48. Boudou, A.; Maury-Brachet, R.; Coquery, M.; Durrieu, G.; Cossa, D. Synergic effect of gold mining and damming on mercury contamination in fish. *Environ. Sci. Technol.* **2005**, *39*, 2448–2454.

49. Park, J.-G.; Curtis, R.L. Mercury distribution in sediments and bioaccumulation by fish in two oregon reservoirs: Point-source

and nonpoint-source impacted systems. *Arch. Environ. Contam. Toxicol.* **1997**, *33*, 423–429.

50. Egler, S.G.; Rodrigues-Filho, S.; Villas-Bôas, R.C.; Beinhoff, C. Evaluation of mercury pollution in cultivated and wild plants from two small communities of the Tapajós Gold Mining Reserve, Pará State, Brazil. *Sci. Total Environ.* **2006**,*368*, 424–433.

7
CHAPTER

Sustainable Land Management in Mining Areas in Serbia and Romania

Vesna Popović [1], Jelena Živanović Miljković [2], Jonel Subić [1], Andrei Jean-Vasile [3], Nedelcu Adrian [3] and Eugen Nicolăescu [4]

[1]*Institute of Agricultural Economics, Volgina 15, Belgrade 11 000, Serbia*

[2]*Institute of Architecture and Urban & Spatial Planning of Serbia, Bulevar kralja Aleksandra 73/II, Belgrade 11 000, Serbia*

[3]*Faculty of Economic Sciences, Petroleum-Gas University of Ploiesti, B-dul Bucuresti, No.39, Ploiesti, Prahova 100680, Romania*

[4]*Department of Finance, Banking, and Accountancy, Dimitrie Cantemir Christian University, Bucharest 030134, Romania*

ABSTRACT

The paper analyzes the impacts of mining activities on sustainable land management in mining areas in the Republic of Serbia and Romania and discusses the main challenges related to the management of these issues in legislation and practice. Particular attention is paid to land disturbance, mine waste management and land reclamation, as well as access to land for mining purposes, the transfer of mining royalties and the partnerships of the mining industry, governments, communities and civil society for sustainable mining. Both governments are willing to provide the adequate role to mining in strengthening the national economies, but they face numerous constraints in this matter. Sustainable mining practices and consistent implementation of the mining for the closure planning approach, within an improved legislative framework and in cooperation with stakeholders at all levels, create conditions for the development of creative, profitable, environmentally-sound and socially-responsible management and reuse of mine lands.

Keywords: sustainable land management; mining; mining waste management; land reclamation

1. INTRODUCTION

Land performs many key environmental, economic, social and cultural functions that are vital for life. These functions are generally interdependent, and the extent to which land performs them is strongly related to sustainability [1].

According to the World Commission on Environment and Development definition, sustainable development is the type of development that meets the needs of the present without compromising the ability of future generations to meet their own needs [2]. The objective of sustainable land management (SLM) is to harmonize the complementary goals of providing environmental, economic and social opportunities for the benefit of present and future generations while maintaining and enhancing the quality of land resources [3].

The need for urgent action to reverse land degradation and to achieve a land degradation-neutral world in the context of sustainable development is stressed in the UN Rio+20 conference [4]. The 2011 Roadmap to a Resource Efficient Europe [5], as a part of the Europe 2020 Strategy for smart, sustainable and inclusive growth [6], promotes resource-efficient development and forces the EU policies to consider, by 2020, their direct and indirect impacts on land use, striving for the goal of no net land take by 2050 [7]. These approaches comprise the development of land policy and planning based on inter-sectoral policy coherence, strengthening of institutions and interactive stakeholder participation.

Much of the impact from mining is directly or indirectly linked to land. The exploitation of mineral resources, especially surface mining, causes numerous negative environmental externalities and socio-economic impacts, e.g., land use changes, ecosystem disturbances, watercourse relocation and a decrease in ground water level, changes in infrastructure networks, non-balanced industrial development, resettlement and changes in the economic and social structure of the local population,*etc.* [8]. Positive impacts can also occur, such as increased conservation activities and water quality management [9,10]. Managing the impacts of changes in land use, *i.e.*, mitigating the negative impacts and enhancing the positive impacts, is an important sustainable development goal that poses a challenge for mining companies and regulatory agencies [11].

The key to success in the role of minerals in sustainable development is in the utilization of this resource as a platform for achieving economic diversification, growth and broad development goals, while maintaining social stability and protecting environmental integrity [12].

Regarding sustainable mining practice, the focus is on how mining companies, operating in a responsible and sustainable manner, can contribute to sustainable development at all levels. Mining companies are expected to create and share value and responsibility across a complex set of economic, environmental and social issues in the sustainable world, and good environmental and social practice is considered a part of the company's competitive advantage [13]. Advances in technology help operations to extract lower-grade ores than those just a few decades ago, to increase productivity and to reduce costs. Operations are meeting stricter environmental requirements for lower emissions, energy, waste and water use. Rehabilitation standards are higher and costlier to meet, and management responsibility has longer terms. Companies need to actively plan how to reduce the environmental impacts of their operations to secure and maintain their license to operate [14].

Beyond the legal license, the concept of a social license to operate (ongoing approval within the local community and other stakeholders of both mining companies and their projects) has been widely recognized by the industry as an essential attribute of success [15]. The social impact assessment (SIA), strengthened by the concept of free, prior and informed consent (FPIC) as an overarching philosophy, and the impact and benefit agreement (IBA), as well as the social impact management plan (SIMP) as the outputs, emerges as a powerful tool for assessing, managing and monitoring the social impacts during implementation and as a tool for a proactive response to changes across the lifecycle of developments [16,17,18].

Rather than simply regulating, the governments tend to improve industry behavior by encouraging transparency and boosting civil engagement. Policies are formulated to promote shared value for industry, governments and civil society, and the partnerships of these stakeholders serve to maintain and manage shared-value opportunities [14].

Being highly dependent on the imports of raw materials, the EU in the Athens Declaration on sustainable access to resources (2014) [19] asks the member states to upgrade their minerals policies, land use planning and legal framework to secure sustainable access to their domestic mineral resources. National governments were in particular encouraged to establish adequate regulatory frameworks to secure mineral deposits of public importance that facilitate investments, but also to ensure that mineral property rights are sufficiently protected. In land use planning, the existence of mineral deposits should be taken into account in the development of urban, highly populated or industrial areas, avoiding potential conflicts, on the basis of national/regional mineral strategies.

A sustainable minerals policy needs to: facilitate the transformation of natural mineral capital into built physical, economic, environmental or social capital of equal or greater value; ensure that environmental and negative social impacts of mining are minimized and their costs incorporated into production functions; require transparency and information sharing; reconsider the allocation of rights and the availability of resources across generations; address benefit/risk trade-offs from the perspective of multiple stakeholders and to create contingency plans that will ameliorate the effects of mineral market booms and busts; and be correlated and consistent with other governmental policies [20,21].

The aim of a land use policy for minerals should be, ultimately, to ensure fair and equal consideration of all potential uses of land, including the eventual extraction of raw materials. Land use planning for minerals should be long-term and undertaken at a national or regional level. A national planning framework can help to ensure that minerals are given due weight in the land use planning process; it is therefore recommended as the best practice in appropriate national circumstances [22].

Mining operations, except mining waste, which is regulated at the EU level [23,24], are a matter of concern of the national administrations and, consequently, regulated at the national level. Mining law usually defines general principles and key rules and procedures for mining explorations and exploitation projects (regulatory authorities, property rights, fees and taxes, environmental, health and safety obligations, mine closure and land rehabilitation).

In practice, all of these efforts are far from consistent within each company, let alone across the entire industry. The mining costs have escalated mainly due to poor planning, low investments in technology innovation, inefficient operations and rising prices of water and energy. Disputes among mining companies and local communities and governments have posed a particular challenge, causing about $25bn worth of projects being delayed or suspended worldwide [25].

A new roadmap for collaborative, long-term sustainable mining should be based on three pillars: shared purpose during and after the mine's operation, flourishing ecosystems, with a reduced environmental footprint of every mining activity as the first step, and achievement of long-term sustainable prosperity for the communities and nations in which the mining companies operate, as well as for the companies themselves [26].

2. MINE LAND MANAGEMENT IN LEGISLATION AND PRACTICE: PRODUCTION CAPACITIES AND POTENTIALS IN BRIEF

Mining activities have a long-standing tradition in the Republic of Serbia and Romania. The restructuring of the mining sector, transition and privatization in the last two decades have brought many mines to closure or to operating with reduced capacity, particularly in Romania, where by 2004, production had been terminated in 344 unviable mines, and the process of mine closure is still ongoing [27].

In recent years, with the rising prices of copper and noble metals on the world market and enlargement of the EU list of important and critical raw materials for the EU economy [28], there is growing interest of investors for the mining sectors in both countries.

2.1. The Republic of Serbia

Geological exploration and exploitation of mineral resources in the Republic of Serbia are today taking place in 105 exploration and 250 exploitation fields [29].

Coal mining in Serbia is carried out in: two large lignite basins with surface exploitation: the mining basin (MB) Kolubara (open pit Field B, Field D, Tamnava-West Field and Veliki Crljeni) and MB Kostolac (open pit Drmno), within the public company Electric Power Industry (EPS) of Serbia; underground hard coal mines (seven) and lignite mine Lubnica, within the Public Enterprise for Underground Coal Exploitation Resavica, in restructuring; and underwater lignite mine Kovin at the Danube.

About 100 million m^3 of overburden and over 38 million tons of lignite per year are exploited by surface mining, over 220 thousand tons of lignite by underwater exploitation and up to 660 thousand tons of hard coal and minor quantity lignite in underground mines [30]. According to EURACOAL data for 2012, the resources of hard coal amount to 184 million tons, and reserves amount to 176 million tons, while lignite resources amount to 5363 million tons, while reserves amount to 3660 million tons [31].

Oil and gas production takes place on 42 oil fields with 650 wells, with an annual production of 1.2 million tons of oil and 562 million m^3 of natural gas [32]. The state oil company Petroleum Industry of Serbia (NIS), majority-owned by GazpromNeft (St. Petersburg, Russia), is the only company engaged in the exploration and production of oil and gas.

Shale oil reserves are estimated to be at about 4.8 billion tons and are found mostly in the area of the Aleksinac coal basin (about two billion tons) [30]. The exploitation is not done, but it is envisaged in the Strategy of Mineral Resources Management until 2030 (SMRM).

The exploitation of copper and associated metals in copper ores—lead, zinc, gold and silver—takes place in majority state-owned Mining and Smelting Combine Bor (RTB Bor Group), in copper mines in Bor and Majdanpek. According to RTB Bor Group data, during century-long mining activities (up to 2011), more than 750 million tons of copper ore have been excavated; approximately 3.35 million tons of copper, 155.5 tons of gold and 971 tons of silver have been produced; and more than 1820 million tons of overburden have been removed. Certified geological reserves amount to 1.38 billion tons of ore containing about 5.1 million tons of copper, 202 tons of gold and 1477 tons of silver. Confirmed reserves, with the average annual exploitation of 26 million tons of ore, guarantee copper production for the next 50 years [33].

Four lead and zinc mines are active; they are privately owned and produce 10%–12% of the volume of two decades ago, when there were over ten active lead and zinc mines [34]. The balance ore reserves of lead and zinc amount to about 6.22 million tons and off-balance ore reserves about 4.2 million tons, containing about 0.34 million tons of lead and about 0.35 million tons of zinc [30]. In 2010, the production of zinc was 2600 tons and the production of lead 1800 tons [35].

There is also significant potential for the exploitation of antimony, lithium, nickel, cobalt and molybdenum [30].

Non-metallic minerals in exploitation are clays, quartz raw materials, barite, gypsum and anhydrite, zeolites, feldspars, diabase, ornamental and building stone. Borates, phosphates, magnesite and fluorite also have good prospects for exploitation [30]. The privatization process gave the best results in this segment of mining, and the future production growth will depend on the development of the industry and construction in the country.

The SMRM predicts that the production of the mining sector will increase from the current 2% to over 5% of GDP in 2020, with investments of over 3.5 billion euros, which would bring revenues of three billion euros.

2.2. Romania

Romania has significant mineral resource potentials, especially in oil and natural gas, including shale gas deposits, lignite and hard coal, gold, silver, copper, lead, zinc, uranium and salt, as well as a larger number of other non-metallic minerals.

According to the draft of Energy Strategy 2015–2030, with an average annual production of 4.2 million tons, Romania is the fourth-largest oil producer in the EU and the fifth-largest in Europe. The annual production of natural gas totals to 11 billion m^3. Proven oil and gas reserves amount to 60 million tons and 150 billion m^3 [36]. OMV Petrom (Bucharest, Romania), a member of Austria's OMV Group, is the major domestic crude oil producer and one of the two leading natural gas producers in the country, after the majority state-owned Romgaz (Mediaş, Romania).

Romania has one of the biggest shale gas reserves in Eastern Europe in the amount of 1.4 trillion m^3 [37]. After five years of public protests, ban and re-approval, Chevron withdrew from exploration in Romania in 2015, as drilling results fell short of expectations, despite large investments. The recent Energy Watch Group study (2015) indicates that the costs and the environmental damage produced by fracking are out of proportion to the amount of raw materials extracted [38].

Coal production has almost halved during the 20 years of the transition and restructuring period [39]. In 2012, it amounted to 34 million tons. Hard coal reserves and resources are estimated at 2446 million tons, of which 252.5 million tons are commercially exploitable, but only 11 million tons might be economically recoverable. Reserves of lignite total to 280 million tons, with a further 9640 million tons of resources [31].

Hard coal is exploited within two separate operating units under the majority state-owned Hunedoara Energy Complex (CNH SA Petrosani). One unit produces about 40 thousand tons monthly in three coal mines in the Jiu Valley (Uricani, Paroseni and Petrila) that will be closed by 2018, following the EU Council Decision on state aid to facilitate the closure of uncompetitive coal mines [40]. The other unit operates four coal mines in MB Petrosani (Lonea, Livezeni, Vulcan and Lupeni) without state aid, with an annual production of 1.5 million tons. The annual production of lignite in 2012 amounts to 32.1 million tons. The majority state-owned Oltenia Energy Complex (Targu Jiu, Romania) is the leader in lignite production [39].

Despite significant reserves of non-ferrous metals, Romania currently exploits relatively small quantities of copper, gold, silver, lead and zinc ores. Besides silver, the production is highly reduced compared to the beginning of the previous decade. According to World Mining data, in 2012, the production of copper amounted to 9.5 thousand tons, gold 500 kg, silver 18 thousand tons, lead 5.5 thousand tons and zinc 8.4 thousand tons [35].

The state-owned company Cupru Min holds the largest part of Romania's copper reserves (the Rosia Poieni open pit mine with estimated mineral reserves of over one billion tons) [41]. Lead and zinc (associated with copper, antimony, bismuth, cadmium, gold and silver) are exploited at underground mines in Baia Mare, Baia Borsa, Certej and Rodna counties [42]. Romania has the largest gold reserves in Europe (760 million tons of gold-silver ores, of which 40 million tons are in 68 gold deposits of Transylvania's Golden Quadrilateral, in the Apuseni Mountains, comprising the Rosia Montana, Bucium, Baia de Aries, Almas, Brad and Sacaramb mining areas) [43].

Non-metallic industrial minerals in exploitation (2012 data) are bentonite, feldspar, gypsum, kaolin, graphite, salt and talc [35]. The mines of these minerals are located throughout the country, and the privatization in this part of the mining sector was quite successful.

The uranium deposit in Crucea, in Suceava County, has been exploited for 30 years and will be closed in 2015 after being exhausted [44].

According to [45], mining and quarrying accounted for 3.5% of the value of industry production in 2011 (coal and lignite 0.6%, oil and natural gas 1.8%, metal ores 0.1% and other mining and quarrying and support services 1.0%).

3. MINING LAND MANAGEMENT IN THE REPUBLIC OF SERBIA AND ROMANIA: AN OVERVIEW OF SELECTED ISSUES

Mining companies are making efforts to minimize the footprint of their activities throughout the mine life cycle, including post-mining site rehabilitation. Mining revival and development need updated legislation and clear political support, as well as better understanding and cooperation between the government, investors and local communities and civil society.

The mining for closure planning approach includes related environmental, social and economic issues, as well as the participation of stakeholders, particularly in consideration of land rehabilitation objectives [46]. Mine rehabilitation and closure plans are required by regulatory agencies around the world and are often a component of the environmental impact assessment process practiced in over 100 countries [47,48].

Land disturbance, waste disposal and post-mining site rehabilitation, as well as access to mineral deposits and land acquisition for mining purposes, the transfer of mining royalties and public participation are the main issues faced by the industry, governments, investors, communities, civil society and other stakeholders. The status and challenges related to the management of these issues in legislation and practice in Serbia and Romania are discussed below.

3.1. Land Use Change

Despite the fact that mining has a small share of a country's total land use, conflict can arise at the local level where mining is perceived as competing with agriculture and livestock grazing or other land uses [49,50].

The Serbian Strategy of Mineral Resources Management until 2030 (SMRM) envisages an increase of demand for coal to 48 million tons by 2025. In order to increase the production of lignite, EPS expands the existing and opens new open pits.

MB Kolubara is surrounded by dense population, fertile agricultural land and developed infrastructure. Termination of the exploitation in this basin is expected for the period 2055–2060, and a rough estimation is that in the period 2020–2060, production will be conducted on the three open pits (Radljevo, Tamnava-West Field and Field E). There is a plan for the acquisition of 4694 ha for mining and an energy-industrial complex, while 3411 hectares of degraded land will be reclaimed until 2020 [51].

MB Kostolac is also located in an agricultural area and includes Viminacium, one of the most significant archaeological sites in Serbia. For the expansion of the open pit Drmno, in the period of 2010–2022, there are plans for the acquisition of 953.4 ha, which includes 833 ha of fertile agricultural land and about 1180 ha throughout the open pit life cycle (up to 2054). The exploitation of the western lignite deposit Dubravica will require over 2000 ha, resettling parts of three villages and reconstruction of the infrastructure systems. The loss of agricultural land will be partially offset by reclamation (1200 ha by 2022) [52] (Figure 1 and Figure 2).

The SMRM envisages a rise in copper production by three times, and the production of lead and zinc will double by 2020. The government seeks strategic partners to open two new open pits and to activate the underground copper mine "Borska reka" in Bor for the new smelter capacity of 80 thousand tons of copper, as well as the underground gold mine "Čoka Marin" in Majdanpek.

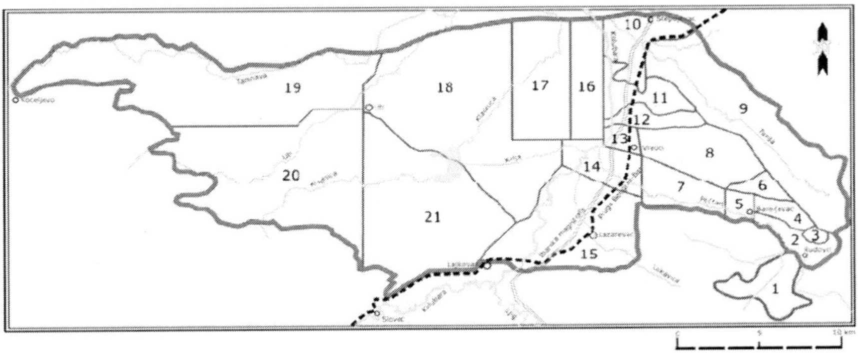

Figure 1. Mining basin (MB) Kolubara: open pits and potential areas for exploitation. 1, Field Kruševica; 2, Field Rudovci; 3, Field A; 4, Field B; 5, Field C; 6, Field Baroševac; 7, Field E; 8, Field D; 9, Field Turija; 10, Field Stepojevac; 11, Field Veliki Crljeni; 12, Field Volujak-Vreoci; 13, Field G; 14, Field F; 15, Field Šopić-Lazarevac; 16, Tamnava-East Field; 17, Tamnava-West Field; 18, Field Radljevo; 19, Field Trlić; 20, Field Zvizdar; 21, Field Ruklade. Source: [53].

In July 2013, the government of Romania adopted the national plan for strategic investment and job creation in five investment fields, including seven projects in the mining sector (among them Rosia Montana and Certej open pit gold mines) [54].

The Rosia Montana mining project comprises 1346 ha to be exploited, of the 2388 ha for which the company owns a license, and 300 ha covered by a tailings management pond of heavy metals and cyanide, which will permanently endanger more than 6000 people from the neighboring areas, 2064 private properties relocated, 975 houses torn down, out of which 41 houses are considered national heritage sites, as well as seven churches, 11 cemeteries relocated and only 634 jobs created over a period of 17 years [55].

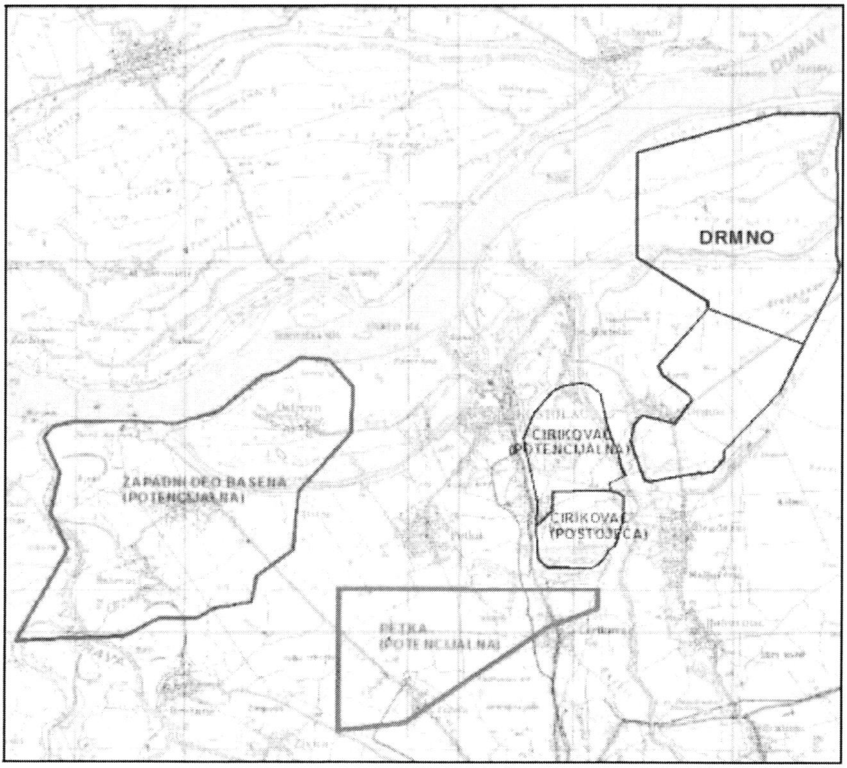

Figure 2. MB Kostolac: open pits (marked with the blue line) and potential areas for exploitation (red, magenta and upper green line). Source: [52].

The Certej mining project, within a 16-year duration period, involves deforestation, soil uncovering, an open pit, two huge waste dumps and the use of cyanides to obtain gold, as well as building two tailings dams over 63.6 ha. The project area of 456.2 ha is currently covered by: forests in 187 ha, meadows in 30.7 ha, arable fields in 5.3 ha and residential areas in 18.7 ha. Certej is located within the Natura 2000 Special Protection Area (SPA) site ROSPA 0132 [55].

There are also considerations about small-scale cyanide gold mining. According to Marincea, in the new 50 deposits identified, the use of the cyanide-based preparation technology "would generate no less than 2900 ha with cyanide waste" [43]. The mining area is usually located on forest lands, pastures and meadows, often in protected areas and/or areas dedicated to agriculture and rural tourism [43].

A new uranium exploitation site is planned for opening in Neamt County, with annual exploitation of 124 thousand tons for the needs of the Cernavoda nuclear plant [44].

Land use change associated with the direct and offsite footprints of mining causes environmental and social impacts [11]. Although systematic analysis of mining land use dynamics is not included in this overview, it is clear that land use acquisition for mining purposes reduces the area of agricultural and forest land in both countries, but also leads to intensive land abandonment.

The reasons for leaving agricultural production in the vicinity of large open pit coal mines in Serbia are changes in the socio-economic structure of the population due to higher earnings in the mining sector, resettlement and the migration to the nearby big urban centers and abroad [56,52]. In Romania, the exploitation of gold-silver and polymetallic ores in the former mining regions discourages the development of agriculture and rural tourism, bearing in mind the serious environmental legacies (see below) and the negative perception of food consumers and tourists regarding new potential pollution of soil, water and air and disruption of ambient values.

3.2. Mining Waste Management

Overburden and waste rock piles and dumps, tailing ponds, oil and gas wells and produced water ponds occupy considerable areas of land, especially large open pit mines. Most mining wastes, such as overburden, are inert solid materials; they do not represent a significant pollution threat to the environment, but they do cause erosion, fertile land loss and ecosystem and landscape disruption. However, other fractions, in particular those generated by the non-ferrous metal mining industry, may contain large quantities of dangerous substances (inherently present in the ore or added during extraction) and generate acid or alkaline drainage contaminating soil and waters [57].

The management of tailings refers to serious risks (dam failures) and environmental, often cross-border, impacts (dust, tailings dam seepage, dam wall failure or direct disposal of tailings into waterways). Technological advances and changes in regulations have resulted in significant improvement in tailings and overall mining waste management practices over the last two decades [58,59,60].

Due to the lack of budget funds, a large number of abandoned mines have been waiting for rehabilitation for decades, as well as mines owned by private business firms in bankruptcy (and unable to take on appropriate rehabilitation measures). In the Vojvodina province alone, according to the recently established cadastre of abandoned mines, there are 217 abandoned mine and quarries [61]. According to UNEP studies of reducing environment and security risks from mining in the Western Balkans, the hotspot sites with the potential transboundary risk via waterways (Drina, Sava, Danube) in Serbia were principally associated with antimony mining, processing and smelting, lead and zinc mining, processing and smelting and copper mining, processing and smelting [46,62].

Rehabilitation of old open pits in the Bor mine was done by making the open pit "RTH" backfilled with tailings and open pit Bor backfilled with overburden from open pit "Veliki Krivelj". In 2012, the government cancelled the remediation of overburden dumps and tailing ponds in Bor and Veliki Krivelj, provided by the World Bank Project P092999, since national experts advocate re-mining, having in mind that mine waste has a relatively high concentration of metals due to earlier applied inefficient metal extraction processes [63].

In 2012, the Petroleum Industry of Serbia (NIS) (Novi Sad, Serbia) began implementing the plan of oil- and gas-produced water ponds' rehabilitation by 2016, with a special focus on 50 of those with more than 5% of hydrocarbons, which, not being repaired for decades, pollute the air and threaten the land and waters. During 2011–2012, NIS remediated and reclaimed 80 produced water ponds in Vojvodina province and returned the reclaimed land to its owners [64].

In the period 1999–2010, the Romanian government, together with the World Bank, implemented two projects of mine closure and environmental and socioeconomic regeneration, which resulted in the rehabilitation of 51 closed mine sites: remediation of contaminated soils, stabilization of tailings dams and waste dumps and reclamation of more than 600 ha of mine land. Given that hundreds of mines were closed during the restructuring of the mining sector and that rehabilitation is a long process, the main objective of these projects was to build national capacities for similar operations on a large scale and in the long term.

There is no doubt that significant progress has been made in this regard, but there is still a lot of work, especially on hotspots requiring urgent action (e.g., Bosneag and Geamana tailings ponds), having in mind the accidents of the cyanide spill into the Tisa River in Baia Mare, 2000 [65]. In October 2014, the EC took Romania to the Luxembourg Court for failure to comply with EU mine waste legislation regarding the toxic Bosneag tailings pond in Moldova Noua. The Geamana tailings pond of Rosia Poieni copper mine threatens the water of the river Aries and through the rivers Mures, Tisa and Danube, risks cross-border pollution [66].

Pollution with dumps, decantation ponds, flotation tailings deposits, garbage deposits, *etc.*, is the biggest threat to the environment and human health, both in the country and across borders. This type of pollution affects 6639 ha, mainly in Oltenia, of which 5773 ha is excessively affected. In the Bacau, Covasna, Gorj and Prahova counties, oil pollution is registered on 751 ha, of which 278 ha are heavily affected. Salt water (from oil extraction) pollution is registered on 2654 ha and strongly-excessive on 1205 ha. The areas affected by radioactive pollution amount to 566 ha and are located mainly in Brasov (500 ha) [67].

New mining projects draw particular attention regarding the environmental risks they cause. The Rosia Montana mining project comprises 300 ha covered by a tailings pond of cyanide, which will be used in the amount of 240 thousand tons in 17 years of exploitation. The Certej mining project requires two waste dumps, two tailings dams over 64 ha and the use of 26,448 tons of sodium cyanide and 15,280 tons of copper sulfate during the 16 years of operation.

Mine site rehabilitation has been integrated into the early planning process, even preceding the start of mine operations. Rehabilitation is an ongoing consideration throughout the mine's lifecycle, both from a technical and a financial standpoint [68].

The costs of land rehabilitation are classified as the mine's operating costs. According to Directive 2004/35/EC on environmental liability with regard to the prevention and remedying of environmental damage, the prevention and remedying of environmental damage should be implemented by applying the "polluter pays" principle, in line with the principle of sustainable development. Directive [24] on the management of waste from extractive industries and amending Directive [23] obliges operators to provide (and periodically adjust in size) a financial guarantee for waste facility maintenance and post-closure site restoration, including land rehabilitation.

The Serbian Law on Mining and Geological Research [69] only partially transposed Directive [24] on the management of waste from extractive industries. Disposal and management of mining waste is based on a permit for waste management, which is issued in accordance with the waste management plan. The operator is obliged to perform remediation and reclamation of degraded land during exploitation and after mining operations. Permanent cessation of mining operations is regulated by the main mining project for a permanent mine closure. The public mining companies are obliged to prepare a mine closure program, approved and financed by the government, which includes measures to address environmental and socio-economic consequences of the mine closure.

The draft of the Law on Amendments to the Law on Mining and Geological Research [70], which is under public debate, introduces the obligation for operators to ensure financial guarantees for rehabilitation and reclamation of mining fields.

The draft of the Law on Land Protection (2015) [71] obliges the land operator to remediate and/or reclaim degraded land (including land degraded by exploitation of minerals) using its own funds, according to the remediation and reclamation project, approved by departmental ministry. If the operator is unknown, inaccessible or does not comply with the legal provisions, the government finances the project, with the right of recourse of the costs from the operator. The law stipulates the obligation of land monitoring and keeping a register of contaminated sites. The analysis of the national legislative related to the transposition of Directive 2004/35/EC is still expected.

According to the Romanian Mining Law [72], the license holder has the following obligations related to land use and protection: to provide environmental agreements as one of the prerequisites for a building permit; to regularly update the mine closure plan; to set up and maintain the financial guarantee for environmental rehabilitation; and to execute and finalize the environmental rehabilitation of affected land in the mining site, according to the mine closure plan, including the post-closure monitoring program implementation and financing (in the case of state mining companies, these works, including abandoned mines, are financed by the state).

Apart from the Mining Law, the Government Decision (GD) [73], which transposes EU Directive [24] on the management of waste from extractive industries, as well as Government Emergency Ordinance [74], which implements the requirements of EU Directive 2004/35/CE on environmental liability, requests financial guarantees for waste facilities maintenance and for environment restoration in the case of pollution, respectively. In practice, there are problems in the calculation of the financial guarantee and the development of financial security instruments and markets as required by Directive [23], due to the lack of expertise in financial, economic and liability matters [75]. Mining companies are usually not required to set up a special guarantee for the waste facilities, but only to set up and maintain the financial guarantee regulated under the Mining Law [76].

It is obvious that both countries (especially Romania because of the structure of its mining sector) have serious environmental legacies, a lack of funds for their restoration and the need to strengthen the administrative capacity in this area, as well as the important tasks on harmonization and/or implementation of the EU mining waste legislation. Cooperation and joint engagement in applying for EU funds for addressing the most urgent mining environmental legacy issues would be of great benefit for sustainable development in the region.

3.3. Mine Land Reclamation

Mine sites, particularly large coal open pits, are usually returned to their pre-mine uses, primarily when prime farmland is mined, but where conditions for land restoration are poor, an adaptive management approach allows reclamation of disturbed land in line with different surrounding land uses and conditions [77,78,79,80]. The choice of reclamation method depends on geological, soil and climate conditions, the planned use of land (crop and ornamental plant production, forestry, wildlife habitats, water accumulations, infrastructural networks, sport and recreation objects, tourism and culture events,*etc.*), the phase of the mine lifetime and available technological solutions, equipment and financial resources [8]. There is evidence that the benefits of sustainable, multi-functional use of natural and semi-natural landscapes exceed the gains from their conversion to single-purpose land use types [81,82]. Rehabilitated mine sites should be able to take advantage of the existing infrastructure and contribute to the local economy after the mine has closed [83].

The largest area of land degraded by mining in Serbia and in Romania is located in lignite open pits. MB Kolubara covers 7.3 thousand ha, and it is planned to expand to a total of 13.4 thousand ha. The open pits and waste dumps of MB Kostolac cover 2.1 thousand ha [84]. In 2014, the total area of reclaimed land in MB Kolubara amounted to 915.5 ha, of which 790 ha is under forest, including lakes and ponds, and 125.5 ha under field crops and orchards. The cultivation of fast growing energy crops is also considered [85]. In MB Kostolac, until 2010, technical and biological reclamation (forest, field crops and vineyards) was accomplished on over 500 ha of degraded land. A nursery of trees and shrubs was also planted, and it is planned to raise a modern apple plantation [52].

According to the evaluation of land reclamation contained in the above-mentioned MB spatial plans, the results so far are modest. Reclamation did not lead to restoration of the previous agricultural and economic potentials of the land, which can be partly attributed to the complex issues of land ownership and land use of reclaimed land [86]. Reclaimed land is managed by a public company. Organizational weakness and a lack of financial resources result in high costs of reclamation and low profitability of production [85].

In the Romanian lignite open pits, according to preliminary data, a total of 24,432 ha was affected in 2013, most of it in the counties of Gorj (12,093 ha), Cluj (3915 ha) and Mehedinti (2315 ha). The reclamation of mine land with agricultural crops is considered the primary goal in most mine land reclamation programs [87]. In Gorj County, a total of 3333 ha was reclaimed, and 12,093 ha are arranged to be reclaimed; in Valcea County and Mehedinti County, 318 ha and 94 ha, respectively, are arranged, while 1074 ha and 466 ha, respectively, shall be reclaimed [67].

Land reclamation takes place in changed environmental, economic and social conditions in the local communities in both countries. It should, therefore, encourage multifunctional use of reclaimed land, adjusted to local conditions, in order to ensure full respect for complex ecosystem, economic and social functions of land as integrated natural resources [88].

3.4. Land Access and Partnerships for Sustainable Mining

Mineral resources and other geological resources in the Republic of Serbia, as well as in Romania are under state property. Mining rights in Serbia are transferred from the state to investors through approvals issued by the ministry upon the request of the commercial entity, *i.e.*, another legal entity and entrepreneur. Mining rights in Romania are transferred from the state to investors through concession licenses (administration rights for public institutions). License holders acquire land for mining activities using general legal mechanisms, which include purchase, exchange, lease, concession, association with the owners and expropriation.

Land use conflicts between miners and private landholders reflect the problems in the way property rights to land and minerals are specified. Land ownership may theoretically assume the absolute title, but according to [89], Mattsson (2004, p. 143) argues that, in practice, absolute ownership rarely exists due to restrictions of various kinds. Explorers and miners may have certain rights of access to private land under state mining legislation [89].

Both the Serbian and Romanian government seek to establish adequate regulatory frameworks to enable more efficient use of mineral resources, respecting the needs to protect the public interest and attracting investments into the sector.

The Serbian draft of the Law on Amendments to the Law on Mining and Geological Research [70] nominates mineral resources of strategic importance for the Republic of Serbia (oil and natural gas, coal, copper and gold ores, ores of boron and lithium, oil shale and other mineral resources determined by a special act of the government). Detailed geological exploration and exploitation of these resources represent an activity of public interest. For the needs of the holders of exploration and mining rights, the competent authority may perform expropriation of property in accordance with the regulations governing the expropriation.

Expropriation in Romania has to be made in accordance with the Expropriation Law [90] for a public utility cause of national interest. As private projects are usually not considered in this sense, the application of this mechanism to obtain property rights over real estate for mining purposes is only theoretical [76]. However, in the draft of Law PL-x 573/2013 on amending the mining law, geological prospecting and exploration, as well as exploitation and processing of mineral resources are declared as public utility works. Such categorization of mining works, which entails expropriation and the provision of some other extraordinary powers to investors (the exemption of some permitting procedures related to works on archaeological sites, reissuing of permits cancelled by courts, *etc.*), was sharply criticized by the public, up to charges of the violation of constitutional rights.

The distribution of mining royalties is another issue that causes anxiety and resistance in the local community, but also controversy among experts from civil society.

The revenues from mining royalties in Serbia are directed to the state budget (40%), the budget of local authorities (40%), the provincial budget 10% (if the exploitation is performed on the territory of the province) and to the departmental ministry funds (20% or 10% if the exploitation takes place on the territory of the province). These funds are used for the elimination of the harmful impacts of mining and mining development promotion (national level) and for improving living conditions and infrastructure (local community) [69].

Mining royalties in Romania are directed to the state budget. One of the amendments to the Mining Law (draft of Law PL-x 573/2013) refers to the transfer of a part of mining royalties to the local authorities in the mining areas for infrastructure development and environmental rehabilitation. According to critics, such a mechanism would undermine the application of the "polluter pays" principle, transposed in the environmental protection legislation and in the Mining law [72].

As stated in the Introduction, the partnerships of industry, governments, local communities and civil society are essential to maintain and manage shared-value opportunities [14]. Through a public review process, the local communities in Serbia are present in the development of spatial and urban plans for the areas with the possibility for exploitation of mineral resources. Compliance with these documents is part of the exploration and mine permitting process. However, a broad public debate, which was recently conducted in Serbia related to nickel exploration, has exposed their dissatisfaction with the role they have in the decision-making process [91].

The procedures for providing environmental agreements (Environmental Impact Assessment—EIA, Strategic Environmental Assessment—SEA) in Romania envisage public debate and cross-border consultations, but public opinion is not mandatory for the environmental authorities [76]. However, the above-mentioned draft law PL-x 573/2013 failed to receive a majority vote in the parliament, as well as some other special laws with most of these provisions, after strong public resistance [43]. Foreign investors are still waiting for a more transparent, simplified and predictable regulatory framework for mining operations, and the government is looking for the best legal updates that will ensure majority support on these sensitive issues.

As Cutifani and Bryant [26] stressed, mining companies must understand that they can no longer operate in isolation; rather, the mining industry can only thrive into the long term if local communities and the local environment thrive. They have to anticipate their challenges into the future and move away from a short-term reactive approach embedded in "survival" to a long-term philosophy beyond the life of a mine that articulates clear, common goals to creating shared value and that provides the infrastructure to make this a reality.

4. CONCLUDING REMARKS

Both analyzed countries, Serbia and Romania, have experienced a mineral policy, regulation and business models of state-owned mining companies in socialism, when strengthening the national economy had precedence over environmental protection and property rights in local communities. Even more devastating ecological and social consequences have occurred in the period of transition, burdened by financial crisis, undeveloped institutions, non-harmonized legislation and corruption. Two decades later, both governments are willing to provide the adequate role to mining in strengthening the national economies, but new challenges place new sustainability demands before all of the stakeholders in the mining sector.

There must be clear legal relations between the state as the owner of natural resources, the investors and the land owners and other stakeholders affected by the exploration and mining. Sustainable spatial, regional and mineral policies, taking the environmental and economic evaluation of space, must be able to provide answers where and when it is or is not appropriate to mine. A transparent business policy of mining companies and their acceptance of the role of a regional development partner, ready to maintain and manage shared value, will contribute towards a consensus in mutual benefit.

Particular attention should be devoted to land expropriation and rehabilitation in order to help create a positive public opinion on the acceptance of mining activities, particularly in densely-populated areas with developed agriculture, respecting the FAO SLM principles of land user-driven and participatory approaches, integrated the use of natural resources at the ecosystem and farming system levels and targeted policy and institutional support to income generation at the local level.

Sustainable mining practice and consistent implementation of the mine closure planning approach require a developed partnership of mining companies, spatial planners, investors, institutions and local communities to identify creative, profitable, environmentally-sound and socially-responsible management and reuse of mine lands.

Transnational and, in particular, cross-border cooperation between countries in the Danube River Basin is extremely important, bearing in mind the transnational risks of large mining environmental hotspots in both countries. The accident of the cyanide spill into the river Tisa from the gold mines in northern Romania (2000) left behind an ecological disaster and long-term consequences for the ecosystems in Romania, Hungary and Serbia. Continuous inventory and risk assessment, information exchange, joint control and early warnings are priorities in this area.

Since both countries are signatories of the Danube River Protection Convention (DRPC), Romania is an EU member and Serbia a candidate for membership in the EU, there are significant opportunities for cooperation and joint engagement in applying for EU funds for addressing the most urgent mining environmental legacy issues.

ACKNOWLEDGMENTS

The paper was prepared within the scientific projects: Sustainable agriculture and rural development in the function of accomplishing strategic objectives of the Republic of Serbia in the Danube region (III 46006) and Sustainable spatial development of the Danube region in Serbia (TR 36036), financed by the Ministry of Education, Science and Technological Development of the Republic of Serbia in the period 2011–2015. Furthermore, the authors would like to thank the reviewers for their valuable comments and suggestions, which helped strengthen this paper.

AUTHOR CONTRIBUTIONS

Vesna Popović contributed to the chapters about mine land management in legislation and practice in Serbia, land access and partnerships for sustainable mining and concluding remarks; Jelena Živanović Miljković contributed to the chapters about mine land change, mining waste management and mine land reclamation in Serbia, Jonel Subić wrote Introduction chapter and contributed to concluding remarks; Andrei Jean-Vasile, Nedelcu Adrian and Eugen Nicolăescu wrote about mine land management in legislation and practice in Romania, mine land change, mining waste management, mine land reclamation and partnerships for sustainable mining in Romania. All authors read and approved the final manuscript.

CONFLICTS OF INTEREST

The authors declare no conflict of interest.

REFERENCES AND NOTES

1. FAO. *Land Evaluation. Towards a Revised Framework*; Land and Water Discussion Paper 6; FAO: Rome, Italy, 2007.
2. Brundtland, G.H. *Our Common Future. Report of the World Commission on Environment and Development*; Oxford University Press: Oxford, UK, 1987.

3. Smyth, A.J.; Dumanski, J. *FESLM: An International Framework for Evaluating Sustainable Land Management. World Soil Resources Report*; FAO: Rome, Italy, 1993; p. 73.

4. UN. The Future We Want: Outcome document adopted at Rio+20, 2012. Available online: http://www.uncsd2012.org/ content/ documents/727The%20Future%20We%20Want%2019%20June%201230pm.pdf (accessed on 12 April 2015).

5. European Commission. Communication from the Commission to the European Parliament, the Council, The European Economic and Social Committee and the Committee of the Regions. Roadmap to a Resource Efficient Europe. Brussels, 20.9.2011, COM(2011) 571 final

6. European Commission. Communication from the Commission. EUROPE 2020—A strategy for smart, sustainable and inclusive growth. Brussels, 3.3.2010, COM(2010) 2020 final.

7. European Commission. *Study Supporting Potential Land and Soil Targets under the 2014 Land Communication. Final Report*; EU: Luxembourg, 2014.

8. Spasić, N.; Stojanović, B.; Nikolić, M. Environmental impact of the mining activity and revitalization of degraded space. *Arhit. Urban.* **2005**, *16–17*, 75–85.

9. Sonter, L.J.; Barrett, D.J.; Soares-Filho, B.S. Offsetting the impacts of mining to achieve no-net-loss of native vegetation.*Conserv. Biol.* **2014**, *28*, 1068–1076.

10. Sonter, L.J.; Moran, C.J.; Barrett, D.J. Modeling the impact of revegetation on regional water quality: A collective approach to manage the cumulative impacts of mining in the Bowen Basin, Australia. *Resour. Policy* **2013**, *38*, 670–677.

11. Sonter, L.J.; Morgan, C.J.; Barrett, D.J.; Soares-Filho, B.S. Processes of land use change in mining regions. *J. Clean. Prod.***2014**, *84*, 494–501.

12. Hobbs, J.C.A. Enhancing the contribution of mining to sustainable development. *Geol. Soc.* **2005**, *250*, 9–23.

13. Buxton, A. *MMSD+10: Reflecting on a Decade. IIED Discussion Paper*; International Institute for Environment and Development: London, UK, 2012.

14. Bartels, R.; Garcia, J.; Reynolds, O.; Thimmiah, S.; Tost, M.; Wong, A. Scoping Paper: Mining and Metals in a Sustainable World. In Proceedings of the World Economic Forum Mining & Metals

Industry Partnership in Collaboration with Accenture, Geneva, Switzerland, 12–18 February 2014.

15. International Council on Mining & Metals—ICMM. Mining's contribution to sustainable development—An overview. 2012. Available online: https://www.icmm.com/document/3716 (accessed on 4 July 2015).

16. Esteves, A.M.; Franks, D.; Vanclay, F. Social impact assessment: The state of the art. *Impact Assess. Proj. Apprais.* **2012**,*30*, 34–42.

17. Franks, D. Social impact assessment of resource projects. International Mining for Development Centre Mining for Development: Guide to Australian Practice. 2012. Available online http://im4dc.org/wp-content/uploads/012/01/UWA_1698_Paper-02_Social-impact-assessment-of-resource-projects1.pdf (accessed on 28 June 2015).

18. Franks, D.M.; Vanclay, F. Social Impact Management Plans: Innovation in corporate and public policy. *Environ. Impact Assess. Rev.* **2013**, *43*, 40–48.

19. Athens Declaration on sustainable access to resources—Critical Raw Materials, Minerals Policy, Land-Use Planning and Permitting. Available online: http://www.euromines.org/files/news/importance-extractive-industry-growth-eu/final-version-athens-declaration-20june2014.pdf (accessed on 3 May 2015).

20. Shields, D.J.; Šolar, S.V. *Sustainable Mineral Resource Management and Indicators: Case Study Slovenia*; Geological Survey of Slovenia: Ljubljana, Slovenia, 2004.

21. Shields, D.J.; Šolar, S.V.; Martin, W.E. The role of values and objectives in communicating indicators of sustainability.*Ecol. Indic.* **2002**, *2*, 149–160.

22. Grantam, J.; Owens, C.; Davies, E. Improving framework conditions for extracting minerals for the EU: Exchanging best practice on land use planning, permiting and geological knowledge sharing. Available online: http://ec.europa.eu/enterprise/policies/raw-materials/files/best-practices/sust-full-report_en.pdf (accessed on 28 June 2015).

23. Directive 2004/35/CE of the European Parliament and of the Council on environmental liability with regard to the prevention and remedying of environmental damage, OJ L 143, 30.4.2004.

24. Directive 2006/21/EC of the European Parliament and of the Council on the management of waste from extractive industries and amending Directive 2004/35/EC, OJ L 102, 11.4.2006.

25. Wang, U. Sustainable mining: An inherent contradiction in terms? Available online: http://www.theguardian.com /sustainable-business/2015/jan/05/sustainable-mining-business-poverty-environment-new-framework (accessed on 10 July 2015).

26. Cutifani, M.; Bryant, P. Reinventing Mining: Creating Sustainable Value—Introducing the Development Partner Framework. Kellogg Innovation Network. 2014. Available online: http:// sustainbusiness.org/PDF/DevelopmentPartnerFramework.pdf (a ccessed on 10 July 2015).

27. UN. *Environmental Performance Reviews—Romania. Second Review. Environmental Performance Reviews Series*; UNECE: New York, NY, USA; Geneva, Switzerland, 2012; p. 37.

28. European Commission. *Report on Critical Raw Materials for the EU*; DG-ENTR: Brussels, Belgium, 2014.

29. Stanojević, J. To more foreign investors by the new law. Available online: http://www.rtb.rs/media/kolektiv/2258.pdf(accessed on 11 August 2015).

30. Government of the Republic of Serbia. *Strategy of Mineral Resources Management by 2030 (Draft)*; National Assembly: Belgrade, Serbia, 2012. Available online: http://www.srbija.gov .rs/vesti/dokumenti_sekcija.php?id=45678 (accessed on 11 August 2015).

31. Ricketts, B., Ed.; *Coal Industry across Europe*, 5th ed.; European Association for Coal and Lignite—Euracoal: Brussels, Belgium, 2013.

32. Official Gazette of the Republic of Serbia. Decision on setting the Energy Balance of the Republic of Serbia for 2015. Available online: http://www.slglasnik.info/sr/147-31-12-2014.html (accessed on 27 May 2015).

33. RTB Bor Group. Resources. Available online: http: //rtb.rs/en/ (accessed on 7 July 2015).

34. Magdalinović, N. Situation and prospects of mineral processing in Serbia. In Proceedings of the Roundtable Condition and Prospects of Mineral Processing in Serbia, Belgrade, Serbia, 17–22 April 2011.

35. Reichl, C.; Schatz, M.; Zsak, G. *Minerals Production, 29*; World-Mining-Data: Vienna, Austria, 2014.

36. Radut, C. Romania's oil and gas reserves. Available online: http://www.nineoclock.ro/romania%E2%80%99s-oil-and-gas-reserves/ (accessed on 24 February 2015).

37. U.S. Energy Information Administration—EIA. Technically Recoverable Shale Oil and Shale Gas Resources. Available online: http://www.eia.gov/analysis/studies/worldshalegas/pdf/fullreport.pdf?zscb=60747969 (accessed on 29 May 2015).

38. Zittel, W. *Fracking: Interim Report*; Energy Watch Group: Berlin, Germany, 2015; Available online: http://energywatchgroup.org/new-energy-watch-group-study-end-fracking-business/ (accessed on 29 May 2015).

39. Nutu, O.; Ivan, V.; Ondrich, J.; Bebiak, M. *Cost of Coal to Romania. The Cost of Subsidies Received by Romanian Coal Industry in Comparison with Renewable Industry*; Candole Partners: Bucharest, Romania, 2014; Available online: http://www.candole.com/research (accessed on 28 May 2015).

40. Council Decision 2010/787/EU on State aid to facilitate the closure of uncompetitive coal mines, OJ L 336, 21.12.2010.

41. Cupru Min. The main activity of S.C. CUPRU MIN S.A. Abrud. Available online: http://www.cuprumin.ro/ en/despre.htm (accessed on 11 June 2015).

42. UN. *Environmental Performance Reviews—Romania*; UNECE: New York, NY, USA; Geneva, Switzerland, 2001.

43. Pencea, R.; Bradatan, T.; Simion, Ş. *Transylvania—Undermined Territory*; Mining Watch Romania: Cluj-Napoca, Romania, 2013; Available online: http://www.academia.edu/10230358/Romania_-_a_mined_field (accessed on 28 May 2015).

44. Stroe, D. Romania to start a new uranium extraction project. Available online: http://www.balkaneu.com/romania-start-uranium-extraction-project/#sthash.NNf2qgf1.dpuf (accessed on 22 May 2015).

45. National Institute of Statistics of Romania. Romanian Statistical Yearbook 2012. Available online: http://www.insse.ro/cms/files/Anuar%20arhive/serii%20de%20date/2012/pdf/ASR%202012_rom.pdf (accessed on 24 May 2015).

46. Peck, P.; Stuhlberger, C.; Simonett, O.; Williams, S. Mining in the Western Balkans. In *Mining and Environment in the Western*

Balkans; Stuhlberger, C., Ed.; UNEP: Vienna, Austria, 2010; pp. 41–90.

47. Noble, B.F. *Introduction to Environmental Impact Assessment: A Guide to Principles and Practice*; Oxford University Press: Oxford, UK, 2006.

48. Sassoon, M. *Financial Surety: Guidelines for the Implementation of Financial Surety for Mine Closure*; Extractive Industries for Development Series; World Bank, Oil, Gas, and Mining Policy Division: Washington, DC, USA, 2009; p. 7. Available online: http://siteresources.worldbank.org/INTOGMC/Resources/7_eifd_financial_surety.pdf (accessed on 12 May 2015).

49. Mitchell, P. Mining and economic growth: The case for Ghana and Tanzania. *South Afr. J. Int. Aff.* **2006**, *13*, 55–67.

50. Schueler, V.; Kuemmerle, T.; Schroder, H. Impacts of Surface Gold Mining on Land Use Systems in Western Ghana.*Ambio* **2011**, *40*, 528–539.

51. Institute of Architecture and Urban&Spatial Planning of Serbia (IAUS). *Amendments on the Spatial Plan for the Exploitation Area of Kolubara Lignite Basin. Internal Documentation Material*; IAUS: Belgrade, Serbia, 2015.

52. Official Gazette of the Republic of Serbia. Spatial plan for the Kostolac mining basin Special-Purpose Area. Available online: http://www.rapp.gov.rs/rudarski-baseni/cid296-83227/prostorni-plan-podrucja-posebne-namene-kostolackog-ugljenog-basena (accessed on 11 May 2015).

53. University of Belgrade, Faculty of Mining and Geology. *Strategy of Coal Mineral Resources Management in Kolubara i Kostolac Basins for the Period until 2017, Documentation Materials*; Faculty of Mining and Geology: Belgrade, Serbia, 2012.

54. Mining Journal. Better times to come. Focus: East-Central Europe. Available online: http://www.euromines.org/files/publications/mining-journal-feature-july-2014-east-central-europe.pdf (accessed on 14 May 2015).

55. Mining Watch Romania. Under Action: Certej, Baita Craciunesti, Baia Mare, Rosia Montana. Available online: http://miningwatch.ro/certej/?lang=en (accessed on 7 July 2015).

56. Official Gazette of the Republic of Serbia. Spatial plan of Kolubara lignite basin area. Available online: http://www rapp.gov.rs/rudarski-baseni/cid296-83173/prostorni-plan-

podrucja-eksploatacije-kolubarskog-lignitskog-basena (accessed on 11 May 2015).

57. European Commission—DG Environment. Mining Waste. 2015. Available online: http://ec.europa.eu/environment/waste/mining/ (accessed on 7 June 2015).

58. Lottermoser, B. *Mine Wastes: Characterization, Treatment and Environmental Impacts*; Springer: New York, NY, USA, 2012.

59. Edraki, M.; Baumgartl, T.; Manlapig, E.; Bradshaw, D.; Franks, D.M.; Moran, C.J. Designing mine tailings for better environmental, social and economic outcomes: A review of alternative approaches. *J. Clean. Prod.* **2014**, *84*, 411–420

60. Franks, D.M.; Boger, D.V.; Cote, K.M.; Muligan, D.R. Sustainable development principles for the disposal of mining and mineral processing wastes. *Resour. Policy* **2011**, *36*, 114–122.

61. Blic. In Vojvodina, 217 abandoned mining pits. 2015. Available online: http://www.blic.rs/Vesti/Vojvodina/570603/U-Vojvodini-217-napustenih-rudarskih-kopova (accessed on 7 July 2015).

62. Peck, P. Reducing Environment & Security Risks from Mining in South Eastern Europe. Desk-assessment study for the Environment and Security Initiative Project. Available online: http://www.envsec.org/publications/Reducing%20environment%20and%20security%20risks%20from%20mining%20in%20SEE.%20Desk%20assessment%20study.pdf(accessed on 11 June 2015).

63. Bogdanović, G.; Trumić, M.; Antić, D.V. Mining waste management: Genesis and possibility of processing. *Reciklaža Održivi Razvoj* **2011**, *4*, 37–43.

64. Živanović, M. Vojvodina's municipalities and villages are complaining about the NIS. Vojvodina Research and Analytical Center—VOICE. Available online: http://voice.org.rs/voice-vojvodanske-opstine-i-sela-se-zale-na-nis/(accessed on 22 March 2015).

65. Pencea, R.; Bradatan, T.; Simion, S. Baia Mare—Avanpremiera dezastrului provocat de industria miniera. Mining Watch Romania. Available online: http://miningwatch.ro/wp-content/uploads/2014/01/RaportBaiaMare-MiningWatchRomania.pdf (accessed on 7 July 2015).

66. Besliu, R. Romania's unsolved communist ecological disaster. OpenDemocracy. Available online: https://www.opendemocracy.net/countries/romania (accessed on 19 March 2015).

67. Ministry of Environment and Climate Changes of Romania. *National Environmental Protection Agency, National Report on the State of Environment in 2013*; ANPM: Bucharest, Romania, 2014. Available online: http://www.anpm.ro/raport-de-mediu (accessed on 30 May 2015).

68. Minalliance. *100 Innovations in the Mining Industry*; PDAC and The Mining Association of Canada: Montreal, QC, Canada, 2012. Available online: http://www.oma.on.ca/en/ontariomining/resources/Minalliance_100_innovations_en.pdf (a ccessed on 17 May 2015).

69. Official Gazette of the Republic of Serbia. Law on Mining and Geological Research. Available online: http://www.en.terragold.co.rs/Laws/LAW-ON-MINING-AND-GEOLOGICAL-EXPLORATION.pdf (accessed on 28 May 2015).

70. Ministry of Mining and Energy of the Republic of Serbia. Draft of Law on Amendments to the Law on Mining and Geological Research. 2015. Available online: http://www.mre.gov.rs/javne-rasprave.php (accessed on 18 May 2015).

71. Ministry of Agriculture and Environmental Protection of the Republic of Serbia. Draft of Law on Land Protection. 2015. Available online: http://www.eko.minpolj.gov.rs/dokumenti/ (accessed on 18 May 2015).

72. Official Gazette of Romania. Mining Law no. 85/2003. Available online: http://www.mpo.cz/kalendar/download/71546/priloha003.pdf (accessed on 5 June 2015).

73. The Government of Romania. Decision no. 856/2008 on the management of waste from extractive industries. Available online: http://lege5.ro/en/Gratuit/geytgobygi/hotararea-nr-856-2008-privind-gestionarea-deseurilor-din-industriile-extractive (accessed on 4 June 2015).

74. The Government of Romania. Emergency Ordinance no. 68/2007 on environmental liability with regard to the prevention and remedying of environmental damage. Available online: http://lege5.ro/en/Gratuit/geydqobvgm/ordonanta-de-urgenta-nr-68-2007-privind-raspunderea-de-mediu-cu-referire-

la-prevenirea-si-repararea-prejudiciului-asupra-mediului (accessed on 4 June 2015).

75. Permanent Representation of Romania to the EU. Report on the implementation of the Directive 2004/35/EC on environmental liability, Brussels, 2013, 3508. Available online: http://ec.europa.eu/environment/legal/liability/pdf/eld_ms_reports/RO.pdf (accessed on 29 May 2015).

76. Dragomir, C.; Halcu, B. Romania. In *The Mining Law Review*, 3rd ed.; la Fleche, E.R., Ed.; Law Business Research Ltd.: London, UK, 2014; pp. 242–252.

77. Bowman, B.; Baker, D. *Mine Reclamation Planning in the Canadian North. Northern Minerals Program*; Working Paper no.1; Canadian Arctic Resources Committee: Ottawa, ON, Canada, 1998.

78. Doley, D.; Audet, P. Adopting novel ecosystems as suitable rehabilitation alternatives for former mine sites. *Ecol. Process.* **2013**, *2*, 1–11.

79. Limpitlaw, D.; Briel, A. Post-mining land use opportunities in developing countries–A review. *J. South. Afr. Inst. Min. Metall.* **2014**, *114*, 899–903.

80. Minerals Education Coalition—MEC. Reclamation Stories—Coal. 2015. Available online: https://www.mineralseducationcoalition.org/reclamation-stories/coal (accessed on 2 July 2015).

81. De Groot, R. Function-analysis and valuation as a tool to assess land use conflicts in planning for sustainable, multi-functional landscapes. *Landsc. Urban Plan.* **2006**, *75*, 175–186.

82. Zhang, J.; Fu, M.; Hassani, F.P.; Zeng, H.; Geng, Y.; Bai, Z. Land Use-Based Landscape Planning and Restoration in Mine Closure Areas. *Environ. Manag.* **2011**, *47*, 739–750.

83. Dutta, S.; Rajaram, R.; Robinson, B. Mineland Reclamation. In *Sustainable Mining Practices—A Global Perspective*; Rajaram, V., Dutta, S., Eds.; A.A. Balkema Publishers: Leiden, The Netherlands, 2005; pp. 179–191.

84. Serbian Environmental Protection Agency. *The Report on the Status of Land in the Republic of Serbia for 2011*; SEPA: Belgrade, Serbia, 2012.

85. Kolubara, M.B. *Report on the State of the Environment in 2014*; Sector for Environmental Protection and Enhancement:

Lazarevac, Serbia, 2015. Available online: http://www.rbkolubara.rs/index.php?option=com_content&view=article&id=470&Itemid=225&lang=en (accessed on 12 May 2015).

86. Popović, V.; Živanović Miljković, J. Key issues of land policy in Serbia in the context of spatial development—Case study of Danube basin area. In Proceedings of 2nd International Scientific Conference Regional development, spatial planning and strategic governance (RESPAG 2013), 22–25 May 2013; Vujošević, M., Milijić, S., Eds.; IAUS: Belgrade, Serbia, 2013; pp. 271–297.

87. Dumitru, E.; Popescu, A. Reclamation on Land Disturbed by Surface Mining in Romania. Available online: http://www.asmr.us/Publications/Conference%20Proceedings/2002/0115%20Dumitru.pdf (accessed on 23 May 2015).

88. Popović, V.; Nikolić, M.; Katić, B. *Agricultural Land Use and Protection in Serbia*; Institute of Agricultural Economics: Belgrade, Serbia, 2011.

89. Jonson, L.E. Mineral Rights. Legal Systems Governing Exploration and Exploitation. Ph.D. Thesis, School of Architecture and the Built Environment, Royal Institute of Technology (KTH), Stockholm, Sweden, 2010. Available online: http://nickelmountain.se/wp-content/uploads/2014/01/Mineral-rights-Legal-Systems-Governing-Exploration-and-Exploitation.pdf (accessed on 15 July 2015).

90. The Parliament of Romania. Law no. 33/1994 on the expropriation for public utility. Available online: http://lege5.ro/en/Gratuit/gi2tqnbtgi/legea-nr-33-1994-privind-exproprierea-pentru-cauza-de-utilitate-publica (accessed on 14 May 2015).

91. Center for Democracy and Germany's Friedrich Ebert Foundation. Nickel: Economic benefits and environmental risks. Debate. Available online: http://www.dc.politickiforum.org/index.php?vrsta=tribina&naredba=prikaz&tekst=80(accessed on 8 November 2012).

8

CHAPTER

An Integrated Assessment Approach to Address Artisanal and Small-Scale Gold Mining in Ghana

Niladri Basu [1,2,*], Elisha P. Renne [3,4] and Rachel N. Long [2]

[1]*Faculty of Agricultural and Environmental Sciences, McGill University, 21, 111 Lakeshore Rd., Ste. Anne de Bellevue, Quebec H9X 3V9, Canada*

[2]*Department of Environmental Health Sciences, School of Public Health, University of Michigan, 1415 Washington Heights, Ann Arbor, MI 48109, USA*

[3]*Department of Anthropology, University of Michigan, 101 West Hall, Ann Arbor, MI 48109, USA*

[4]*Department of Afro-American and African Studies, University of Michigan, 4700 Haven Hall, Ann Arbor, MI 48109, USA*

ABSTRACT

Artisanal and small-scale gold mining (ASGM) is growing in many regions of the world including Ghana. The problems in these communities are complex and multi-faceted. To help increase understanding of such problems, and to enable consensus-building and effective translation of scientific findings to stakeholders, help inform policies, and ultimately improve decision making, we utilized an Integrated Assessment approach to study artisanal and small-scale gold mining activities in Ghana. Though Integrated Assessments have been used in the fields of environmental science and sustainable development, their use in addressing specific matter in public health, and in particular, environmental and occupational health is quite limited despite their many benefits. The aim of the current paper was to describe specific activities undertaken and how they were organized, and the outputs and outcomes of our activity. In brief, three disciplinary workgroups (Natural Sciences, Human Health, Social Sciences and Economics) were formed, with 26 researchers from a range of Ghanaian institutions plus international experts. The workgroups conducted activities in order to address the following question: What are the causes, consequences and correctives of small-scale gold mining in Ghana? More specifically: What alternatives are available in resource-limited settings in Ghana that allow for gold-mining to occur in a manner that maintains ecological health and human health without hindering near- and long-term economic prosperity? Several response options were identified and evaluated, and are currently being disseminated to various stakeholders within Ghana and internationally.

Keywords: Minamata Convention; implementation; policy; mercury; public health; anthropology; history; social sciences; economics; gender; small-scale mining; artisanal mining; gold mining

"Be contented and turn your eyes to what I shall present you concerning the Gold itself.....This Illustrious Metal is generally found in three sorts of places: First, the best in found in or betwixt particular Hills; and apprehending where the Gold, [they] dig Pits; and separate it from the Earth which comes out with it. The second place is in, at, and about some Rivers and Water-falls; whose violence washeth down great quantities of Earth, with carry the Gold with it. The third is on the Sea shore; where (as at Elmina and Axim) there are little Branches or Rivulets into which the Gold is driven from Mountainous Places, as well as Rivers.... these places are visited by hundreds of women [who are] furnished with large and small Troughs or Tray[s], which they first fill full of Earth and Sand, which they wash with repeated fresh Water;...if there is any Gold...it is thrown into the small Tray..."

From Willem Bosman, 1705. *A New and Accurate Description of the Coast of Guinea...*, translated from the original Dutch. London: J. Knapton, 1705, pp. 80–81 [1].

1. ARTISANAL AND SMALL-SCALE GOLD MINING (ASGM) IN GHANA

Informal gold mining, referred to as artisanal and small-scale gold mining (ASGM), has proliferated worldwide due to contemporary and powerful economic forces and widespread poverty across low- and middle-income countries. It is estimated that upwards of 15 million people worldwide may directly be involved in ASGM, and another 100 million people may be reliant upon the sector [2,3]. Gold from small-scale mines may represent 20%–30% of the world's output [4].

West Africa, particularly Ghana, is one of the world's most important gold mining regions. Gold has been mined in Ghana for over 1000 years [5] and, as Bosman has discussed [1], gold mining was well-developed in the Gold Coast by the 16th century. Its production has increased 700% since 1980 and in 2012 gold accounted for 43% of the country's national exports [6,7]. Ghana is second only to South Africa in gold production from Africa [8]. The ASGM sector is estimated to have increased from 6% of Ghana's gold production in 2000 to 23% in 2010 [9]. The sector may employ an estimated 1.1 million people (with estimates suggesting that most work illegally) and represent nearly two-thirds of the country's total mining labor force [10,11]

The ASGM sector is faced with numerous challenges. Many mining community residents are impoverished and live in rural settings that lack basic resources such as health care services and clean potable water [12]. Mining activities can destroy landscapes, grossly contaminate water resources, and stress water tables [13]. Mercury contamination of water, soil, and food is well documented in ASGM communities, and newer studies have documented that exposures to other toxic elements such as arsenic and lead may be widespread [14]. Many residents have expressed deep concern about access to clean water, but are willing to accept compromised water quality in exchange for income and sustenance [15]. In Ghana, while unlicensed small-scale gold mining is illegal, the government is unable to monitor the activities of *galamsey* miners [16]. *Galamsey* is the term commonly used for illegal small-scale miners [17]. While some *galamsey* are working in areas where they have long resided and practiced traditional gold mining as their legitimate right [18], since they are not registered and do not pay taxes, government officials cannot provide resources to these communities. Small-scale miners' fear of prosecution exacerbates the situation, fueling this illicit market economy. Furthermore, women, who have long been engaged in small-scale gold mining, face different problems from their male counterparts, particularly concerning prevailing gender roles and reproductive health [19].

The problems in ASGM communities are multi-faceted and complex, and so improvements to the sector can only be realized by considering different approaches and disciplinary perspectives. Here we outline the use of an Integrated Assessment approach to increase understanding of policy and response options concerning ASGM in Ghana. As elaborated upon in the following sections, Integrated Assessments have been used in the realms of environmental science and sustainable development [20], but their use to address specific matters in public health, and in particular, environmental and occupational health, is quite limited. This is despite the numerous benefits they offer as discussed below. This paper also serves as an introduction to a special issue of the International Journal of Environmental Research and Public Health (http://www.mdpi.com/journal/ijerph/special_issues/asgm) entitled "Integrated Assessment of Artisanal and Small-Scale Gold Mining (ASGM) in Ghana".

2. A PRIMER ON INTEGRATED ASSESSMENTS

Integrated Assessments have been defined as:

"a way of bringing together knowledge of ecosystems, people, and policy in order to find solutions for particularly challenging problems and to develop tools and information that policy makers can use." [21];

"a collective, deliberative process by which experts review, analyze, and synthesize scientific knowledge in response to user's information needs relevant to key questions, uncertainties or decisions" [22]; "policy-motivated research to develop an understanding of the issue, not based on disciplinary boundaries but based on boundaries defined by the problem" [23].

A major activity of an Integrated Assessment is to identify and assess best available information using diverse and multi-disciplinary approaches, so as to enable consensus-building and effective translation of scientific findings to stakeholders, help inform policies, and ultimately improve decision making [20]. An Integrated Assessment aims to bring together key aspects and goals of other assessment methods including Process Assessments (which describe a process and evaluate status, trends, and underlying causes), Impact Assessments (which identify and assess the possible consequences of an environmental problem towards human health, ecosystems, and socioeconomic systems), and Response Assessments (which identify and evaluate the possible responses, including policies, to an environmental problem).

There are several benefits to tackling complex problems, particularly those lacking a clear cause or solution, using an Integrated Assessment framework. According to a document published by NOAA [24], Integrated Assessments can help to produce outcomes or decisions that have higher likelihood of success. The process includes a number of steps in which response options and policy actions, as well as their forecasted impacts, are evaluated iteratively. Often they are conducted in the context of a specific management of policy issue, and thus they have a clear end user. By bringing together groups normally separated along disciplinary (e.g., natural sciences and social sciences) or sectoral (e.g., industry and government) lines, they can facilitate multi-disciplinary dialogue and work towards building consensus. For example, as depicted inFigure 1, social scientists and natural sciences are often disconnected from policies, yet informed decisions require the involvement of those with a policy background. By engaging with diverse stakeholders, and especially the public, the Integrated Assessment process helps enable broad support due to its consultative nature, facilitation of local knowledge into decision-making practices, and advancement of public understanding of the issue. The process also creates new relationships amongst participants who may not normally interact.

Figure 1. An Integrated Assessment brings together groups normally separated along disciplinary or sectoral lines to facilitate multi-disciplinary dialogue and work towards building consensus. Figure is reproduced with permission from the Michigan Sea Grant.

3. INTEGRATED ASSESSMENT TO ADDRESS ASGM IN GHANA

Conducting the activity in Ghana made sense for many reasons. Foremost is that the Co-Principal Investigators have been working in Ghana since 2006, and ASGM was amongst the various topics under investigation [15,19,25,26]. Over time relationships had been formed with stakeholders including community residents, mine concession owners, government officials, and researchers from across academia, NGOs, and the government. Ghana also represented an excellent platform for this work owing to a relatively large group of in-country researchers who have published widely on ASGM in their own country. A simple bibliometric analysis revealed that Ghanaian researchers are amongst the most prolific worldwide on the topic of ASGM. A search in Scopus on 19 March 2015 for the term "gold mining" resulted in 2531 papers, and research hailing from Ghana was in the top 10 of all countries (Figure 2). The Integrated Assessment process afforded an unprecedented opportunity to bring together several of these active researchers, many of whom had not previously met.

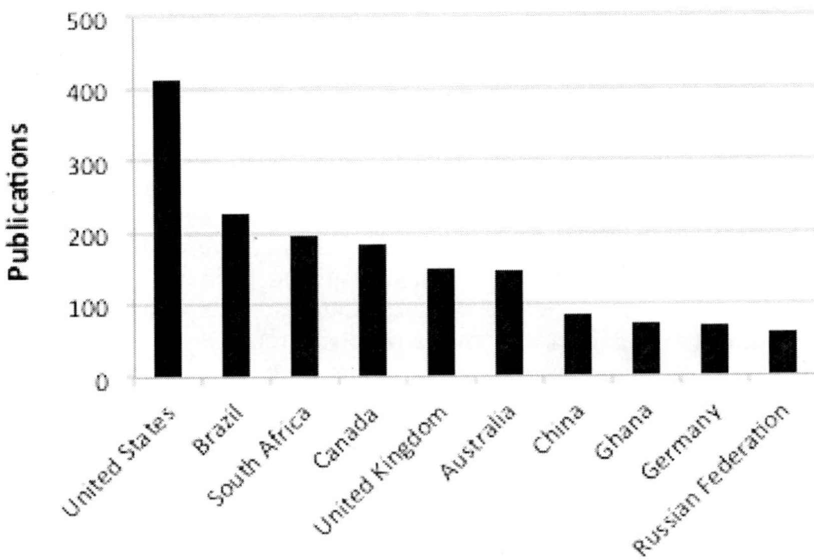

Figure 2. A simple bibliometric analysis using Scopus reveals that Ghanaian researchers are amongst the most prolific worldwide. On 19 March (2015), the term "gold mining" was searched resulting in 2531 papers. Research hailing from Ghana was in the top 10 of all countries.

4. APPROACH AND ACTIVITIES

From its conception the project was carefully designed to bring together interested Ghanaian ASGM researchers. To ensure that critical perspectives were captured and to align with the Integrated Assessment framework, three disciplinary workgroups were formed: Natural Sciences, Human Health, and Social Sciences and Economics. For each workgroup, co-leaders were identified and immediately engaged. The co-leaders consisted of at least one researcher from Ghana paired with at least one researcher from the University of Michigan. All of the co-leaders had prior collaborative history and thus had developed collegial relationships.

The co-leaders had two initial tasks: first, they developed and refined the policy relevant question that would guide the Integrated Assessment. Following several discussions over email and meetings in-person, the group decided to organize the Integrated Assessment to tackle the question: *What are the causes, consequences and correctives of small-scale gold mining in Ghana?* More specifically: *What alternatives are available in resource-limited settings in Ghana that allow for gold-mining to occur in a manner that maintains ecological health and human health without hindering near- and long-term economic prosperity?*

The second initial task of the co-leaders was to identify workgroup members. It was necessary to ensure that each group consisted of at least five members who hailed from different institutions and/or sectors government, academia, NGO), and to also ensure that they were mainly Ghanaian researchers. As mentioned earlier, a great benefit of pursuing research activities in Ghana is the great in-country expertise concerning ASGM (Figure 2). In total, 26 workgroup members were recruited, hailing from a range of institutions including: the Ministry of Lands and Resources; the Ghana Health Service; the Water Research Institute (at the Council for Scientific and Industrial Research); the Centre for Energy, Environment, and Sustainable Development; Kwame Nkrumah University of Science and Technology; University of Ghana; University for Development Studies; University of Cape Coast; Institute for Development Studies; Kumasi Polytechnic; University of Michigan; and McGill University. In the initial stages of the work we purposefully limited membership to active scientific researchers and thus did not include other important stakeholders such as miners and policy makers. This was to ensure that we focused on the scientific evidence, but also had confidence that other perspectives were infiltrating into the initial phases given that all the researchers involved work intimately with other stakeholder groups. Further, the latter stages of the project (such as

Meeting #4) involved dissemination and exchange with relevant stakeholders.

The overall project was designed to be 40 months in duration (May 2012 through August 2015) with several milestones pre-identified. The major activities include a series of workshop meetings held in Accra (Table 1) coupled with regular exchanges via email, both carefully orchestrated to shepherd the workgroups through the key Integrated Assessment steps (Table 2). To facilitate open dialogue, workshops included a range of activities including, for example, guided and unstructured brainstorming sessions, individual researcher updates, and iterative polls using paper and electronic surveys. During these meetings researchers mainly from Ghana delivered 37 presentations. The core activities concerned the identification and scrutiny of datasets so as to document the status, trends, causes, and consequences of ASGM in Ghana (Step 2 in Table 2). This was achieved via in person meetings (Meetings #2 and #3) and online exchanges for a period of approximately 1.5 years. Related to this, the process has facilitated the researchers further developing their own datasets and findings into publications that are being published in this special issue of the International Journal of Environmental Research and Public Health (http://www.mdpi.com/journal/ijerph/special_issues/asgm). The second main activity was the identification and evaluation of response options (Steps 3 and 4 in Table 2). As elaborated upon below, this activity entailed several exchanges within Workgroups to develop a list of 12 response options that were justified based on the reviewed evidence. During Meeting #3, the Workgroups presented their options to the entire delegation who then provided feedback and evaluative scores. Meeting #3 ended with a list of the ranked list of response options, of which the top six were presented to decision makers and stakeholders (38 individual representing 28 Institutions) at Meeting #4.

The organization of the meetings was carefully considered so as to promote open discussion and active participation. Meals and break periods enabled free-flowing unstructured discussions and provided networking opportunities, and this dedicated time was critical to establishing and growing relationships. Event scheduling occurred several months in advance to ensure that all participants could be engaged and that dates did not conflict with other events. All participants were afforded an opportunity to help set the agenda (*i.e.*, sent draft agendas prior to the meeting to comment upon), and were provided all materials in advance of the meeting. Time was effectively moderated so as to respect participants' time. Workgroup members, especially those from the global North, were cognizant of the need to listen and enable

Table 1. Outline of meeting activities as part of our integrated assessment.

Meeting #	Dates	Participants (# in Brackets)	Key Activities
1	6–7 August 2012	Organizer (1), Co-Leaders (9)	policy relevant question defined workgroup members and stakeholders identified
2	11–12 April 2013	Organizer (1), Co-Leaders (9), Workgroup Members (11)	22 research presentations delivered by individuals workgroup breakout sessions
3	28–30 April 2014	Organizer (1), Co-Leaders (9), Workgroup Members (13), External Advisors (4), Facilitators (2), Observer from Funding Agency (1)	3 summary presentations on each workgroup's findings and recommended response options Delphi poll used by researchers to refine and evaluate response options
4	21–22 April 2015	Organizer (1), Co-Leaders (3), Facilitators (2), External Advisors & International Stakeholders (5), Decision Makers (25)	summary presentations on findings and recommended response options Delphi poll used by stakeholders to refine and evaluate response options focused discussion on implementation of the Minamata Convention

Table 2. Key steps of the integrated assessment.

	Brief Overview of Integrated Assessment Steps
1	Define policy-relevant question
2	Document status/trends; causes/consequences
3	Identify desired outcomes/policy options
4	Evaluate policy options
5	Provide technical guidance for implementation
6	Assess uncertainty
7	Evaluation
8	Submit findings for peer review and public comment

open discussion, rather than dominate conversation. The meeting was designed to provide a neutral and open forum. This was continuously discussed by the organizers and queried in the yearly evaluations. For the final two meetings neutral facilitators (from Ghana's Science and Technology Policy Research Institute, STEPRI) were brought into the process.

Engagement was also promoted by constantly reminding the workgroup members of the overall Integrated Assessment process, and how their activities both within their workgroups but also as individual researchers (*i.e.*, publication of their individual research papers; involvement in future grant applications) would feed into the larger process. Careful notes were taken during each meeting and promptly shared with all participants within two weeks of each meeting. Between meetings there was constant communication amongst workgroup members and the larger team. A website (asgmresearch.weebly.com) served as a clearinghouse of information, and advertised our work internally and to outside stakeholders.

5. KEY FINDINGS FROM THE INTEGRATED ASSESSMENT

The results of the Integrated Assessment are being published in this special issue of the International Journal of Environmental Research and Public Health (http://www.mdpi.com/journal/ijerph/special_issues/asgm). In addition to this introductory piece, each of the three workgroups has published a detailed assessment report [12,13,14], and there is a paper outlining the application of the Delphi method [27,28] to the identification of response options, and 12 individual research papers written by team members on a variety of ASGM matters. Here we provide very brief summaries of the assessment reports written by the three working groups.

The Human Health report [14] provides evidence from across multiple Ghanaian ASGM sites that documents relatively high exposures to mercury [29,30] and other heavy metals, occupational injuries [31,32,33] and noise exposure [34]. The work also reviews limited data on psychosocial health, nutrition, cardiovascular and respiratory health, sexual health, and water and sanitation. The top three recommended response options from this workgroup were (note that the final three options listed below were tied):

- *That registration of small-scale miners be increased by improving the process by, for example, reducing or eliminating fees and localizing registration.*
- *That universities, the EPA, and the Minerals Commission explore and implement high-yield mercury free alternatives.*

- *That ministries, local governments, and District Assemblies promote diversification of economic opportunities.*
- *That the government provide water, electricity, telecommunications, and sanitation in partnership with enterprises to ASGM communities and other affected communities.*
- *That a national framework for policy and planning implementation be established (i.e., taskforces, workgroups) that considers stakeholder input.*

The Natural Sciences Report [13] documents evidence of mercury contamination in a range of ecological media, including soil, foodstuffs, sediment, and water [29,35,36,37]. Other water quality parameters near ASGM sites show impairment, with some samples exceeding guidelines for acidity, turbidity, and nitrates. Deforestation and land degradation often accompany ASGM activity, potentially decreasing biodiversity, farmland, and soil fertility. While not well-documented, effects from and on climate change as well as contamination from mining waste may add additional stress on environmental quality and ecosystem services. The top three recommended response options from this workgroup were:

- *That the royalties from the proceeds of mining be placed into a central account and directed towards improving health and environment of ASGM communities.*
- *That a national framework for policy and planning implementation be established (i.e., taskforces, workgroups) that considers stakeholder input.*
- *That there be public and private support for education with ASG miners on ecological and human health risks, mercury and metals, mercury reduction strategies, and business practices.*

The Socioeconomics Report [12] discusses sustainable ASGM development in terms of the legalization and formalization of small-scale gold miners' activities. The costs of obtaining mineral rights and permits, as well as the related costs of travel and work time lost, contribute to miners' calculations concerning the perceived value of registering as a legal gold miner. Yet registration is fundamental to miners' investing in sustainable and environmentally-sound ASGM practices. However, the enforcement of regulations has to be consistent and fair, with rules that are simple, unequivocal and broadly accepted. The authors suggest that to improve stability and sustainability of the ASGM industry, the government should enhance capacity to locally certify and track ASGM activities before enforcing sanctions against non-compliant miners [11]. Such certification and tracking is likely to enhance the long-term stability of ASGM economic benefits and human development potential, while minimizing the adverse health and environmental impacts that need to be addressed. The top three recommended response options from this workgroup were:

- *That registration of small-scale miners be increased by improving the process by, for example, reducing or eliminating fees and localizing registration.*
- *That a national framework for policy and planning implementation be established (i.e., taskforces, workgroups) that considers stakeholder input.*
- *That ministries, local governments, and District Assemblies promote diversification of economic opportunities.*

6. IDENTIFICATION OF RESPONSE OPTIONS AND THE POLICY LANDSCAPE

One of the major expectations of the Integrated Assessment was that teams would identify response options, or actions to address the social, economic, health and environmental issues associated with ASGM, that could be acted upon by stakeholders. Groups were provided minimal guidance on the development of response options, and were asked only to ensure that there was some evidence supporting the response options developed.

A detailed description of the entire development and evaluation of the response options can be found in a companion paper [28]. In brief, the response options were developed by workgroups in advance of Meeting #3, and described in the final section of their respective discipline-specific reports. Before the meeting, team leaders reviewed these response options and condensed them into 18 options that were shared with all participants via an online poll (SurveyMonkey.com). At the meeting, workgroups gave brief presentations on their findings, and then elaborated on the response options developed in the reports. Response options were then discussed among the entire group and evaluated via two rounds of polling based on the Delphi technique [28].

The current Integrated Assessment is timely for many reasons. Internationally, in 2013, the member nations of the United Nations Environment Program (UNEP) agreed on the Minamata Convention on

Mercury. The Convention has both a stand-alone article (Article 7) and a dedicated Annex (Annex C) concerning the ASGM sector. Notably, countries with significant ASGM sectors, such as Ghana, are required to develop a public health strategy for ASGM communities, and take specific measures to protect vulnerable populations such as children and women of childbearing age. The Convention also encourages Parties to cooperate in education, outreach and capacity-building initiatives specific to ASGM (7.4B), as well as engage in general strengthening of public health measures to address mercury pollution (Article 16), especially protection of vulnerable groups (women and children; 16.1A) with specific mention of strengthening of institutional and health professional capacities (16.1D). At the national level, ASGM is built into Ghana's Minerals and Mining Act, 2006 (Act 703). While many miners are formally registered, it is estimated that about 80% of ASGM miners are not registered and thus conducting the activity illegally (personal communication with Ghana Minerals Commission, 6 August 2014). To help legitimize the sector, in 2013 the President of Ghana launched an Inter-Ministerial Task Force on Small-Scale Mining. In addition, the National Development Planning Commission, which reports to President of Ghana, has recently identified ASGM as a sector of major economic, social and national security concern that requires swift policy action. Efforts are underway to share results of this Integrated Assessment with the aforementioned programs, and other stakeholders.

7. EVALUATION

A critical feature of the Integrated Assessment was a series of evaluations. The funders of this Integrated Assessment have published a standalone evaluation guideline [37] within which they cite the work of Conley and Moote [38] that state that evaluations "can help: (1) determine when the idealized narrative used to justify collaborative [efforts] holds true; (2) address criticism of these efforts; and (3) assess and refine efforts to institutionalize a movement that has developed largely at the grassroots level." At the outset of this Integrated Assessment a framework of activities was established with expected outputs and outcomes defined. These are discussed below from an evaluation perspective.

Technical Adequacy. To ensure that the work is scientifically sound, it is to be judged via the publication of peer-reviewed papers, generation of factsheets accessible to a range of stakeholders, and the involvement of international experts as external advisors. As mentioned previously a major output of this Integrated Assessment will be the publication of ~18 peer-reviewed papers in this special issue of the International Journal of Environmental Research and Public Health (http://www.mdpi.com/journal/ijerph/special_issues/asgm). Acceptance of these papers by the scientific community will provide strong support of the work's technical adequacy. We are in the process of finalizing a series of fact sheets as well as data resources that, for example, outline the relevant mining laws in Ghana, provide a table of mercury content in nearly 10,000 specimens (*i.e.*, fish, tailings, sediment, plants, soil, water, hair, urine) from across the country, and summarize workgroup findings. These will be made available as hardcopies and also posted on our project website.

Value and Effectiveness. The Integrated Assessment is very timely and relevant to ongoing decision making at the international and national levels. As mentioned earlier, Ghana has signed onto the global UNEP Minamata Convention on Mercury that has special provisions for the ASGM sector. At the national level ASGM is built into various Acts and is a focus of an Inter-Ministerial Task Force on Small-Scale Mining and a topic of concern for the National Development Planning Commission.

Meeting #4 was co-organized by the Principal Investigators along with Ghana's Science and Technology Policy Research Institute (STEPRI) to disseminate findings to officials (including Ministers, Deputy Ministers, Unit Heads) representing 28 organizations including groups such as the Ministry of Environment, Science, Technology and Innovation (MESTI), the Ministry of Health, the Ministry of Lands and Natural Resources, the National Development Planning Commission (NDPC), the Minerals Commission, the Ministries of Local Government and Rural Development, the Water Research Institute, the University of Ghana School of Public Health, the Ghana Health Service, the University of Mines and Technology, the Environmental Protection Agency, the Ghana Mine Workers' Union, the Ghana National Association of Small Scale Miners, and the Ghana National Association of Small Scale Businesses. The participants involved in Meeting #4 provided structured and open-ended feedback on the top six response options presented to them, and this is detailed in the companion paper [28].

In addition, the work is warranted given the recent boom in gold prices, global economic crises, expansion of resource development across Africa, and a dearth of information concerning multiple stressors in mining communities. The work also builds off, and thus mutually reinforces, several research grants and research training grants in Ghana (and across West Africa) held by the Co-Principal Investigators and several workgroup members.

The effectiveness of an Integrated Assessment can be determined by its influence on policy. While it is too early to determine this, there are a number of planned activities in coming months that involve key decision makers and managers, both nationally and internationally. Further, the policy landscape concerning ASGM is currently under scrutiny worldwide, and the countrywide assessment conducted here is expected to inform policies in Ghana and potentially elsewhere.

Legitimacy. To ensure fairness and ensure an assessment that is balanced and objective, a wide range of stakeholders has been engaged from across disciplines and sectors, including external international advisors. As discussed earlier, meetings were carefully designed so as to enable open dialogue and encourage diverse perspectives. Engagement was tracked via responsiveness to emails and online polls, participation in meetings (both in-person and virtual).

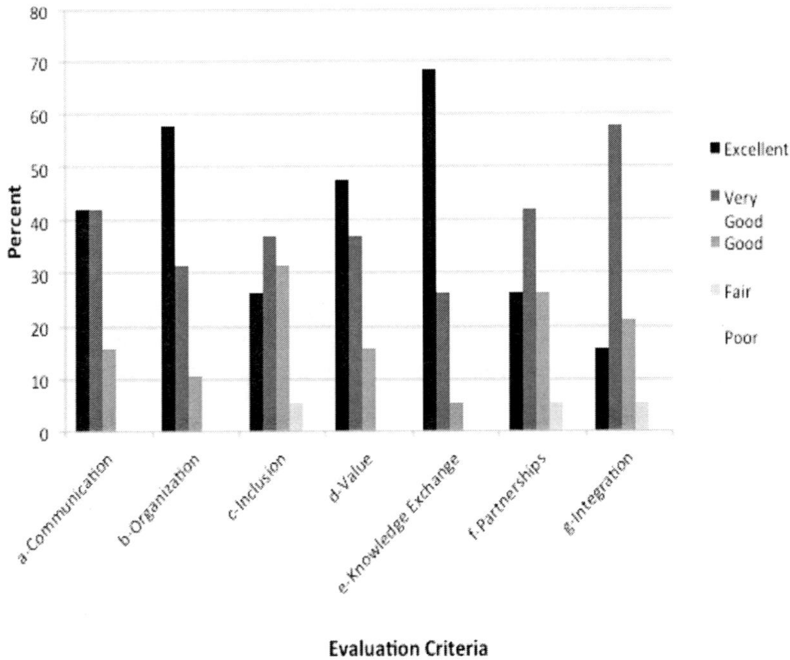

Figure 3. Participant evaluation scores from the third meeting. Scores are representative of other meetings. The evaluation criteria presented here include: A—Clear communication of project goals, activities, and timeline; B—Organization of planning process; C—Inclusion of appropriate stakeholders; D—Value in addressing the problem and research questions; E—Exchange of knowledge; F—Formation of partnerships; G—Integration among diverse stakeholders/experts.

Table 3. Representative comments of the integrated assessment from the evaluations.

Positive Aspects of the Integrated Assessment Process and/or Meetings
Most impressive were the far reach and depth of discussion of this multifaceted, difficult problem among so many disciplinary perspectives.
The poll strategy was very effective in getting the thoughts of individuals synthesized despite the short period available.
People openly discussed ideas and respected each other's opinions. Was a sense that people were open to new ways of looking at things.
The [Delphi] poll strategy was very effective in getting the thoughts of individuals synthesized despite the short period available.
Negative aspects of the Integrated Assessment process and/or meetings:
In my opinion, the IA process can be improved by engaging the local communities.
There is the need to identify all relevant stakeholders who are in a position to support the initiative and get the results to influence policy.
Inclusion of policy makers during the process could improve ownership and hence facilitate policy dialogue and implementation.
Time for each presentation should be increased to elicit extensively detailed information from groups.
The options were quite confusing during the [Delphi] poll. Took a long time during the workshop to re-define the options.

A major scheme by which legitimacy was ascertained was via a series of anonymous surveys of meeting participants. These surveys were administered on paper at the conclusion of each meeting and collected and analyzed by the project manager. The surveys contained both quantitative and open-ended, qualitative questions that addressed both the most recent meeting and the entire Integrated Assessment process. Quantitative questions about the meeting asked about the logistics of the meeting and the meeting's effectiveness in achieving its stated goals. Quantitative questions about the Integrated Assessment process asked about communication, organization, inclusion of appropriate stakeholders, and the value of the process in addressing the problem, the exchange of knowledge, the formation of partnerships, and integration among diverse stakeholders and experts. Qualitative questions asked for open-ended responses about both positive and negative aspects of the meetings and of the Integrated Assessment process. Figure 3 shows the results of the latest quantitative survey on the Integrated Assessment process, and Table 3 provides some representative responses (pro and con) to qualitative questions.

8. CONCLUSIONS

This Project Report aimed to describe how an Integrated Assessment approach was used to study ASGM activities in Ghana. In doing so, the activities and approaches described and implemented here helped increase understanding of problems associated with the sector, particularly within and across disciplinary lines. In addition, the project enabled consensus building and included effective translation of scientific findings to stakeholders including those in decision-making positions.

ACKNOWLEDGEMENTS

We thank the Graham Environmentally Sustainable Institute (GESI) at the University of Michigan (Ann Arbor, USA) for funding this Integrated Assessment project. At GESI we thank John Callewaert and Margaret Allen for their active involvement in the project, as well as their guidance. We also acknowledge the wonderful contributions to the Integrated Assessment by all the workgroup members, stakeholders, and their organizations.

AUTHOR CONTRIBUTIONS

Niladri Basu wrote the paper, with guidance and contributions from Rachel Long and Elisha Renne. Niladri Basu and Elisha Renne served as Principal Investigators of the project, and Rachel Long served as the Project Manager.

CONFLICTS OF INTEREST

The authors declare no conflict of interest.

REFERENCES

1. Bosman, W. *A New and Accurate Description of the Coast of Guinea..., Translated from the Original Dutch*; James Knapton: London, UK, 1705.

2. Hilson, G.; Pardie, S. Mercury: An agent of poverty in Ghana's small-scale gold-mining sector? *Resour. Policy* **2006**, *31*, 106–116.

3. Swain, E.B.; Jakus, P.M.; Rice, G.; Lupi, F.; Maxson, P.A.; Pacyna, J.M.; Penn, A.; Spiegel, S.J.; Veiga, M.M. Socioeconomic consequences of mercury use and pollution. *Ambio* **2007**, *36*, 45–61.

4. *Inter-organizational Programme for the Sound Management of Chemicals*; Global Mercury Assessment: Geneva, Switzerland, 2002.

5. Curtin, P. Africa and the wider monetary system, 1250–1850. In *Precious Metals in the Later Medieval and Early Modern Worlds*; Richards, John F., Ed.; Carolina Academic Press: Durham, NC, USA, 1983; pp. 231–268.

6. Massachusetts Institute of Technology, "The Observatory of Economic Complexity", 2012. Available online: http://atlas.media.mit.edu/explore/tree_map/hs/export/gha/all/show/2012/ (accessed on 26 January 2015).

7. Hilson, G. Harvesting mineral riches: 1000 Years of gold mining in Ghana. *Resour. Policy* **2002**, *28*, 13–26.

8. Tschakert, P.; Singha, K. Contaminated identities: Mercury and marginalization in Ghana's artisanal mining sector.*Geoforum* **2007**, *38*, 1304–1321.

9. Tetteh, K. *An overview of Ghana's Artisanal and Small-Scale Mining (ASM) Sector*; Minerals Commission of Ghana: Accra, Ghana, 2010.

10. Tschakert, P. Recognizing and nurturing artisanal mining as a viable livelihood. *Resour. Policy* **2009**, *34*, 24–31.

11. *Minerals and Africa's Development: The International Study Group Report on Africa's Mineral Regimes*; U.N. Economic Commission for Africa, Economic Commission on Africa: Addis Ababa, Ethiopia, 2011.

12. Wilson, M.L.; Renne, E.P.; Roncoli, C.; Agyei-Baffour, P.; Tenkorang, E.Y. Integrated assessment of artisanal and small-scale gold mining in Ghana—Part 3: Social sciences and economics. *Int. J. Environ. Res. Public Health* **2015**, *12*, 8133–8156.

13. Rajaee, M.; Obiri, S.; Green, A.; Long, R.; Cobbina, S.; Nartey, V.; Buck, D.; Antwi, E.; Basu, N. Integrated assessment of artisanal and small-scale gold mining in Ghana—Part 2: Natural sciences review. *Int. J. Environ. Res. Public Health***2015**, *12*, 8811–8827.

14. Basu, N.; Clarke, E.; Green, A.; Long, R.; Calys-Tagoe, B.; Chan, L.H.M.; Dzodzomenyo, M.; Fobil, J.N.; Neitzel, R.L.; Obiri, S.; *et al.* Integrated assessment of artisanal and small-scale gold mining in Ghana—Part 1: Human health review. *Int. J. Environ. Res. Public Health* **2015**, *12*, 5143–5176.

15. Long, R.; Renne, E.; Robins, T.; Wilson, M.; Pelig-Ba, K.; Rajaee, M.; Yee, A.; Koomson, E.; Sharp, C.; Lu, J.; Basu, N. Water values in a Ghanaian small-scale gold mining community. *Hum. Organ.* **2013**, *72*, 199–210.

16. Hilson, G.; Hilson, C.J.; Pardie, S. Improving awareness of mercury pollution in small-scale gold mining communities.*Environ. Res.* **2007**, *103*, 275–287.

17. Ofosu-Mensah, E.A. Historical overview of traditional and modern gold mining in Ghana. *J. Libr.* **2011**, *1*, 6–22.

18. Abakzaa, T.M.; Seyire, J.S.; Afriyie, K. *The Glittering Façade: Effects of Mining Activities on Obuasi and Its Surrounding Communities*; Third World Network-Africa: Accra, Ghana, 2007.

19. Renne, E.; Basu, N.; Gager, E.; Koomson, E.; Lee, B.; Lee, S.; Leeth, A.; Manigault, D., III; Rajaee, M.; Sajjad, A.; Smith, M.; Yee, A. Women's work, health and the environment in a small-scale mining site in northeastern Ghana. *Women Environ.* **2011**, *2*, 13–16.

20. Lund, K.; Dinse, K.; Callewaert, D.; Scavia, D. Benefits of using integrated assessment to address sustainability challenges. *Int. J. Environ. Stud.* **2011**.

21. Michigan Sea Grant and Graham Environmental Sustainability Institute. Tackling Wicked Problems through Integrated Assessment. Available online: http//www.miseagrant.umich.edu/downloads/research/tackling-wicked-problems.pdf (accessed on 9 September 2015).

22. NRC. *Analysis of Global Change Assessments—Lessons Learned*; The National Academy Press: Washington DC, USA, 2007.

23. Rotmans, J.; Dowlatabadi, H. *Integrated Assessment of Climate Change: Evaluation of Methods and Strategies*; Rayner, S., Malone, E., Eds.; Batelle Press: Washington, DC, 1998.

24. *Introduction to Stakeholder Participation*; NOAA Coastal Services Center: Charleston, SC, USA, 2007.

25. Paruchuri, Y.; Siuniak, A.; Johnson, N.; Levin, E.; Mitchell, K.; Goodrich, J.; Renne, E.; Basu, N. Occupational and environmental mercury exposure among small-scale gold miners in the Talensi-Ndboum district of Ghana's upper east region. *Sci. Total Environ.* **2010**, *408*, 6079–6085.

26. Basu, N.; Nam, D.-H.; Kwansah-Ansah, E.; Renne, E.; Nriagu, J. Multiple metals exposures among small-scale artisanal gold miners. *Environ. Res.* **2010**, *111*, 462–467.

27. Helmer, O. *Analysis of the future: The Delphi method*; Rand Corporation: Santa Monica, CA, USA, 1967.

28. Basu, A.; Phipps, S.; Long, R.; Basu, N. Applying the Delphi technique to evaluate response options to small-scale gold mining in Ghana. *Int. J. Environ. Res. Public Health* **2015**. (In Press).

29. Rajaee, M.; Long, R.N.; Renne, E.P.; Basu, N. Mercury exposure assessment and spatial distribution in a Ghanaian small-scale gold mining community. *Int. J. Environ. Res. Public Health* **2015**, *12*, 10755–10782.

30. Rajaee, M.; Sánchez, B.N.; Renne, E.P.; Basu, N. An investigation of organic and inorganic mercury exposure and blood pressure in a small-scale gold mining community in Ghana. *Int. J. Environ. Res. Public Health* **2015**, *12*, 10020–10038.

31. Calys-Tagoe, B.N.L.; Ovadje, L.; Clarke, E.; Basu, N.; Robins, T. Injury profiles associated with artisanal and small-scale gold mining in Tarkwa, Ghana. *Int. J. Environ. Res. Public Health* **2015**, *12*, 7922–7937.

32. Kyeremateng-Amoah, E.; Clarke, E.E. Injuries among artisanal and small-scale gold miners in Ghana. *Int. J. Environ. Res. Public Health* **2015**, *12*, 10886–10896.

33. Long, R.N.; Sun, K.; Neitzel, R.L. Injury risk factors in a small-scale gold mining community in Ghana's upper east region. *Int. J. Environ. Res. Public Health* **2015**, *12*, 8744–8761.

34. Green, A.; Jones, A.D.; Sun, K.; Neitzel, R.L. The association between noise, cortisol and heart rate in a small-scale gold mining community—A pilot study. *Int. J. Environ. Res. Public Health* **2015**, *12*, 9952–9966.

35. Cobbina, S.J.; Duwiejuah, A.B.; Quansah, R.; Obiri, S.; Bakobie, N. Comparative assessment of heavy metals in drinking water

sources in two small-scale mining communities in northern Ghana. *Int. J. Environ. Res. Public Health* **2015**, *12*, 10620–10634.

36. Bortey-Sam, N.; Nakayama, S.M.M.; Akoto, O.; Ikenaka, Y.; Fobil, J.N.; Baidoo, E.; Mizukawa, H.; Ishizuka, M. Accumulation of heavy metals and metalloid in foodstuffs from agricultural soils around Tarkwa area in Ghana, and associated human health risks. *Int. J. Environ. Res. Public Health* **2015**, *12*, 8811–8827.

37. Grace, A.J.; Callewaert, J. *Integrated Assessment: Stakeholder Perspective Evaluation Guide*; University of Michigan Graham Institute of Environmental Sustainability: Ann Arbor, MI, USA, 2013.

38. Conley, A.; Moote, M.A. Evaluating collaborative natural resource management. *Soc. Natur. Resour.* **2003**, *16*, 371–386.

9

CHAPTER

Application of a Backfilling Method in Coal Mining to Realise an Ecologically Sensitive "Black Gold" Industry

Xiaowei Feng [1,2,†], Nong Zhang [1,2,3,*], Lianyuan Gong [4,†], Fei Xue [1,2] and Xigui Zheng [1,2]

[1] *Key Laboratory of Deep Coal Resource Mining, Ministry of Education, China University of Mining and Technology, Xuzhou 221116, China*

[2] *State Key Laboratory of Coal Resources and Safe Mining, School of Mines, China University of Mining and Technology, Xuzhou 221116, China*

[3] *Hunan Key Laboratory of Safe Mining Techniques of Coal Mines, Hunan University of Science and Technology, Xiangtan 411201, China*

[4] *Key Laboratory of Coal Processing and Efficient Utilization of Ministry of Education, School of Chemical Engineering and Technology, China University of Mining and Technology, Xuzhou 221116, China*

ABSTRACT

China, as the largest coal-producing and -consuming country in the world, is highly dependent on its coal industry, or "Black Gold" industry, for the national energy and economy. The consequent environmental crises, however, have persisted for decades, and the most serious effect is surface subsidence induced by underground mining. Underground coal excavation in China has ignored this problem for thousands of years, even though it causes conspicuous damage to the surface ecosystem and construction projects due to the subsidence of overlying strata. This study recommends paste backfilling to replace the space originally occupied by coal resources to avoid such subsidence and proposes backfilling schemes for two mainstream mining methods used in China's collieries, namely, continuous mining and fully mechanised coal mining. These methodologies have been successfully implemented in some collieries, and the gob area can be backfilled immediately to prevent surface subsidence. To promote an ecological ideology when conflict exists between economic profits and environmental protection, experience from developed countries should be considered, support and appropriate legislation from the government are essential, and the perspective of colliery managers should be taken into account, and further in-depth study on strata subsidence and backfilling material must be pursued.

Keywords: underground mining; surface subsidence; "Black Gold" industry; backfilling method; environmental protection; ecological sensitivity

1. INTRODUCTION

Worldwide, 30.1% of primary energy, 40% of electricity generated, and 70% of steel manufacturing depend on coal, and the total global coal production in 2013 was 7,822.8 Mt. Regional contributions to this total are available from the International Energy Agency and the BP Statistical Review of World Energy [1,2]. The data show that China was the largest coal producer (45.6%) in 2013; thus, China continues to play a leading role in environmental disruption even though the nation resolved to reduce the carbon intensity of its economy (the amount of CO_2 emitted per unit of GDP) by 40%–45% during 2005–2020 [3]. It does not make sense to place greater value on production or economic development than on the protection of our environment, yet the environmental crises resulting from both coal mining and use, including soil erosion, water pollution, surface subsidence, the release of pollutants such as oxides of sulphur (SO_x) and nitrogen (NO_x), and impacts on local biodiversity, persist.

These crises comprise a number of environmental challenges; one such challenge that urgently needs to be addressed is surface subsidence [4]. Many factors may lead to surface subsidence, such as the collapse of columns underground or invisible karst rivers among strata. The most prominent factor causing subsidence is underground mining [5,6]; when a block of sufficient width and length is excavated from a coal field, the original equilibrium state is broken up regardless of whether the mining method used is longwall mining or room and pillar mining. Consequently, rock strata between the immediate roof of the coal seam and the ground surface can be moved. During our research, we investigated coal mine fields in Xuzhou City, where the ground surface has been particularly damaged by underground mining, as shown in Figure 1. It is clear that subsidence around the coal mine field has become a serious concern, and unlike other issues, such subsidence is not only a technical problem but also a socially relevant issue.

Many studies have been carried out around the world to evaluate surface subsidence for the purpose of developing corresponding countermeasures [7,8,9,10,11], but what if we change our approach from the outset, taking measures to prevent this type of subsidence from occurring? A potential solution to these questions may lie in radically preventing subsidence induced by mining activity at its source, *i.e.*, by using some other material to fill the space occupied by the extracted coal. Traditional schemes include waste rock backfilling [12], coal ash backfilling, and paste backfilling, among others. With measures such as these, ground surface subsidence can be reduced to some extent. However, backfilling is not mandatory in China's collieries as this environmentally sound practice has long been perceived as a costly, time-consuming, and low-efficiency engineering method.

(a) (b) (c)

Figure 1. Example of surface subsidence problems induced by underground mining. (**a**) Surface subsidence around Zhangji Colliery in Xuzhou, at latitude and longitude of 34.3667 and 117.0229, respectively; the mining depth is more than 900 metres; (**b**) Building damage associated with Zhangji Colliery in Xuzhou, at latitude and longitude of 34.3846 and 117.0315, respectively; (**c**) Waterlogging around the Pangzhuang Colliery in Xuzhou, at latitude and longitude of 34.3566 and 117.0766, respectively. (Xiaowei Feng and Lianyuan Gong performed investigations around the subsidence areas and obtained these figures by themselves).

Hence, the status since the 1970s has remained the use of backfilling. Currently, a simple, natural caving method in which the roof above the coal seam is allowed to naturally cave in under the effects of gravity is mainly adopted because it incurs no extra cost; this method can easily lead to surface subsidence. It is critical that innovative backfilling methods be developed to create a more ecologically sound "Black Gold" industry, and such methods must provide both economic and technical value.

2. CHARACTERISTICS OF SURFACE SUBSIDENCE INDUCED BY UNDERGROUND MINING

As a result of continuous sedimentation effects and tectonic movements occurring on the scale of billions of years, the Earth's surface shows a general trend toward stability, even though seismic activities and volcanism are spread worldwide. However, within the Earth, artificial underground mining poses enormous hazards to the safety of miners and the environment. Surface subsidence induced by mining has long been ignored because governments focus more on financial gains than environmental protection or simply because most people are more familiar with greenhouse gas emissions and naturally consider emissions to be the greatest problem caused by underground coal excavation.

The extraction of underground coal resources has been undertaken for thousands of years in China, according to the record in *The Tiangong Kaiwu*, or *The Exploitation of the Works of Nature*, a Chinese encyclopaedia compiled by Song Yingxing, initially published in May 1637 (see Figure 2a). Once excavation occurs within the ground, the overlying rock strata experience subsidence, as illustrated in Figure 2b. It is clear that the associated damage compensation is much more expensive than the economic profits from excavation. Lakes, rivers, reservoirs, and other water bodies are easily disturbed, causing mine flooding underground, while the surface is subjected to water erosion/pollution, desertification, and building damage.

Figure 2b also indicates an apparent three-zone division along the vertical direction from the mining level to the surface level—the caving zone, fracture zone, and bending zone. This zone division is based on the degree of fracture of strata overlying the coal seam, and the degree of fracture is in turn associated with the mining method, spatial relationships, geotechnical conditions of the rock strata, and excavation duration [13]. It is generally thought that these three zones can only

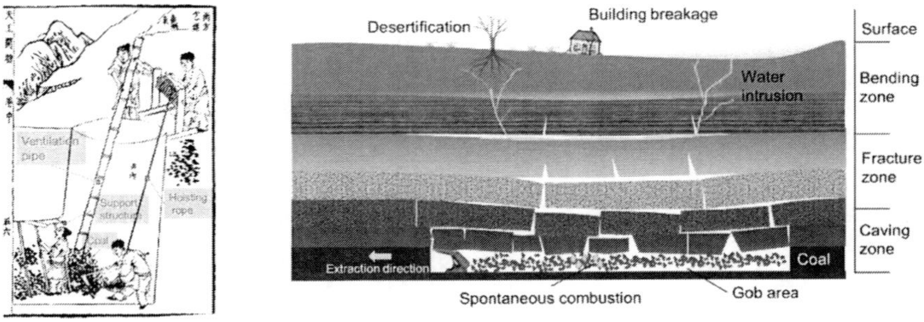

(a) (b)

Figure 2. Comparison of an ancient mining method and current mechanised mining without backfilling. (**a**) The ancient mining method described in *The Tiangong Kaiwu*, published in 1637; (**b**) The current mechanised mining method; the method is characterised by its caving system, which is both economical and efficient to enable high productivity.

develop when the ratio of the thickness of overlying strata to the thickness of the coal seam is greater than 40, but this ratio is surpassed extremely easily in the case of the many 1,000-metre-deep collieries; data show that there are already more than 50 collieries with an excavation depth greater than 1,000 metres in China [14]. Underground fracturing induces varying hazards on the surface, such as water loss and soil erosion, building damage, and ecological disruption, among others. Moreover, the insufficient backfilling produced by single roof caving behind hydraulic supports is inadequate to prevent the deformation of the overlying rock mass. Some unfilled space can remain in the gob area, which causes problems for the excavation of the neighbouring working face because the dead weight of the surrounding rock mass is only partially supported by the caving gangue in the gob area, and the remaining weight must be transferred to the adjacent intact coal body to maintain stability. In other words, this transfer causes a pressure increase on the neighbouring working face, thus increasing the costs required to maintain stability in the adjacent face. Additionally, it is impossible to realise a one hundred percent of coal recovery ratio, which means certain amount of coal can be left over in gob area, then spontaneous combustion may be incurred under the influence of unstable environment and oxygen in the gob area, which presents a danger to both the efficiency and safety of production [15]. Surface ecology and construction are also disturbed by subsidence. Such damage is especially predominant in the north-western coal districts of China, like Qinghai Province, Sinkiang Province, Ningxia Province, *etc.*, where the ecological environment, primarily semi-desert grassland, is extremely fragile. Thus, disturbances underground readily lead to surface water losses and can cause environmental conditions to turn from bad to worse.

Injecting grout into the bed separation in the overburden of coal seams may serve as a solution to this environmental problem [16]. Such an approach is a remedial measure, though it has long been treated as an effective measure to seal off spaces between strata to resist on-going subsidence, it cannot rehabilitate the ground surface to its original landform. Furthermore, injecting grout can only take place when bed separation does occur between strata, and the costs and technical problems can be much greater than an immediate backfilling method following coal mining. Immediate backfilling is conducted from the outset to backfill the space originally occupied by coal resources (see Figure 3). This method is a win-win situation from both an economic and ecological perspective.

Figure 3. Current mechanised mining with backfilling procedure employed (drawing designed by Xiaowei Feng).

Figure 3 demonstrates that timely backfilling prevents the occurrence of the three-zone division above the coal seam, and the ecosystem on the ground is also minimally influenced. In conclusion, mining that protects water resources and is environmentally friendly, or green mining, is a challenge, and it is urgent that we turn this previously economically driven engineering practice into an ecological engineering practice.

3. METHODOLOGY FOR SURFACE SUBSIDENCE PREVENTION

As mentioned in the previous sections, the key to preventing surface subsidence lies in the timely filling of man-made spaces underground with alternative materials after excavation. The alternatives can vary from waste rock backfilling [17], coal ash backfilling and hydraulic sand backfilling to paste backfilling and similar measures. Gob backfilling makes it possible to excavate coal resources trapped under artificial constructions and natural water bodies [18], and this method is thus increasingly used in underground mines all over the world [19]. Currently, there are nearly 13.79 billion tons of coal trapped under the aforementioned conditions in key state-owned collieries in China [20]. To choose the most effective of these alternatives, the following equation can be utilised during practical field trials:

$$\frac{\rho}{\rho'} = k\frac{M}{M'} \tag{1}$$

where

- M is weight of coal excavated, in tons.
- M' is weight of backfilling material to be applied, in tons.
- ρ is the density of coal, in tons per cubic metre (1.3 t/m^3 for coal, in general).
- ρ' is the density of backfilling material, in tons per cubic metre (1.6 t/m^3 for backfilling material, in general).
- k is a backfilling factor (0.6–0.95) that varies by system; the general value for coal mines is 0.8.

Based on the above specific values, it can be found that $M'/M \approx 1$. It is clear that the amount of backfilling material required is approximately equal to the coal production. It is difficult to prepare such a large amount of material and transfer it to the space behind the coal face, especially for measures like waste rock backfilling, coal ash backfilling, and hydraulic sand backfilling as a result of their very one-dimensional technical system. This is the greatest disadvantage to backfilling because of the dramatic increase in cost, which may be twice the production profits from coal excavation. However, among these alternatives, the most effective method is paste backfilling [21,22,23] by mixing waste rock, aeolian sand, or tailings with coal ash and cement. First, a certain amount of chemical admixtures and auxiliary material is added to upgrade the pumpability of the mixture, and then the mixture is pumped to the rear space of the coal excavation line. This approach has the potential to reduce surface subsidence and is a promising method for large amounts of waste reuse [24]. The aforementioned technical procedure is illustrated in Figure 4.

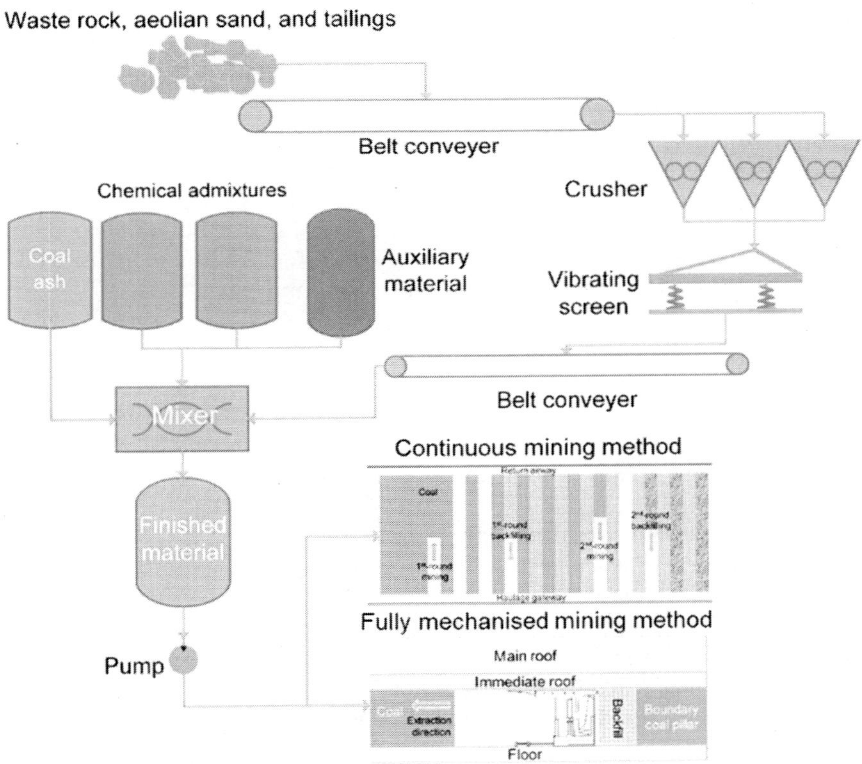

Figure 4. Diagram of underground paste backfilling (drawing designed by Xiaowei Feng).

Generally, the objectives of paste backfilling should be as follows: to geotechnically ensure a safe and efficient mining operation; to increase the extraction ratio and decrease water consumption; to decrease soil contamination and surface subsidence; and most importantly, to decrease cost compared with other measures and let the value of M'/M approach 1 as closely as possible because of the backfill material's expansion ability.

The pumpability of backfill material will dramatically decrease the cost of transportation. Thus, the pumping pressure should be lowered and the hydration time should be increased to allow for longer transportation distances. The backfilling material should possess appropriate expansion and solidification properties to decrease porosity and improve uniformity throughout the gob area when it reaches the excavated site. The material can thus support the roof and prevent the subsidence of overlying strata [25]. Two mainstream methodologies are used: the first addresses continuous mining faces, and the second is designed for fully mechanised coal mining faces. The schematic details for backfilling methodology in a continuous mining face are shown in Figure 5, the coal seam is divided into strips by a skip-mining method. The first round of mining leaves a certain width of coal pillars, this preliminary coal excavation is a type of room and pillar method, where pillars are left to support the roof in case of roof collapse or further subsidence of the overlying strata and the coal is excavated for economic profit. At a certain distance from the first-round mining face, the first-round backfilling face is arranged to fill up the space initially occupied by coal mass. These two faces are capable of excavating a certain amount of coal while simultaneously controlling the potential subsidence of the overlying roof. However, nearly half of the coal is left behind, and it is thus proposed to apply a second-round mining face to excavate the remaining coal pillars among the previous first-round backfilling faces. The same volumes of gob area occur when excavating these remaining coal pillars; hence, to thoroughly prevent roof subsidence, a second-round backfilling face behind the second-round mining face must completely fill the gob area. In this type of backfilling method, these four operational sequences are relatively independent. Furthermore, this method both guarantees a high recovery rate of coal and controls surface subsidence. Such backfilling has been successfully employed in the Yuyang colliery in Shanxi Province, where the total subsidence area before backfilling was 4.6398 square kilometres on the ground [26]. Another successful case is the Xuchang Colliery in the city of Zibo in Shandong Province [27].

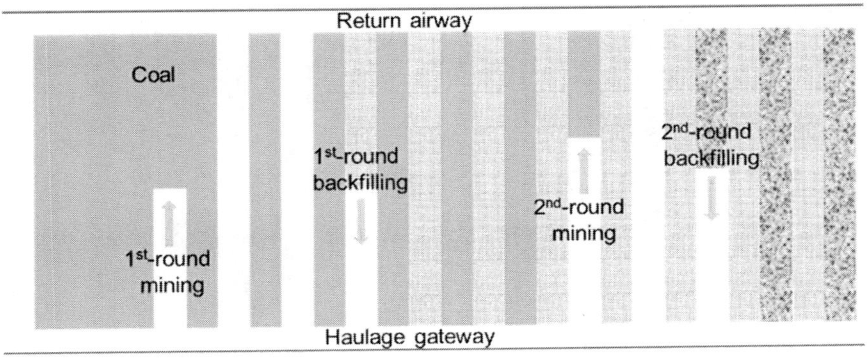

Figure 5. Schematic illustration of backfilling methodology in a continuous mining face (drawing designed by Xiaowei Feng).

For fully mechanised coal mining, a different backfilling method should adopted to address the unique nature of the coal mining face and equipment used. The hydraulic support is characterised by its backfilling function, which is capable of pumping paste backfilling material into the gob area behind the support. The working procedure can be divided into four steps (see Figure 6). The movement of the support along the extraction direction allows the mould bag to stretch automatically (see Figure 6b) thus allowing backfilling material to be pumped in (see Figure 6c). One cycle is finished when the previously pumped material coagulates, and the process is repeated throughout the coal mining process, as shown inFigure 6d. Apparently, this approach can also fully pack the gob area and confine the subsidence of overlying roof strata such as the immediate roof and main roof, thus supporting the concept of "green mining" [28]. The immediate roof generally consists of siltstone, sandy shale, and shale, which can remain stable for a certain period even if the coal seam underneath it has been mined out for some distance, but will eventually sag due to the gap below if no backfilling procedure has been applied. The main roof is much stiffer and thicker than the immediate roof, which will apply greater pressure on the mining face, underlying strata, and related galleries below; this relationship also highlights the importance of utilising backfilling strategies during mining activities. This type of backfilling technology is also widely being used at the Zhucun Colliery in the city of Jiaozuo in Henan Province [29].

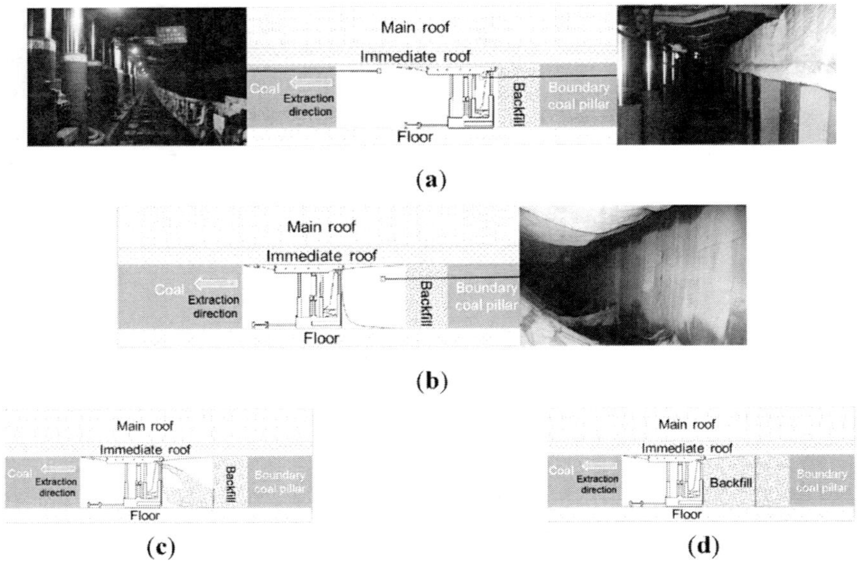

Figure 6. Backfilling procedure for fully mechanised coal mining with self-backfilling hydraulic supports. (**a**) The moment when the first-round backfilling neighbouring the boundary coal pillar is finished; the left and right photos, taken in the Yulin colliery of Shanxi Province, indicate corresponding sections in the centre sketch; (**b**) The left sketch shows the moment when the mould bag stretches as a result of the forward movement of the hydraulic support, which is also indicated in the right photo, also taken in the Yulin colliery of Shanxi Province; (**c**) Backfilling material being pumped into the rear space formed by the mould bag; (**d**) End of the first-round backfilling procedure. (drawings in this figure were designed by Xiaowei Feng and photos were taken by Xigui Zheng and Fei Xue in a related field trial in Shanxi Province).

4. DISCUSSION AND FUTURE OUTLOOK

National legislation must support both economic development and ecological sustainability. Currently, China is still a developing country with an energy economy that is largely dependent on its coal resources. Although China's annual coal production accounts for a large part of the global total, the production method applied in collieries still lags behind those used in developed countries such as the United States, Germany, and Australia. Many workers contribute a lifetime of labour, or even their lives, to the national energy industry. Legislation under these circumstances cannot limit the widespread effects of collieries in a short time, which may even cause a hysteresis phenomenon in technical development in coal production. However, legislation addressing the environmental crises induced by underground mining is imperative, and during this process, China can benefit from other countries' experiences. The Ruhr District in Germany can serve as an example, where land subsidence was extremely serious due to large-scale coal extraction over the past 100 years [30], yet the regional initiatives implemented by the government since the 1970s have turned this district into a world-renowned "Garden Industrial Zone" in view of its current diversification of economy, emerging industry, environment rehabilitation, and other factors. As another example, consider the United States, where the conflict between environmental protection and growing coal excavation has been on-going since the 1970s. In the Yellowstone River Basin of eastern Montana, the abundant, high-quality water and the nation's largest coal reserves presented a great conflict of interest because the extraction of the latter requires large volumes of water [31]. In the late 1970s, south-eastern Montana's Tongue River basin was experiencing rapid development of its extensive coal deposits, which was greatly impacting the basin's hydrologic systems, and the Water Use Act required certain adjustments to relieve this situation despite some unexpected restrictions [32]. Under such a historical background, US president Jimmy Carter signed the Surface Mining Control and Reclamation Act (SMCRA) in 1977. SMCRA is the first mining environmental protection law in the USA, and it provides standards for environmental protection related to mining activity. Simultaneously, the Office of Surface Mining (OSM) was set up to supervise the protection and rehabilitation of environment [33].

Obviously, experience from these countries should be examined and adjusted based on the unique mining conditions in China. A scientifically feasible and ecologically friendly coal mining industry is a key element for constructing green homes in mine fields, though this notion has only gained attention in recent years in China. Whether the coal mining industry, or "Black Gold" industry, can be realised ecologically depends on both the rationality of the economic structure and the feasibility of its technological background. The government should consider that coal is the main resource for and an extremely reliable source of energy in China, but coal represents a raw product in the national industrial chain and the extent of the nation's reliance on this industry is still unreasonable. Hence, green mining technology deserves support and concern from the government, which will undoubtedly urge supervisors of collieries to place more importance on the environment while simultaneously avoiding negative influences on the economy. From a technological perspective, more research and investments should be focused on basic study of strata subsidence arising from underlying coal mass excavation, including cost optimisation of the paste backfilling process and related backfilling materials. Moreover, different collieries have different characteristics; it should be emphasised that collieries in eastern China should focus more on how to protect the surface ground and construction, whereas those in western China should focus more on how to protect their fragile ecosystems and barren water system. Ecosystem protection patterns and economic evaluation systems should also be studied, including their relationship with enterprise cost, which can thus guide government legislation.

5. CONCLUSIONS

The Chinese economic structure is largely dependent on its coal production, which causes inevitable problems to our shared ecosystem. This paper focuses on the surface subsidence caused by underground mining and proposes corresponding possible solutions, which can be generalised as follows:

(1) Underground mining without additional backfilling has an enormous impact on the overlying strata above coal seams. The direct consequence to the surface is subsidence, which then triggers building damage, water resource erosion or pollution, and desertification.

(2) Among alternatives that may mitigate surface subsidence, this paper recommends backfilling with paste, which is characterised by its pumpability, convenient underground transportation, and high expansion as well as low porosity in the gob area. Paste backfilling thus has the multiple advantages of low cost, high efficiency, and technical feasibility compared with other measures.

(3) In view of the two most commonly used mining methods in Chinese collieries, *i.e.*, fully mechanised coal mining and continuous mining, this paper suggests different backfilling processes for each method. Both approaches are capable of stabilising the overlying strata and mitigating surface subsidence due to underground mining.

(4) As to the future of China's coal industrial development and environmental protection, legislation is still imperative and should consider both successful experiences in developed countries and the current unique situations in China. The key to success lies in finding an equilibrium point between government officers and colliery managers. Additional studies pertinent to

these issues should also be carried out to transform this environmentally damaging industry into a green ecological "Black Gold" industry.

ACKNOWLEDGMENTS

Financial support for this paper from the Fundamental Research Funds for the Central Universities (NO. 2014XT01), Program for Changjiang Scholars and Innovative Research Team in University (IRT_14R55), and National Basic Research Program of China (2013CB227904) is gratefully acknowledged.

AUTHOR CONTRIBUTIONS

X.F. and X.Z. proposed the idea. N.Z. provided suggestions and field trials in Shanxi Province. X.F. and L.G. performed investigations around subsidence areas in the city of Xuzhou in Jiangsu Province. X.F. wrote and revised the manuscript. F.X. assisted X.F. in revising the manuscript. X.F. created the Figures. N.Z. supervised the project.

CONFLICTS OF INTEREST

The authors declare no conflict of interest.

REFERENCES

1. Key World Energy Statistics 2014. Available online: http://www.iea.org/publications/freepublications/publication/key-world-energy-statistics-2014.html (accessed on 20 March 2015).
2. Statistical Review of World Energy 2014. Available online: http://www.bp.com/en/global/corporate/about-bp/energy-economics/statistical-review-of-world-energy.html (accessed on 20 March 2015).

3. Guan, D.B.; Klasen, S.; Hubacek, K.; Feng, K.S.; Liu, Z.; He, K.B.; Geng, Y.; Zhang, Q. Determinants of stagnating carbon intensity in China. *Nat. Clim. Change* **2014**, *4*, 1017–1023.

4. Cui, F.; Liu, P.L.; Feng, R.M.; Zhang, H.X. Coal mining present application status of coal backfill in China. In Proceedings of the 30th Annual International Pittsburgh Coal Conference 2013, PCC 2013, Beijing, China, 15–18 September 2013; pp. 616–628.

5. Hu, Z.Q.; Yang, G.H.; Xiao, W.; Li, J.; Yang, Y.Q.; Yu, Y. Farmland damage and its impact on the overlapped areas of cropland and coal resources in the eastern plains of China. *Resour. Conserv. Recy.* **2014**, *86*, 1–8.

6. Bell, F.G.; Stacey, T.R.; Genske, D.D. Mining subsidence and its effect on the environment: Some differing examples.*Environ. Geol.* **2000**, *40*, 135–152.

7. Jung, Y.B.; Song, W.K.; Cheon, D.S.; Lee, D.K.; Park, J.Y. Simple method for the identification of subsidence susceptibility above underground coal mines in Korea. *Eng. Geol.* **2014**, *178*, 121–131.

8. Sheorey, P.R.; Loui, J.P.; Singh, K.B.; Singh, S.K. Ground subsidence observations and a modified influence function method for complete subsidence prediction. *Int. J. Rock Mech. Min.* **2000**, *37*, 801–818.

9. Kim, K.D.; Lee, S.; Oh, H.J.; Choi, J.K.; Won, J.S. Assessment of ground subsidence hazard near an abandoned underground coal mine using GIS. *Environ. Geol.* **2006**, *50*, 1183–1191.

10. Zhao, C.Y.; Lu, Z.; Zhang, Q.; Yang, C.S.; Zhu, W. Mining collapse monitoring with SAR imagery data: A case study of datong mine, China. *J. Appl. Remote Sens.* **2014**, *8*, 083574.

11. Blachowski, J.; Chrzanowski, A.; Szostak-Chrzanowski, A. Application of GIS methods in assessing effects of mining activity on surface infrastructure. *Arch. Min. Sci.* **2014**, *59*, 307–321.

12. Zhang, Q.; Zhang, J.X.; Zhao, X.; Liu, Z.; Huang, Y.L. Industrial tests of waste rock direct backfilling underground in fully mechanized coal mining face. *Environ. Eng. Manag. J.* **2014**, *13*, 1291–1297.

13. Lian, X.G.; Jarosz, A.; Saavedra-Rosas, J.; Dai, H.Y. Extending dynamic models of mining subsidence. *Trans. Nonferrous Met. Soc.* **2011**, *21*, 536–542.

14. China National Coal Association (CNCA); Shandong Energy Group Co., Ltd. *Mining Technology for Thousand-Metre-Deep Pits around*

Countrywide Collieries; China University of Mining and Technology Press: Xuzhou, Jiangsu, China, 2013; pp. 2–23.

15. Zhou, C.C.; Liu, G.J.; Cheng, S.W.; Fang, T.; Lam, P.K.S. The environmental geochemistry of trace elements and naturally radionuclides in a coal gangue brick-making plant. *Sci. Rep.* **2014**, *4*, 6221.

16. Xuan, D.Y.; Xu, J.L. Grout injection into bed separation to control surface subsidence during longwall mining under villages: Case study of Liudian coal mine, China. *Nat. Hazards* **2014**, *73*, 883–906.

17. Zhang, J.X.; Miao, X.X.; Guo, G.L. Development status of backfilling technology using raw waste in coal mining. *Int. Min. Saf. Eng.* **2009**, *26*, 395–401.

18. Fan, G.W.; Zhang, D.S.; Wang, X.F. Reduction and utilization of coal mine waste rock in China: A case study in Tiefa coalfield. *Resour. Conserv. Recy.* **2014**, *83*, 24–33.

19. Li, L. A new concept of backfill design-Application of wick drains in backfilled stopes. *Int. J. Min. Sci. Technol.* **2013**,*23*, 763–770.

20. Zhang, J.X.; Zhou, N.; Huang, Y.L.; Zhang, Q. Impact law of the bulk ratio of backfilling body to overlying strata movement in fully mechanized backfilling mining. *J. Min. Sci.* **2011**, *47*, 73–84.

21. Mez, W.; Schauenburg, W. Backfilling of caved-in goafs with pastes for disposal of residues. In Proceedings of the 6th International Symposium on Mining with Backfill, Brisbane, Australia, 14–16 April 1998; Volume 98, pp. 245–248.

22. Wang, X.M.; Zhao, J.W.; Xue, J.H.; Yu, G.F. Features of pipe transportation of paste-like backfilling in deep mine. *J. Cent. South Univ. Technol.* **2011**, *18*, 1413–1417.

23. Zhang, X.G.; Jiang, N.; Shang, X.L. Phy-chemical properties experiment research on coal mine paste backfilling. In Proceedings of the 3rd International Workshop on Mine Hazards Prevention and Control, Brisbane, Australia, 19–21 November 2013; Guo, W., Shen, B., Tan, Y., Cheng, W., Yan, S., Eds.; 2013; Volume 94, pp. 531–538.

24. Bian, Z.F.; Miao, X.X.; Lei, S.G.; Chen, S.E.; Wang, W.F.; Struthers, S. The challenges of reusing mining and mineral-processing wastes. *Science* **2012**, *337*, 702–703.

25. Li, L.; Aubertin, M. An improved method to assess the required strength of cemented backfill in underground stopes with an open face. *Int. J. Min. Sci. Technol.* **2014**, *24*, 549–558.

26. Ma, J.G. Current situation of environment and rehabilitation research on Yuyang colliery in Shanbei. *Land Resour. Infor.* **2013**, *9*, 39–44.

27. Zhang, J.X.; Miao, X.X.; Mao, X.B.; Chen, Z.W. Research on waste substitution extraction of strip extraction coal-pillar mining. *Chin. J. Rock Mech. Eng.* **2007**, *26*, 2687–2693.

28. Liu, Z.; Zhou, N.; Zhang, J.X. Random gravel model and particle flow based numerical biaxial test of solid backfill materials. *Int. J. Min. Sci. Technol.* **2013**, *23*, 463–467.

29. Sun, X.G.; Zhou, H.Q.; Wang, G.W. Digital simulation of strata control by solid waste paste-like body for backfilling. *J. Min. Saf. Eng.* **2007**, *24*, 117–121.

30. Harnischmacher, S.; Zepp, H. Mining and its impact on the earth surface in the Ruhr District (Germany). *Z. Geomorphol.* **2014**, *58*, 3–22.

31. Thomas, J.L.; Anderson, R.L. Water-energy conflicts in Montana's Yellowstone River Basin. *J. Am. Water Resour. Assoc.* **1976**, *12*, 829–842.

32. Hickcox, D.H. Water rights, allocation, and conflicts in the Tongue River Basin, Southeastern Montana. *J. Am. Water Resour. Assoc.* **1980**, *16*, 797–803.

33. Plotkin, S.E.; Gold, H.; White, I.L. Water and energy in the Western coal lands. *J. Am. Water Resour. Assoc.* **1979**, *15*, 94–107.

10

CHAPTER

Assessing Landscape Ecological Risk in a Mining City: A Case Study in Liaoyuan City, China

Jian Peng [1,*], Minli Zong [2,3], Yi'na Hu [1], Yanxu Liu [1] and Jiansheng Wu [2]

[1]*Laboratory for Earth Surface Processes, Ministry of Education, College of Urban and Environmental Sciences, Peking University, Beijing 100871, China*

[2]*Key Laboratory for Environmental and Urban Sciences, School of Urban Planning and Design, Shenzhen Graduate School, Peking University, Shenzhen 518055, China*

[3]*Shanghai Urban Planning and Design Research Institute, Shanghai 200040, China*

ABSTRACT

Landscape ecological risk assessment can effectively identify key elements for landscape sustainability, which directly improves human wellbeing. However, previous research has tended to apply risk probability, measured by overlaying landscape metrics to evaluate risk, generally lacking a quantitative assessment of loss and uncertainty of risk. This study, taking Liaoyuan City as a case area, explores landscape ecological risk assessment associated with mining cities, based on probability of risk and potential ecological loss. The assessment results show landscape ecological risk is lower in highly urbanized areas than those rural areas, suggesting that not only cities but also natural and semi-natural areas contribute to overall landscape-scale ecological risk. Our comparison of potential ecological risk in 58 watersheds in the region shows that ecological loss are moderate or high in the 10 high-risk watersheds. The 35 moderate-risk watersheds contain a large proportion of farmland, and the 13 low-risk watersheds are mainly distributed in flat terrain areas. Our uncertainty analyses result in a close range between simulated and calculated values, suggesting that our model is generally applicable. Our analysis has good potential in the fields of resource development, landscape planning and ecological restoration, and provides a quantitative method for achieving landscape sustainability in a mining city.

Keywords: ecological disturbance; ecological vulnerability; ecological importance; uncertainty analysis; Liaoyuan City; China

1. INTRODUCTION

Landscape sustainability is defined as the capacity of a landscape to consistently provide long-term, landscape-specific ecosystem services, which is essential for maintaining and improving human wellbeing [1]. Sustainability is an object that not only meets the demands of humans nowadays, but also ensures future benefits. Thus, the forecast for future benefit and security is critical in sustainability research. Risk assessment, which focuses on future damages, is appropriate for evaluating sustainability as both of the concepts concern the future benefits and security of system. In other words, ecological security framework for ecosystem health protection and ecological risk control are essential for the improvement of ecological sustainability in regions and landscapes [2]. Therefore, ecological risk assessment is one effective way to determine regional and landscape-scale ecological sustainability [3]. In 1989, Hunsaker defined regional ecological risk assessment as the evaluation of regional-scale risk faced by environmental resources, or the risk caused by regional-scale pollution and natural disturbance [4], and he proposed to apply ecological risk assessment to regional and landscape scales in 1990 [5]. In the decade that followed, a series of ecological risk assessments were conducted in watersheds and other large areas [6,7,8,9,10]. However, growing recognition of the interplay of factors that might influence risks faced by ecosystems, such as global urbanization, land use change, and climate change, has led to the realization that unilateral risk management is unlikely to be useful in the management of complex systems. The need for a multilateral approach that incorporates the roles of various landscape factors in influencing risk and, therefore, sustainability was addressed with the first landscape ecological risk assessments that were conducted in the 21st century [11].

Landscape ecological risk refers to the possibility of harm to the structure and function of ecosystems from disturbances, such as human activities and natural disasters [12]. Landscapes are typically impacted by multiple disturbances that operate at different spatial and temporal scales [13,14]. Therefore, in contrast to general ecological risk assessment, landscape ecological risk assessment comprehensively assesses various types of potential ecological impacts, and their cumulative effects. It explores the effects of a variety of hazards for large-scale units, and is the complement and expansion of general ecological risk assessment. Landscape ecological risk assessment also accounts for differences in ecological characteristics and risks between different landscapes and assessment units via spatial heterogeneity, and time-series analyses. For example, Graham *et al.* used contagion index to evaluate the regional ecological risk considering the terrestrial and aquatic linkages, and the landscape pattern [15]; Kapustka *et al.* suggested risk management countermeasures grounded in landscape ecology [16]; Liu *et al.* drew attention to potential ecological risks caused by the intensification of soil erosion, and ecological vulnerability [17].

Within the last decade, landscape ecology work in China has produced relatively independent quantitative evaluation models which took landscape patterns and function into consideration [18]. One such model provides a landscape ecological risk index developed from spatial patterns and several indices, such as disturbance and vulnerability indices [19,20], landscape exposure, stability, and external pressure indices [21], as well as threat and intensity indices [22]. However, models of these type tend to calculate risk directly from probability values by the superimposition of landscape pattern indices, ultimately predicting only the probability of the occurrence of an adverse ecological event, and paying little attention to the probability of ecological losses following the event. Models based on threat and intensity indices focus on ecological risk caused by external threats, and ignore the inherent features of landscapes, including their vulnerability, resilience, stability, and the value of certain landscape features. Some other models, that characterize risk loss (the loss of risk), by calculating ecosystem services, are often limited by the direct conversion of land use, and fail to consider the effects of landscape patterns on it [23]. In addition, uncertainty analysis methods, highlighted in ecological risk assessment, is essential for the reliability of the results, and it has drawn increasing attention from scholars [24].

Mining area is a special man-land system, where people are engaged in complex interactions with land, for the exploitation of mineral resources, which has caused serious impacts on the environment [25,26]. Ecological risk assessment plays an important role in both theoretical support, and practical guidance for the implementation of regional sustainable development, and ecological restoration in mining cities [27]. Environmental problems in mining cities result primarily from mining itself, and ecological risk assessment in these cities should be built on the specific social, economic, and natural environments of the mining city in question. When combined with traditional ecological problems and risk type analysis, landscape ecology can form an important component of ecological risk assessment. Current work on ecological risk assessment in mining cities in China largely address the ecological effects of mining on soils, such as heavy metal pollution [28], land use, vegetation and landscape patterns [29,30]. However, quantitative assessments of integrated ecological risks stemming from a variety of risk sources, targeting a number of risk receptors, and driving varied ecological effects, are scarce.

Mining landscape ecological risk assessment entails the analysis of the direct and indirect risks posed by mining on a macro-scale in view of overall regional sustainability. It will be an important warning, as well as a practical guide for the planning and sustainable development of mining landscapes, including mining cities. In this study, using Liaoyuan known as the coal capital of Jilin Province as a case study, a mining landscape ecological risk assessment was quantitatively explored. Liaoyuan belongs to the first batch of China's resource exhausted cities, and urban development and environmental protection have been significantly influenced by mining activities. In the transformation of economic development, the urban ecological problem and regional ecological risk can probably be stimulated. As a result, ecological risk assessment at landscape scale is in great need to provide a spatial approach for ecological security and sustainability. Specifically, the goals of this study were to (1) quantify the disturbance and vulnerability degrees of the landscape using multiple indicators developed on the basis of traditional landscape pattern indices for mining cities; (2) calculate the risk of loss based on the landscape pattern index, in order to explore a new model for the landscape ecological risk assessment of mining cities; (3) construct landscape ecological risk zoning based on risk assessment results to provide direction for the sustainable development of mining cities; and (4) conduct landscape ecological risk uncertainty analysis to verify the reliability of our risk evaluation results.

2. MATERIALS AND METHODS

2.1. Study Area and Data Source

Liaoyuan Prefecture City (5140 km^2; 42°17′40″N to 43°13′40″N, 124°51′22″E to 125°49′52″E), which contains 33 towns, lies in the south-central part of Jilin Province, China. It is upstream of the Dongliao and Huifa Rivers, within the transition zone between the Changbai Mountains and the Songliao Plain, and across Liaohe River and Songhua River (Figure 1). Liaoyuan enjoys a semi-humid, temperate, and continental monsoon climate, with abundant water, and forest resources. Liaoyuan is a coal resource-based city. By the end of 2007, 33 types of minerals had been discovered in Liaoyuan, of which coal and building stone comprise a large proportion. There are 152 mineral ore fields in Liaoyuan, most of which are small, and their distribution is concentrated. Mining has boosted economic development in the region, but has also resulted in severe damage to regional eco-environment. The environmental problems such as ground subsidence, air pollution, and excessive heavy metal waste production, have had serious impacts on local people's life and social development [31]. In March 2008, the National Development and Reform Commission (NDRC, China) listed Liaoyuan in the first batch of resource-exhausted cities that were in urgent need of transformation in economic development.

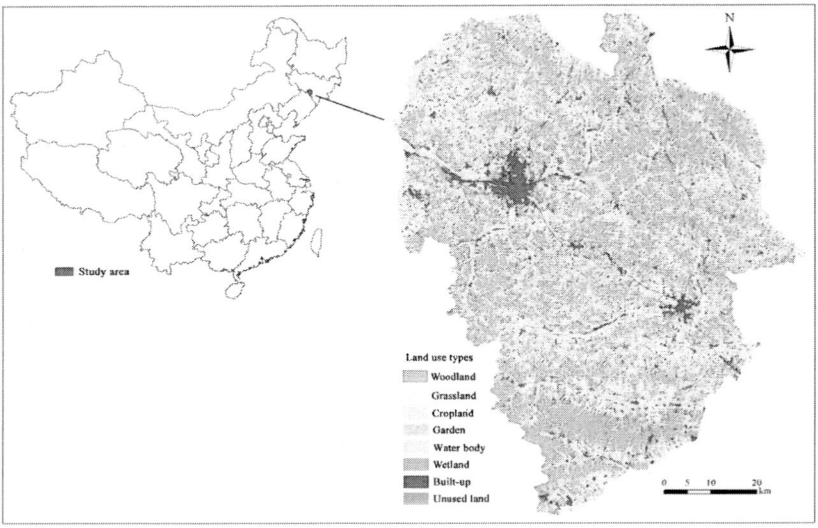

Figure 1. Location of Liaoyuan City, China.

The land use data is from the second national land survey of Liaoyuan. According to land use classification and planning criteria published by China Land Resource Bureau, the landscape of Liaoyuan may be divided into 8 types, *i.e.*, woodland, grassland, cropland, garden, water body (including rivers, surface water ponds and reservoirs), wetland (including inland beaches and marshes), built-up, and unused land (including bare, sandy and saline land; Figure 1). Among these types, wetlands and unused land refer to natural reserves referring to the secondary indicators of classification system.

In this study, cities, rural settlements, mining sites, and other construction sites are treated as ecological risk sources, and natural and semi-natural landscapes are treated as risk receptors, whose risk is assessed. The location of mines is provided by Liaoyuan City Land Bureau, which is used for evaluating the disturbance of mines. Data on gross regional product, population, industry, and other socio-economic data is taken from the "Liaoyuan Statistical Yearbook (2009)" and "Liaoyuan Yearbook (2010)" to measure the development of Liaoyuan [32,33]. Based on a 30 m × 30 m Digital Elevation Model (DEM) image, and 1:200,000 topographic maps of Liaoyuan, in 2009, the study area is divided into 58 small watershed basins (Figure 2). Watershed, which in relation to hydrology process and topography, can determine ecosystem processes in a relatively integrated region without being subjectively sliced. Thus, since watershed unit contains more ecological meanings than town unit, it is set as assessing units for landscape ecological risk assessment. The hydrological analysis module (Hydrology) of ArcGIS10.0 is used for zoning the study area.

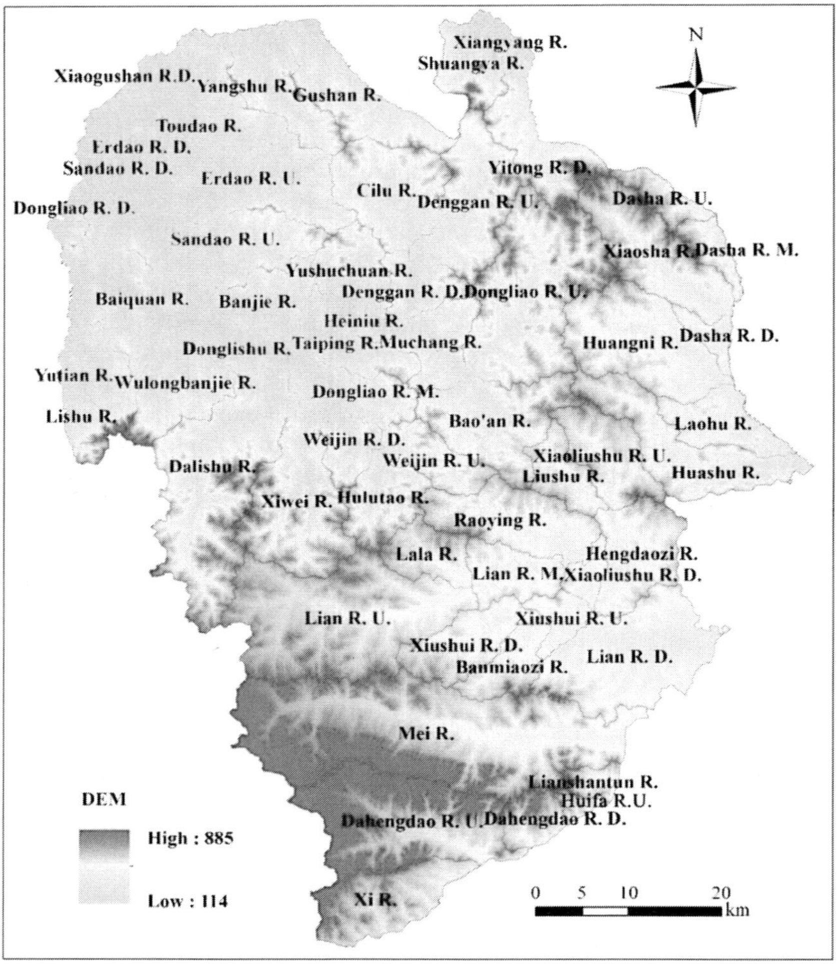

Figure 2. Extracted watersheds based on DEM in Liaoyuan City, China. (Note: R-River, U-Upstream, M-Midstream, D-Downstream).

2.2. Calculation of Landscape Ecological Risk

The formula for risk measurement is the methodology basis of regional landscape ecological risk assessment [34]. In this paper, overall landscape ecological risk (*R*) is calculated as a function of the probability of ecological risk (*P*), and risk loss (*D*), wherein *P* is the product of ecological disturbance (*E*) and ecological vulnerability (*V*), and risk loss represents ecological importance (*S*). However, the final probability (between 0 and 1), sometimes, may be too small to have a meaningful impact on the partitioning of ecological risk. In order to avoid such insignificant values, the cube root result of proposed function is extracted to arrive at final landscape ecological risk.

$$R = \sqrt[3]{P \times D} = \sqrt[3]{E \times S \times V} \qquad (1)$$

In order to clarify the management approach most appropriate to a specific landscape ecological risk, such as risk control guidelines, the calculated risk is "zoned" after the risk assessment. First, landscape ecological risk is divided into three grades (high, moderate and low), and then, using the "Natural Break" function in ArcGIS10.0, the risk is stacked onto ecological risk probability, together with risk loss, to produce a map of landscape ecological risk zones.

2.2.1. Ecological Disturbance

Ecological disturbances are relatively discrete events that alter ecosystems, communities, or demographic structure, and result in changes in resources, substrates, or the physical environment [35]. It is the external cause of regional ecological risk, and one of the main sources of landscape heterogeneity [36]. Land use degree is a representation of how broad and deep landscapes are utilized under the influence of a combination of human activities and social development factors [37]. It reflects not only the natural attributes of the land, but also the comprehensive effect of human activity on the land [38]. However, land use degree represents only current patterns of disturbance, and does not incorporate, in its measure, the probability of future human disturbance. Thus, on the basis of previous researches on land use classification, in this study, a model of landscape ecological disturbance (E), which attempts to incorporate settlement, mining, and road disturbances into the overall measure of human disturbance, is constructed,

$$E = aU + bM + cR + dT \qquad (2)$$

where *U* is landscape disturbance; *M*, *R*, and *T* are mining, settlement and road disturbances, respectively, and a, b, c, d are weights (of 0.4, 0.3, 0.2, and 0.1, respectively) assigned to each disturbance parameter such that a + b + c + d = 1.

Mining, settlement, and road disturbances (M, R, and T) are calculated using the "Buffer Analysis" modules in ArcGIS 10.0. Based on the characteristics of the disturbance and the decreasing relationship between the magnitude of influence of the disturbance and distance from its source, where different buffers are set to measure the influence of different disturbances (Table 1). On the other hand, the influence of mines and settlements are assigned maximum values in area where they overlapped, and the influence of roads is assigned summation due to its additive effects. Depending on the intensity of human activities, different landscapes are assigned different levels to quantify the influence of human activities on landscape disturbance. These levels are as follows: unused (0.2), natural renewable (woodland, grassland, water body, and wetland; 0.4), half natural renewable (cropland, and garden; 0.6), and artificial non-renewable (built-up; 0.8).

Table 1. Influence values of various disturbances based on the size of the disturbance and the distance from the disturbance source.

Disturbance	Mining			Settlement			Road			
Level	Large mining (>10 ha)	Medium mining (1–10 ha)	Small mining (<1 ha)	City	Designated Town	Rural settlement	National road	Provincial road	County road	Highways & railways
Number of Buffer	4	3	2	2	2	2	3	3	2	2
Distance/meter (influence value)	300(0.8) 600(0.6) 1000(0.4) 2000(0.2)	300(0.6) 600(0.4) 1000(0.2)	300(0.4) 600(0.2)	600(0.8) 1200(0.4)	400(0.6) 1000(0.3)	200(0.4) 500(0.1)	50(0.8) 250(0.5) 500(0.1)	50(0.7) 100(0.5) 500(0.1)	50(0.6) 100(0.1)	30(0.5) 50(0.1)

2.2.2. Ecological Vulnerability

Ecological vulnerability (*V*), an integral part of ecological risk assessment, refers mainly to the vulnerability of ecosystems to the strong external disturbance [39], and is a function of vulnerability of landscape type, and vulnerability of landscape structure. Ecological vulnerability of landscape type denotes the probability of the landscapes to deviate from their steady states, or suffer enormous damage from outside interference. Landscapes are classified into seven kinds and given different weights: unused land (7), wetland (6), water body (5), cropland (4), garden (3), grassland (2) and woodland (1), then normalized, and finally multiplied by the area ratio of that type of landscape to the surrounding landscape, to obtain the ecological vulnerability of landscape types.

Based on the pattern-process feedback mechanism, in landscape ecology, landscape patterns can also affect ecological vulnerability, and thereby, landscape ecological risk. As an intrinsic attribute of ecosystems, ecological vulnerability is closely related to ecosystem sensitivity, resilience, and stability [40]. Ecosystem sensitivity refers to the ecosystem's internal adaptive capacity to external pressure or external interference; ecosystem resilience is its ability to recover from these interferences; and ecosystem stability represent the ability to maintain the normal dynamic ecological system. Different landscape types, with different sensitivity and resilience, may play different roles in maintaining biodiversity, protecting species, improving landscape structure, and promoting the overall functioning of the landscape [41]. In this study, such two indicators as landscape fragmentation and area ratio of various landscapes whose slope is greater than 15°, are used to measure ecological sensitivity. The fragmentation of a landscape by natural or human factors is a consequence of changes in landscape patterns, from a continuous structure to patches, and more fragmented landscapes with greater slopes are assumed to be more ecologically sensitive. Ecological resilience, assumed to be positively related to landscape connectivity (the degree to which the landscape facilitates or impedes movement among resource patches), and landscape dominance (the degree to which one or a few land cover types predominate the landscape in terms of area proportion), is calculated as a function of these terms based on the formula described by Wu *et al.* [31]; weights were assigned using the "Analytic Hierarchy Process" in Matlab R2010a (Table 2).

Table 2. Indicators system for structural vulnerability evaluation.

		Woodland	Grassland	Cropland	Garden	Water Body	Wetland	Unused Land
Ecological sensitivity	Landscape fragmentation	0.9001	0.7286	0.9821	0.4864	0.9518	0.8936	0.4563
	Area ratio (slope > 15°)	0.0999	0.2714	0.0179	0.5136	0.0482	0.1064	0.5437
Ecological resilience	Landscape connectivity	0.9359	0.2309	0.9768	0.3165	0.2951	0.2897	0.2500
	Landscape dominancy	0.0641	0.7691	0.0232	0.6835	0.7049	0.7103	0.7500

Finally, ecological vulnerability (V) is calculated based on the following model:

$$V_j = \sum_{i=1}^{7} k_{ij} \times L_j = \sum_{i=1}^{7} k_{ij} \times \frac{F_{ij}}{C_{ij}} \qquad (3)$$

where i is landscape type, j is basin unit, k is ecological vulnerability of landscape type, L is structural vulnerability, F is ecological sensitivity, and C is ecological resilience.

2.2.3. Ecological Importance

Ecological importance refers, fundamentally, to the intrinsic value of an ecosystem. Thus, the greater the ecological importance of a landscape (as measured by the importance of various constituent landscapes or units in the region), the greater the ecological loss associated with adverse impacts on it. In this study, the value of ecosystem services is used as a measure of ecological importance, as suggested by Costanza *et al.* [42] and Xie *et al.* [43].

Ecosystem services are vital functions of the life-support system [44]. They contribute to human welfare, both directly and indirectly, and therefore represent a portion of the total economic value of the planet [42]. Ecosystem services value is the product of interactions between nature and humans, and is directly affected by human activities via changes in landscape patterns [45]. It is also closely related to the spatial distribution of landscape, whose impact may be negative. Landscape fragmentation is a driver of biodiversity loss [46,47]. Increase in fragmentation, implying corresponding increases in patches, leads to reduction in ecosystem services, and therefore decrease in ecosystem services value. In this study, landscape fragmentation is incorporated into the calculation of the ecosystem service value of each watershed, and ecological importance (S) is modeled as follows:

$$S_j = \sum_{i=1}^{7} V_i' \times A_{ij}/A_j = \sum_{i=1}^{7} \frac{V_i}{F_{ij}/F_i} \times A_{ij}/A_j \qquad (4)$$

where i is type of landscape, j is the assessing unit, V_i' is the ecosystem services value of the original ecosystems, V_i is the ecosystem services value after fragmentation disturbance, F_{ij} is the landscape fragmentation of type i in watershed j, F_i is the fragmentation of landscape type i in the whole study area, A_{ij} is the area of type i in watershed j, and A_j is the total area of watershed j.

2.3. Monte Carlo Analysis of Assessment Uncertainty

Uncertainty, in the form of incomplete information and data, and diversity of risk sources, is inevitable in ecological risk assessment. As a part of results of probabilistic uncertainty analysis, sensitive analysis reflects dynamic disturbance of the evaluation results of each parameter [48]. It judges the impact of each parameter by its correlation with evaluation results. If the correlation is high, the parameter has great influence on the result and it is sensitive. Especially for a multi parameter model, sensitive analysis can identify the parameters with higher influence on the evaluation results, so that in the further analysis, we can take effective measures to reduce the uncertainty.

In order to make the results more robust, it is necessary to carry out uncertainty analyses using Monte Carlo methods, which explores the uncertainty associated with the ecological risk assessment process, and its possible impact on results. Based on a Monte Carlo analysis, uncertainty arising from land use classification assignments, and their associated ecological vulnerability, are explored using the program Crystal ball 11.1.2.2.000. Specifically, the simulation computing sets two possible uncertainty distributions (low 20%, high 40%), and carries out 10,000 simulative iterations to calculate the simulated value (Table 3 and Table 4). Outliers are excluded by working within a 95% confidence interval; *i.e.*, simulation results above the maximum 2.5%, and minimum 2.5% were excluded before the final simulated value is calculated.

Table 3. Probability distribution of land use degree under low (20%)/high (40%) uncertainty scenarios.

Land USE degree	Possible Land Use Degree			
	Level-1 (0.2)	**Level-2 (0.4)**	**Level-3 (0.6)**	**Level-4 (0.8)**
Level-1 (0.2)	0.8/0.6	0.2/0.4	0/0	0/0
Level-2 (0.4)	0.1/0.2	0.8/0.6	0.1/0.2	0/0
Level-3 (0.6)	0/0	0.1/0.2	0.8/0.6	0.1/0.2
Level-4 (0.8)	0/0	0/0	0.2/0.4	0.8/0.6

Table 4. Probability distribution of landscape ecological vulnerability under low (20%)/high (40%) uncertainty scenarios.

Vulnerability of Landscape Types	Possible Vulnerability of Landscape Types						
	Level-1 (0.0357)	**Level-2 (0.0714)**	**Level-3 (0.1429)**	**Level-4 (0.1071)**	**Level-5 (0.1786)**	**Level-6 (0.2143)**	**Level-7 (0.2500)**
Level-1 (0.0357)	0.8/0.6	0.2/0.4	0/0	0/0	0/0	0/0	0/0
Level-2 (0.0714)	0.1/0.2	0.8/0.6	0.1/0.2	0/0	0/0	0/0	0/0
Level-3 (0.1429)	0/0	0.1/0.2	0.8/0.6	0.1/0.2	0/0	0/0	0/0
Level-4 (0.1071)	0/0	0/0	0.1/0.2	0.8/0.6	0.1/0.2	0/0	0/0
Level-5 (0.1786)	0/0	0/0	0/0	0.1/0.2	0.8/0.6	0.1/0.2	0/0
Level-6 (0.2143)	0/0	0/0	0/0	0/0	0.1/0.2	0.8/0.6	0.1/0.2
Level-7 (0.2500)	0/0	0/0	0/0	0/0	0/0	0.2/0.4	0.8/0.6

Setting Table 3 as an example, we define that the uncertain values should be attributed to the adjacent class. On one hand, the low (20%) uncertainty means 80% of the values is unchanged, while the other 20% should change the class. Namely in Level-1, 80% is still in Level-1, 20% will change to the adjacent Level-2. On the other hand, the high (40%) uncertainty means 60% of the values is unchanged, while the other 40% should change the class. Namely in Level-2, 60% is still in Level-2, and

40% will change to the adjacent class, where the proportion of adjacent Level-1 and Level-3 should either be 20%, respectively.

3. RESULTS

3.1. Probability of Landscape Ecological Risk

Based on the data of Liaoyuan City, maps showing the spatial distribution of the four types of disturbances—landscape, mining, settlement, and road—are generated (Figure 3). The distribution of ecological disturbances in each watershed (Figure 4), shows that, of the 58 watersheds included in the study, the Banjie River, Donglishu River, and Xiaoliushu River Downstream watersheds have the highest ecological disturbance, and the Dahengdao River Upstream, Dasha River Midstream, and Dasha River Upstream watersheds, have the lowest ecological disturbance. Whereas, in general, ecological disturbance is strongly positively correlated with urbanization and human activities, and in those areas with high ecological disturbances, mines, settlements and roads all strongly influence the landscape types and land uses. In addition, ecological disturbance is significantly correlated with population density ($R = 0.76$); watersheds with high population densities, have high ecological disturbances. For example, in the Banjie River Watershed, where population density is 2657 person/km^2, ecological disturbance is 0.4608; whereas in the Muchang River Watershed, where population density is 78 person/km^2, ecological disturbance is 0.2573.

Ecological vulnerability across watersheds (Figure 4) reveals that the Muchang River Watershed has the highest vulnerability, with Erdao River Upstream Watershed for the lowest. Located in the north-central part of the Liaoyuan City, the Muchang River Watershed has the largest reservoir in the study area and is surrounded by mountains, and other landscapes, which isolate it from the surrounding landscapes. Its high landscape topographic index and low landscape connectivity, are likely responsible for its high ecological vulnerability. On the contrary, Erdao River Upstream Watershed, located in northwest Liaoyuan, is a low-lying area with unbroken rivers, large and centrally located croplands, and woodland cover along the river, mainly adjoined by croplands. Spatially, ecological vulnerability increases substantially from the middle to the edges, where the peak appears in the middle, and the higher areas concentrated in the southeast. It is worth noting that regions in relatively flat terrain have high ecological vulnerability, those located in mountainous areas have intermediate levels of vulnerability, and regions located in transitional zones between mountains and plains have the lowest ecological vulnerability.

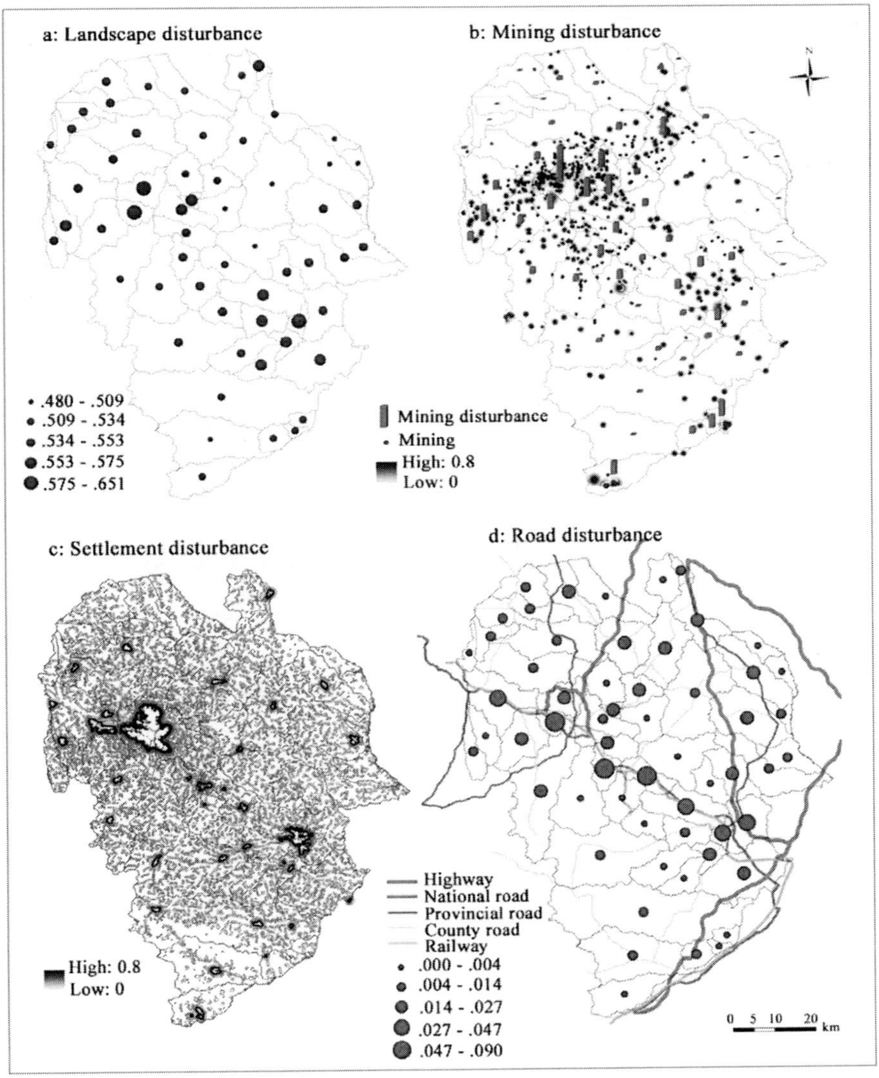

Figure 3. Spatial distribution of four types of disturbance: (**a**) Landscape disturbance; (**b**) Mining disturbance; (**c**) Settlement disturbance; (**d**) Road disturbance.

Ecological risk probability, which is calculated based on ecological disturbance and ecological vulnerability, regardless of ecological losses, is mapped at three levels (low, moderate, and high), for the 58 watersheds (Figure 5). The Banjie River and Xiaoliushu River Downstream watersheds have the highest ecological risk probability, while the low-probability regions, such as the Mei River and Xiaosha River watersheds, mostly concentrated in the southwest, west, northeast and north of Liaoyuan. The overall ecological vulnerability of mountains falls within the moderate and low levels. In summary, the ecological risk probability of mountains (low mountains, hills, and terraces) is lower than that of relatively flat areas.

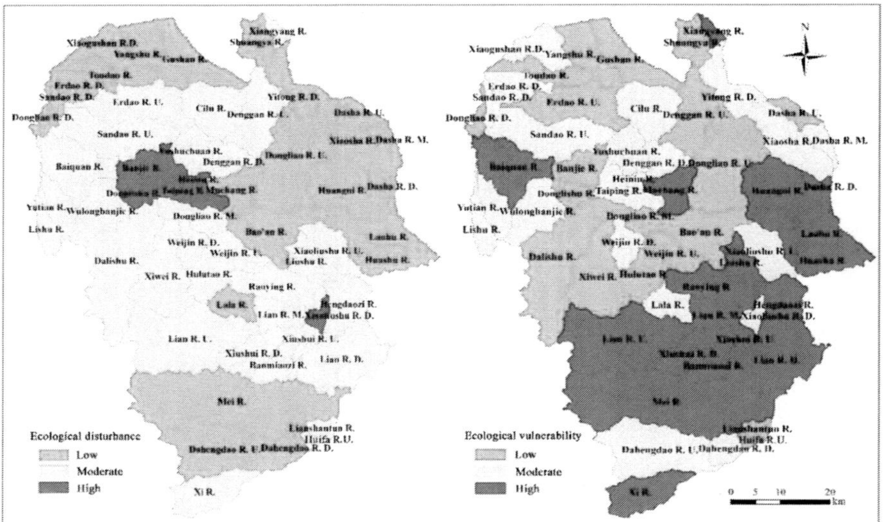

Figure 4. Ecological disturbance and vulnerability of watersheds in Liaoyuan City, China.

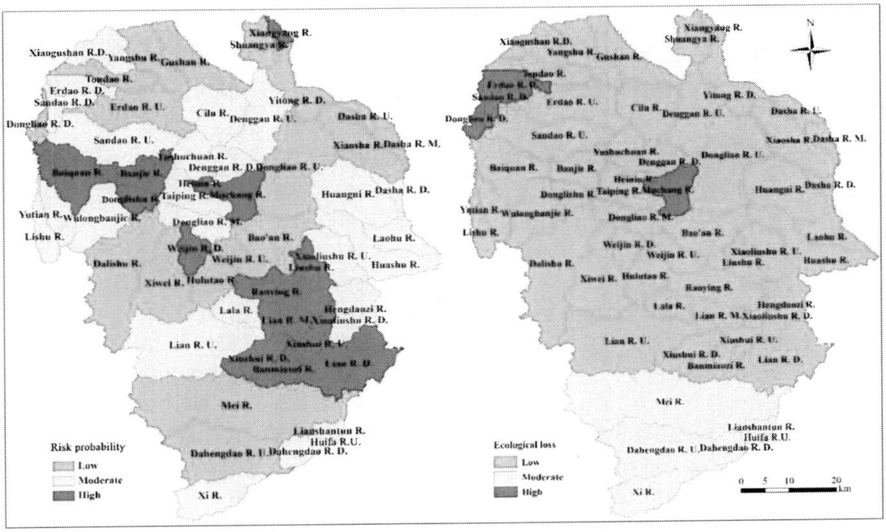

Figure 5. Landscape ecological risk probability and ecological loss in Liaoyuan City, China.

3.2. Ecological Loss and Landscape Ecological Risk

A map of the ecological importance and thus the ecological loss due to landscape ecological risk for 58 watersheds, divided into three levels (high, moderate, low; Figure 5), reveals significant spatial differences across the Liaoyuan City. Sandao River Downstream, Dongliao River Downstream, and Erdao River Downstream watersheds have the highest ecological importance, and Xiaoliushu River Downstream, Banjie River and Donglishu River watersheds, have the lowest ecological importance. The top three most important watersheds are adjacent to one another, and located in the northwest of Liaoyuan, and have a high proportion of little fragmented water bodies, which play an important role in ecosystem services in the watershed. These watersheds also have low proportions of built-up land and cropland, and large proportions of woodland resulting in over 95% of the area of these watersheds consisting of natural and semi-natural landscapes.

The three "lowest loss" watersheds all have experienced high urbanization. Due to urban sprawl, natural and semi- natural landscapes are being continuously converted to built-up land. Consequently, the proportion of construction land in these watersheds is higher than that in others, and resulting in the decline of ecosystem services, and thus their ecological importance. Overall, the ecological importance of watersheds with superior natural resources endowment is higher than that of the watersheds with construction land expansion. That is to say, human activities have a negative impact on the ecological importance of watersheds.

Ecological risk, calculated based on landscape ecological risk probability, and ecological loss, mapped for the watersheds of Liaoyuan City (Figure 6), shows that, except for the Muchang River Watershed, regions with high ecological risk (e.g., Dahengdao River Upstream, Mei River, and Xi River watersheds) are concentrated in northwestern and southern Liaoyuan, whereas regions with low ecological risk (e.g., Lala River, Xiwei River, and Banjie River watersheds) are along the southeast-northwest direction. What is more, cities and counties almost belong to low ecological risk areas, which is due to the natural/semi-natural landscapes, as the object of study, have low ecological loss in these areas. Regions with medium ecological risk constitute the majority of the study area, and are concentrated in the northeast, east, and the west.

3.3. Zoning of Landscape Ecological Risk

The results of our zoning (Figure 7) show that 10 watersheds face high ecological risk, 35 watersheds face moderate risk, and 13 watersheds are at low risk. Each of these risk classes is addressed as follows: (1) High-risk areas can be divided into five types, accounting for 18.39% of the total area. These areas are mainly concentrated in watersheds where there are large water bodies, high coverage of grassland and woodland (Figure 7), such as Muchang River, Mei River and Dahengdao River watersheds. The ecosystem services offered by water, woodland and grassland and other landscapes, such as water conservation, windbreak and sand-fixation, and biodiversity conservation, have important significance for regional eco-environment. Thus, it is important to keep maintaining and strengthening the protection of woodland, grassland through reforestation; (2) Moderate-risk area includes three types, accounting for 62.02% of the total area. In these areas, the area ratio of cropland is larger than that in high-risk areas. Cropland is a semi-natural and man-made landscape which is vulnerable to human activities. During the long winter in the Northeast China, as the fallow period is long, cropland may be frequently destroyed by human activities and weather disasters. Therefore, it is necessary to strengthen the management of cropland and prevent soil erosion; (3) Low-risk area comprises three types, mainly in relatively flat terrain in the northwestern, accounting for 19.59% of the total area. The area ratio of built-ups in these areas is higher than the other two, while the area ratio of natural and semi-natural landscape is the lowest, as it is deeply influenced by human activities.

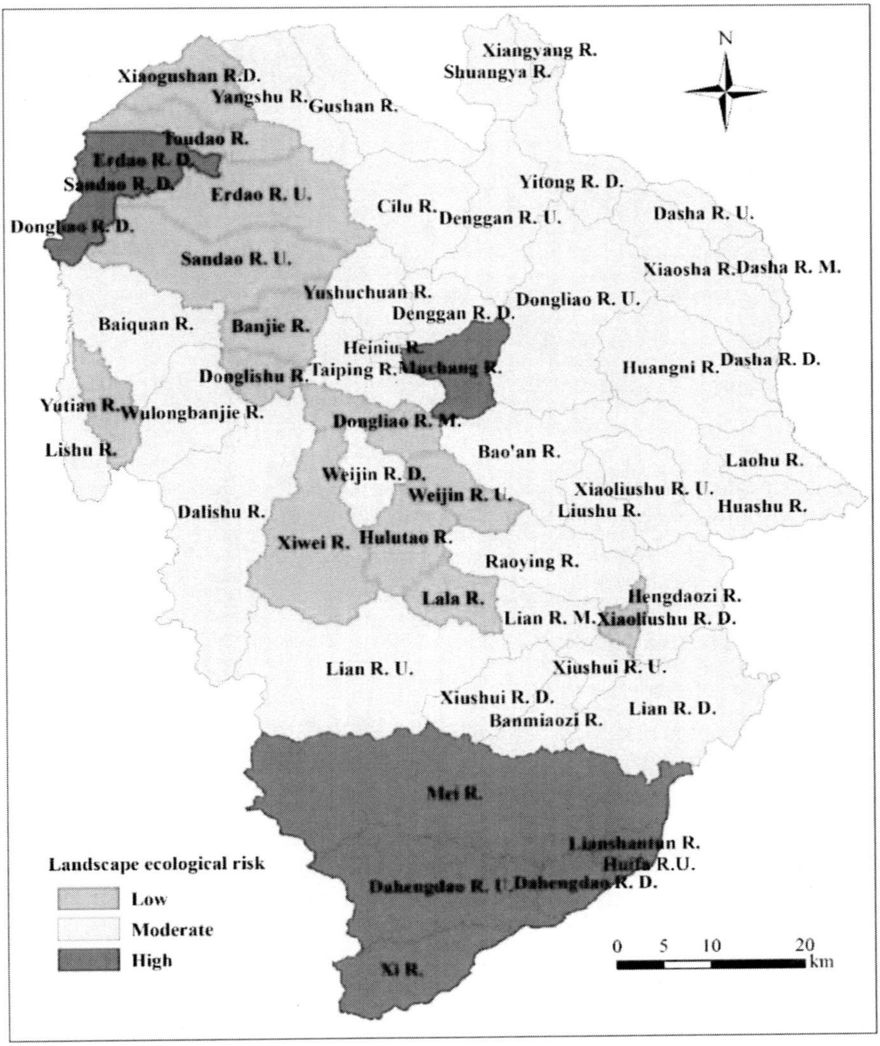

Figure 6. Landscape ecological risk of watersheds in Liaoyuan City, China.

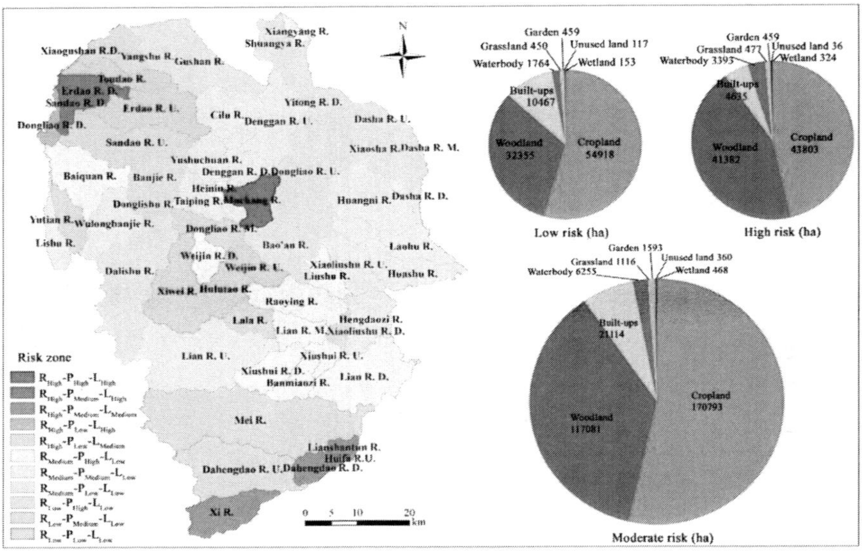

Figure 7. Landscape ecological risk zoning and total area of land use types in each grade of landscape ecological risk in Liaoyuan City, China (Note: R means ecological risk, P means ecological risk probability, and L means ecological loss).

4. DISCUSSION

4.1. Implications of Assessment Results of Landscape Ecological Risk

In the view of ecological risk probability, the high probability may be accounted for by the fact that these watersheds are relatively flat, and adjoin the city and county centers (which are suitable for human habitation, and have higher human disturbances than other areas). The low-probability regions are exposed to the lowest amount of human activity. However, although human interference in mountainous regions is lower, the ecological sensitivity of these regions is high due to the effects of terrain and slope. Nevertheless, the high or complete vegetation coverage of mountainous lands results in high ecological resilience, such that the overall ecological vulnerability of mountains falls within the moderate or low levels. This phenomenon results in a relatively high ecological risk probability in flat areas.

In the view of landscape ecological risk, the areas with high urbanization are faced with lower risk than rural areas with good ecological conditions, a pattern that is consistent with the findings of Guo *et al.* for Beijing [49]. Despite this, we recommend that that watersheds with low ecological risk should not be damaged, but rather their protection must be strengthened, and attention should be paid to the reduction in their extant, albeit low, levels of risk. In the zoning of landscape ecological risk, the internal city and the surroundings are of high possibility to be transformed into built-ups due to urbanization. Therefore, the planning and construction of these areas should pay more attention to protecting the integrity and connectivity of natural and semi-natural landscapes, and changing the growth mode to the approach of sustainable development.

4.2. Uncertainty of Landscape Ecological Risk Assessment

The results of the Monte Carlo analysis (Figure 8) show that the simulated and calculated risk values are similar under both high uncertainty (40%) as well as low uncertainty (20%). This implies that the graded assignment of land use degree and vulnerability of landscape types, is reasonable, and the model is generally applicable in the study area. Whereas under low uncertainty, the ratio of simulated risk to calculated risk is concentrated between 0.99 and 1.05, while under high uncertainty it is concentrated between 0.99 and 1.1; exceptions are the Dasha River Downstream and Xiangyang River watersheds. Results under low uncertainty are slightly better than those under high uncertainty, suggesting that the reliability of calculation values is higher under low uncertainty, and the possibility of bias increases as uncertainty increases [50].

Moreover, under both uncertainties, the spatial distribution of simulated and calculated risk values, in the 58 watersheds, are essentially the same. Under low uncertainty, simulated values, in the east and north of the study area, are closer to calculated values, than they are in the west and southwest. Under high uncertainty, although results are more scattered, the difference between simulated and calculated risk values of the west and southwest is larger than that of the east and north. In general, simulated values under low uncertainty are more geographically concentrated than those under high uncertainty.

Sensitivity analysis, which is important for making recommendations for future monitoring and research, is conducted for two cases. The land use degree is divided into four levels (U-level), and the vulnerability of landscape type is divided into seven levels (V-level). The intervals are equidistant, and a higher level represents to a higher risk value. A variance contribution rate of 1% is set as the threshold in the sensitivity analysis, and under low uncertainty there are five major factors, with four under high uncertainty (Figure 9). Two additional factors, the sixth factor under low uncertainty (V-level 2; level 2 in vulnerability of landscape type), and the fifth factor under high uncertainty (U-level 4; level 4 in land use degree), have a combined variance contribution rate of 0.76%, less than the threshold.

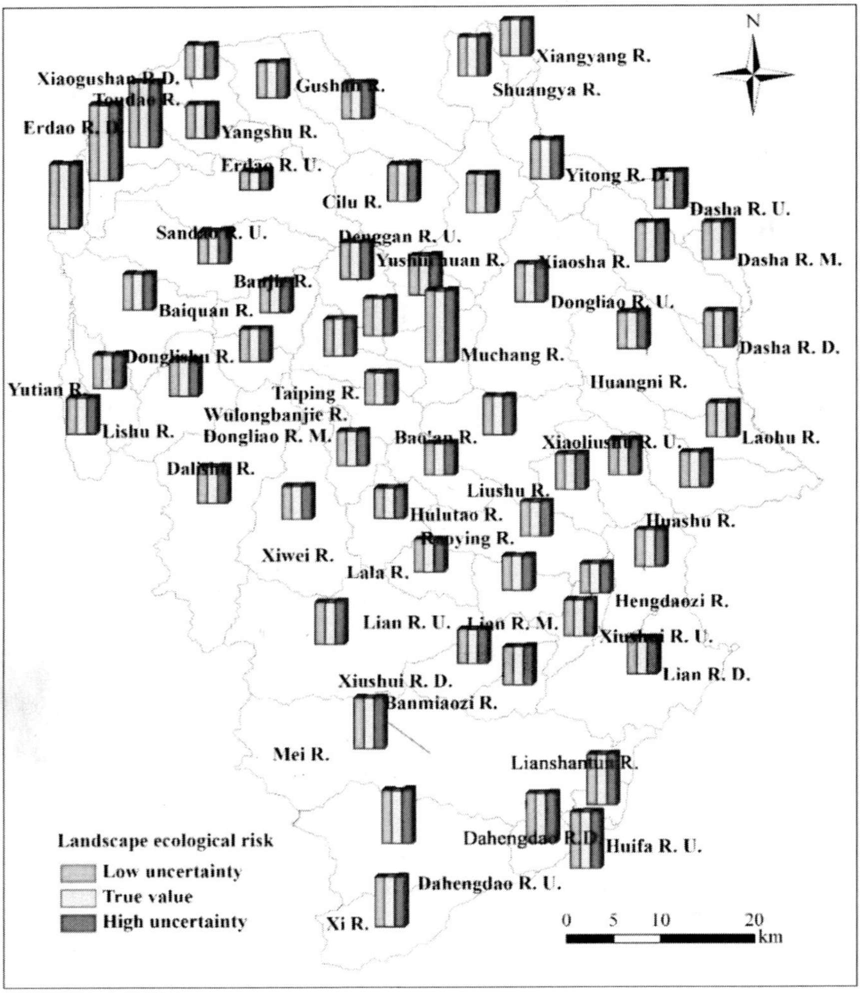

Figure 8. Simulated landscape ecological risk under low or high uncertainty of watersheds in Liaoyuan City, China.

A variance contribution threshold of 85% results in four major factors (with the highest sensitivity) under low uncertainty, and three major factors under high uncertainty. This suggests that the most sensitive factors are similar, showing only slight differences between low and high uncertainty. For example, under high uncertainty, the variance contribution rate of V-level 4 is 43.17%, nearly 5% different from that under low uncertainty; other factors show only approximately 2% difference in variance contribution rates between low and high uncertainties. Under both uncertainty scenarios, U-level 3 has a lower average variance contribution rate (24.69% for low uncertainty, and 26.40% for high uncertainty) than V-level 4 (Figure 9). In addition, V-level 1, U-level 2, and U-level 3 all affect the simulated values to varying degrees. Overall, the impact of vulnerability of landscape type on simulated value is greater than that of land use degree. Considering that landscape types corresponding with V-level 4, V-level 1, and U-level 2 account for high area ratio in each watershed, it testifies that when grading assignments, ensuring which landscape types are most dominated in the watershed is essential for the accuracy of the simulated results.

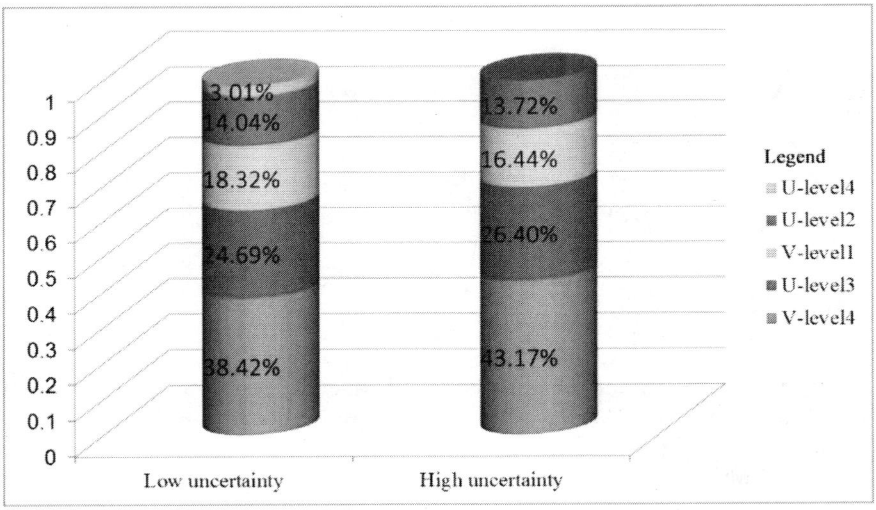

Figure 9. Variance contribution rate of sensitive factors under low or high uncertainty.

Sensitivity analyses also shows that under high or low uncertainty, the most sensitive factors change in only five watersheds, *i.e.*, Yushuchuan River, Dongliao River Upstream, Weijin River Upstream, Lian River Upstream, and Dalishu River watersheds (Figure 10). The most sensitive factors of over 90% of the watersheds are stable, indicating that the assignments of land use degree, and vulnerability of landscape types are reliable. From a regional perspective, in both uncertainty scenarios, the most sensitive factors change the most in the western region, such as Dalishu River and Lian River Upstream watersheds, whereas the factors are stable in the southern, northwestern, and eastern regions, implying that the assignments in these areas is the most reliable.

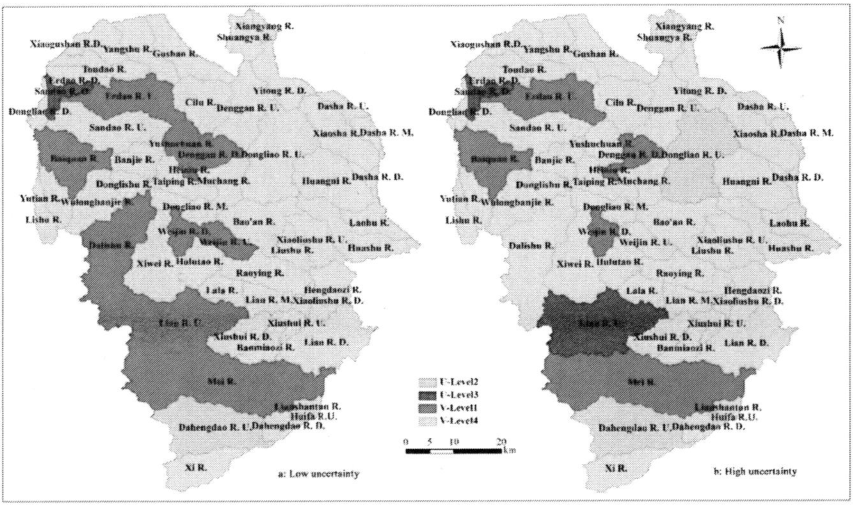

Figure 10. Most sensitive factors in assessing landscape ecological risk under low or high uncertainty of watersheds in Liaoyuan City, China.

5. CONCLUSIONS

Landscape ecological risk assessment is useful in representing landscape ecological security, which is a critical domain of landscape sustainability [50,51,52]. In the assessment process of this study, the higher landscape ecological risk means the more vulnerability and disturbance on the important landscape units relate to less benefit in the future. The result with high risk may suffer more damages in the future which threatens the sustainability of the landscape. Finally, a zoning to identify the spatial risk should be useful to prevent risks and support a sustainable landscape planning. In the former landscape ecological risk assessment studies, ecological disturbance and vulnerability have attracted much attention [53]. However, the concern for ecological importance is less than those two domains. In the study, on one hand, fragmentation of landscape has been introduced to modify the ecosystem services in the assessment of ecological importance, which is an improvement in the evaluation; on the other hand, the uncertainty of risk was often ignored in the former assessment, and the sensitivity analysis in this study may be a novelty in regional ecological risk study.

Mining areas are susceptible to damage through the economic development of mining cities, and the ecological risk posed by mining has gradually drawn increasing attention. Liaoyuan City, once the coal city of Jilin, is a typical mining city in that it has become a resource-exhausted city because of serious damage to its environment, caused by mining. In the mining process, the sustainability of landscapes is disturbed, and the specific ecological risks and uncertainties associated with this disturbance needs to be clarified. Our study of landscape ecological risk in Liaoyuan has found that areas with high probability of risk are mainly concentrated in the relatively flat terrain of the landscape. Not only cities, but also natural and semi-natural landscapes, contribute to ecological risk. Woodland and water dominated watersheds are high-risk areas, cropland-dominated watersheds are moderate-risk areas, and the watersheds with high proportion of built-up land are low-risk areas. Although the similarity of simulated and calculated values is higher under the low uncertainty scenario, in both scenarios, the spatial distribution of both sets of values among the watersheds are the same. Overall, vulnerability of landscape type has a higher impact on the simulated values than land use degree.

Landscape ecological risk assessment, based on the risk probability and ecological loss, directly guides the planning and restoration of landscapes by recognizing spatial structure, by contributing to regional risk management. One of the gaps in our study is that it is quantified through the relative value of ecological services for several landscape types within a watershed, rather than the calculation of an overall absolute value which may be compared among various regions. Another gap is the absence of information on how landscape patterns affect the per unit area value of ecosystem services. Thus, future research in the field of landscape sustainability needs to focus on quantifying landscape ecological risk in ecologically sensitive and fragile areas. It also needs to incorporate further robust analyses of uncertainty, which, although common in regional ecological risk assessment, is still relatively rare in the field of landscape ecological risk assessment. In addition, the directions for improving uncertainty analyses in landscape ecological risk assessment include: (1) analyzing multiple sources of uncertainty; (2) putting forward new methods that are particularly applicable for landscape ecological risk assessment [54]; (3) using a grid system to carry out evaluation at grid scale, in order to improve the accuracy of assessing results [55]; and (4) investigating more specifically the sensitivity of each parameter in the assessment model [56].

ACKNOWLEDGMENTS

This research is financially supported by National Natural Science Foundation of China (41271195).

AUTHOR CONTRIBUTIONS

Jian Peng, Minli Zong, and Jiansheng Wu conceived and designed the study. Jian Peng, Minli Zong, Yi'na Hu and Yanxu Liu made substantial contributions to acquisition, analysis, and interpretation of the data. Jian Peng and Minli Zong wrote the first draft of the article. Yi'na Hu and Yanxu Liu reviewed and edited the first draft. All authors read and approved the submitted manuscript, agreed to be listed, and accepted the version for publication.

CONFLICTS OF INTEREST

The authors declare no conflict of interest.

REFERENCES

1. Wu, J.G. Landscape sustainability science: Ecosystem services and human well-being in changing landscapes. *Landsc. Ecol.* **2013**, *28*, 999–1023.

2. Peng, J.; Wu, J.S.; Pan, Y.J.; Han, Y.N. Evaluation for regional ecological sustainability based on PSR model: Conceptual framework. *Prog. Geogr.* **2012**, *31*, 933–940. (In Chinese).

3. Xu, E.G.B.; Leung, K.M.Y.; Morton, B.; Lee, J.H.W. An integrated environmental risk assessment and management framework for enhancing the sustainability of marine protected areas: The Cape d'Aguilar Marine Reserve case study in Hong Kong. *Sci. Total Environ.* **2015**, *505*, 269–281.

4. Hunsaker, C.T.; Graham, R.L.; Suter, G.W.; O'Neill, B.L.; Jackson, B.L.; Barnthouse, L.W. *Regional Ecological Risk Assessment: Theory and Demonstration*; Oak Ridge National Lab: Oak Ridge, TN, USA, 1989.

5. Hunsaker, C.; Graham, R.; Suter, G.; O'Neill, R.; Barnthouse, L.; Gardner, R. Assessing ecological risk on regional scale. *Environ. Manag.* **1990**, *14*, 325–332.

6. Adams, S.M.; Bevelhimer, M.S.; Greeley, M.S.; Levine, D.A.; Teh, S.J. Ecological risk assessment in a large river-reservoir: 6. Bioindicators of fish population health. *Environ. Toxicol. Chem.* **1999**, *18*, 628–640.

7. Baron, L.A.; Sample, B.E.; Suter, G.W. Ecological risk assessment in a large river-reservoir: 5. Aerial insectivorous wildlife. *Environ. Toxicol. Chem.* **1999**, *18*, 621–627.

8. Cook, R.B.; Suter, G.W.; Sain, E.R. Ecological risk assessment in a large river-reservoir: 1. Introduction and background. *Environ. Toxicol. Chem.* **1999**, *18*, 581–588.

9. Jones, D.S.; Barnthouse, L.W.; Suter, G.W.; Efroymson, B.A.; Field, J.M.; Beauchamp, J.J. Ecological risk assessment in a large river-reservoir. 3. Benthic invertebrates. *Environ. Toxicol. Chem.* **1999**, *18*, 599–609.

10. Sample, B.E.; Suter, G. Ecological risk assessment in a large river-reservoir: 4. Piscivorous wildlife. *Environ. Toxicol. Chem.* **1999**, *18*, 610–620.

11. Landis, W.G. Twenty years before and hence: Ecological risk assessment at multiple scales with multiple stressors and multiple endpoints. *Hum. Ecol. Risk Assess.* **2003**, *9*, 1317–1326.

12. Peng, J.; Liu, Y.X.; Pan, Y.J.; Zhao, Z.Q.; Song, Z.Q.; Wang, Y.L. Study on the correlation between ecological risk due to natural disaster and landscape pattern-process: Review and prospect. *Adv. Earth Sci.* **2014**, *29*, 1186–1196. (In Chinese).

13. Hemstrom, M.; Mezenich, J.; Reger, A.; Wales, B. Integrated analysis of landscape management scenarios using state and transition models in the upper Grande Ronde River Subbasin, Oregon, USA. *Landsc. Urban Plan.* **2007**, *80*, 198–211.

14. Lange, A.; Siebert, R.; Barkmann, T. Sustainability in land management: An analysis of stakeholder perceptions in rural northern Germany. *Sustainability* **2015**, *7*, 683–704.

15. Graham, R.L.; Hunsaker, C.T.; O'Neil, R.V. Ecological risk assessment at the regional scale. *Ecol. Appl.* **1991**, *1*, 196–206.

16. Kapustka, L.A.; Galbraith, H.; Luxon, M. Using landscape ecology to focus ecological risk assessment and guide risk management decision-making. *Toxicol. Ind. Health* **2001**, *17*, 236–246.

17. Liu, S.L.; Cui, B.S.; Dong, S.K.; Yang, Z.F.; Yang, M.; Holt, K. Evaluating the influence of road networks on landscape and regional ecological risk a case study in Lancang River Valley of Southwest China. *Ecol. Eng.* **2008**, *34*, 91–99.

18. Yin, H.; Wang, Y.L.; Cai, J.L.; Lv, X.F.; Liu, X.Q. Regional ecological risk assessment: Its research progress and prospect. *Chin. J. Ecol.* **2009**, *28*, 969–975. (In Chinese).

19. Xie, H.L. Regional eco-risk analysis of based on landscape structure and spatial statistics. *Acta Ecol. Sin.* **2008**, *28*, 5020–5026. (In Chinese).

20. Jing, Y.P.; Zhang, S.W.; Li, Y. Ecological risk analysis of rural-urban ecotone based on landscape structure. *Chin. J. Ecol.***2008**, *27*, 229–234. (In Chinese).

21. Li, J.G.; He, C.Y.; Li, X.B. Landscape ecological risk assessment of natural/semi-natural landscape in fast urbanization regions—A case study in Beijing. China. *J. Nat. Resour.* **2008**, *23*, 33–47. (In Chinese).

22. Wang, C.Y.; Zhang, J.; Xi, H.M.; Fu, J. Ecological risk assessment of island exploitation based on landscape pattern.*Acta Ecol. Sin.* **2008**, *28*, 2811–2817. (In Chinese).

23. Li, Y.F.; Luo, Y.C.; Liu, G.; Ouyang, Z.Y.; Zheng, H. Effects of land use change on ecosystem services: A case study in Miyun reservoir watershed. *Acta Ecol. Sin.* **2013**, *33*, 726–736. (In Chinese).

24. Lanis, W.G.; Wiegers, J.K. Ten years of the relative risk model and regional scale ecological risk assessment. *Hum. Ecol. Risk Assess.* **2007**, *13*, 25–38.

25. Liu, L.; Liu, J.; Zhang, Z.G. Environmental justice and sustainability impact assessment: In search of solutions to ethnic conflicts caused by coal mining in Inner Mongolia, China. *Sustainability* **2014**, *6*, 8756–8774.

26. Himley, M. Global mining and the uneasy neoliberalization of sustainable development. *Sustainability* **2010**, *2*, 3270–3290.

27. McLellan, B.C.; Corder, G.D. Risk reduction through early assessment and integration of sustainability in design in the minerals industry. *J. Clean. Prod.* **2013**, *53*, 37–46.

28. Chang, Q.; Qiu, Y.; Xie, M.M.; Peng, J. Theory and method of ecological risk assessment for mining areas based on the land destruction. *Acta Ecol. Sin.* **2012**, *32*, 5164–5174. (In Chinese).

29. Li, Z.Y.; Zhang, N.; Tang, J.; Ji, Y.; Liu, J.L. Analysis on the landscape ecological risk of Jilin coal mining area. *J. Jilin Univ. (Earth Sci. Ed.)* **2011**, *41*, 207–214. (In Chinese).

30. Li, L.X.; Wang, B.; Zhou, L.B.; Yu, X.B. Ecological landscape risk evaluation in mineral resources exploitation. *Conserv. Util. Miner. Resour.* **2011**, *2*, 1–5. (In Chinese).

31. Wu, J.S.; Zong, M.L.; Peng, J. Assessment of mining area's ecological vulnerability based on landscape pattern: A case study of Liaoyuan, Jilin Province of Northeast China. *Chin. J. Ecol.* **2012**, *31*, 3213–3220. (In Chinese).

32. Wang, E.H. *Liaoyuan Statistical Yearbook (2009)*; Liaoyuan Bureau of Statistics: Liaoyuan, China, 2010. (In Chinese)

33. Chang, Y.J. *Liaoyuan Yearbook (2010)*; Jilin People's Publishing House: Liaoyuan, China, 2010. (In Chinese)

34. Chen, Q.Y.; Liu, J.L. Development process and perspective on ecological risk assessment. *Acta Ecol. Sin.* **2014**, *34*, 239–245. (In Chinese).

35. Pickett, S.T.A.; White, P.S. Chapter 21—Patch Dynamics: A Synthesis. In *The Ecology of Natural Disturbance and Patch Dynamics*; White, S.T.A., Pickettp, S., Eds.; Academic Press: San Diego, CA, USA, 1985; pp. 371–384.

36. Chen, P.; Pan, X.L. Ecological risk analysis of regional landscape in inland river watershed of arid area—A case study of Sangong River Basin in Fukang. *Chin. J. Ecol.* **2003**, *22*, 116–120. (In Chinese).

37. Deal, B.; Pallathucheril, V. Sustainability and urban dynamics: Assessing future impacts on ecosystem services.*Sustainability* **2009**, *1*, 346–362.

38. Scott, C.; Lake, P.; Sergi, S.; John, M.; John, S. The effects of land use changes on streams and rivers in Mediterranean climates. *Hydrobiologia* **2013**, *719*, 383–425.

39. Song, G.B.; Li, Z.; Yang, Y.G.; Semakula, H.M.; Zhang, S.S. Assessment of ecological vulnerability and decision-making application for prioritizing roadside ecological restoration: A method combining geographic information system, Delphi survey and Monte Carlo simulation. *Ecol. Indic.* **2015**, *52*, 57–65.

40. De Lange, H.J.; Sala, S.; Vighi, M.; Faber, J.H. Ecological vulnerability in risk assessment—A review and perspectives.*Sci. Total Environ.* **2010**, *408*, 3871–3879.

41. Wayne, C.Z.; Robert, L.B.; Ralph, D.N. Patterns of deforestation and reforestation in different landscape types in central New York. *For. Ecol. Manag.* **1990**, *36*, 103–117.

42. Costanza, R.; D'Arge, R.; Groot, R.; Farber, S.; Grasso, M.; Hannon, B.; Limburg, K.; Naeem, S.; O'Neill, R.V.; Paruelo, J.; *et al.* The value of the world's ecosystem services and natural capital. *Nature* **1997**, *386*, 253–260.

43. Xie, G.D.; Lu, C.X.; Leng, Y.F.; Zheng, D.; Li, S.C. Ecological assets valuation of the Tibetan Plateau. *J. Nat. Resour.***2003**, *18*, 189–196. (In Chinese).

44. Banzhaf, H.S.; Boyd, J. The architecture and measurement of an ecosystem services index. *Sustainability* **2012**, *4*, 430–461.

45. Hessburg, P.F.; Reynolds, K.M.; Salter, R.B.; Dickinson, J.D.; Gaines, W.L.; Harrod, R.J. Landscape evaluation for restoration planning on the Okanogan-Wenatchee national forest, USA. *Sustainability* **2013**, *5*, 805–840.

46. Manuela, D.G.; Rolf, H.; Silvia, T. Effects of habitat and landscape fragmentation on humans and biodiversity in densely populated landscapes. *J. Environ. Manag.* **2009**, *90*, 2959–2968.

47. Matteo, M.; Duccio, R.; Francesco, G.; Giovanni, B.; Valerio, A. Biodiversity, roads, & landscape fragmentation: Two Mediterranean cases. *Appl. Geogr.* **2013**, *46*, 63–72.

48. Campolongo, F.; Saltelli, A.; Tarantola, S. Sensitivity analysis as an ingredient of modeling. *Stat. Sci.* **2000**, *15*, 377–395.

49. Guo, X.R.; Hu, H.W.; Chen, D.S.; Wei, W.; Cheng, S.Y. Urban ecological risk assessment based on land utilization analysis. *J. Beijing Univ. Technol.* **2012**, *38*, 1114–1120. (In Chinese).

50. Gao, Y.; Wu, Z.F.; Lou, Q.S.; Huang, H.M.; Cheng, J.; Chen, Z.L. Landscape ecological security assessment based on projection pursuit in Pearl River Delta. *Environ. Monit. Assess.* **2012**, *184*, 2307–2319.

51. Zhou, K.H.; Liu, Y.L.; Tan, R.H.; Song, Y. Urban dynamics, landscape ecological security, and policy implications: A case study from the Wuhan area of central China. *Cities* **2014**, *41*, 141–153.

52. Cen, X.T.; Wu, C.F.; Xing, X.S.; Fang, M.; Garang, Z.M.; Wu, Y.Z. Coupling Intensive Land Use and Landscape Ecological Security for Urban Sustainability: An Integrated Socioeconomic Data and Spatial Metrics Analysis in Hangzhou City. *Sustainability* **2015**, *7*, 1459–1482.

53. Peng, J.; Dang, W.X.; Liu, Y.X.; Zong, M.L.; Hu, X.X. Review on landscape ecological risk assessment. *Acta Geogr. Sin.* **2015**, *70*, 664–677. (In Chinese).

54. Li, Y.F.; Sun, X.; Zhu, X.D.; Cao, H.H. An early warning method of landscape ecological security in rapid urbanizing coastal areas and its application in Xiamen, China. *Ecol. Model.* **2010**, *221*, 2251–2260.

55. Ma, L.Y.; Xu, X.G.; Xu, L.F. Uncertainty Analysis of integrated ecological risk assessment of China. *Acta Sci. Nat. Univ. Pekin.* **2011**, *47*, 893–900. (In Chinese).

56. Li, Y.F.; Shi, Y.L.; Qureshi, S.; Bruns, A.; Zhu, X.D. Applying the concept of spatial resilience to socio-ecological systems in the urban wetland interface. *Ecol. Indic.* **2014**, *42*, 135–146.

11
CHAPTER

Mountaintop Removal Mining and Catchment Hydrology

Andrew J. Miller and Nicolas P. Zégre *

Division of Forestry & Natural Resources, West Virginia University, Morgantown, WV 26506-6125, USA

ABSTRACT

Mountaintop mining and valley fill (MTM/VF) coal extraction, practiced in the Central Appalachian region, represents a dramatic landscape-scale disturbance. MTM operations remove as much as 300 m of rock, soil, and vegetation from ridge tops to access deep coal seams and much of this material is placed in adjacent headwater streams altering landcover, drainage network, and topography. In spite of its scale, extent, and potential for continued use, the effects MTM/VF on catchment hydrology is poorly understood. Previous reviews focus on water quality and ecosystem health impacts, but little is known about how MTM/VF affects hydrology, particularly the movement and storage of water, hence the hydrologic processes that ultimately control flood generation, water chemistry, and biology. This paper aggregates the existing knowledge about the hydrologic impacts of MTM/VF to identify areas where further scientific investigation is needed. While contemporary surface mining generally increases peak and total runoff, the limited MTM/VF studies reveal significant variability in hydrologic response. Significant knowledge gaps relate to limited understanding of hydrologic processes in these systems. Until the hydrologic impact of this practice is better understood, efforts to reduce water quantity and quality problems and ecosystem degradation will be difficult to achieve.

Keywords: mountaintop removal mining; valley fills; streamflow; hydrology; Appalachians; surface coal mining

1. INTRODUCTION

Mountaintop removal mining and valley fill (MTM/VF) coal extraction, practiced in the Central Appalachian region of the eastern United States, represents a dramatic change to the landscape. Post mining topography, vegetation, landuse, soils, and runoff pathways can be severely altered during the mining process and subsequent reclamation. While surface mining is used to mine metals, minerals, and oil, it is often associated with coal, accounting for 40% of coal mined globally and accounts for two-thirds of coal production in the US and Australia [1]. Surface mining represents the largest landuse/land cover change in the Central Appalachian region [2] and by 2012, the US Environmental Protection Agency (EPA) estimates that MTM/VF will have impacted approximately 6.8% of the predominately forested Appalachian coalfield region of West Virginia, Kentucky, Tennessee, and Virginia with nearly 4000 km of headwater streams buried under valley fills [3] (Figure 1). While the U.S. Energy Information Administration projects a reduction in Appalachian coal production through 2035, this decline is relatively minimal (−0.6%) indicating that the low-sulfur Appalachian coal extracted by cost effective MTM/VF practices will continue to be a significant component to the energy future of the United States [4]. Furthermore, the large-scale alterations of catchment structure (e.g., topography, slopes, organization, volume, *etc.*), water chemistry, and land cover are likely permanent, having important health implications for aquatic and human communities linked to these streams.

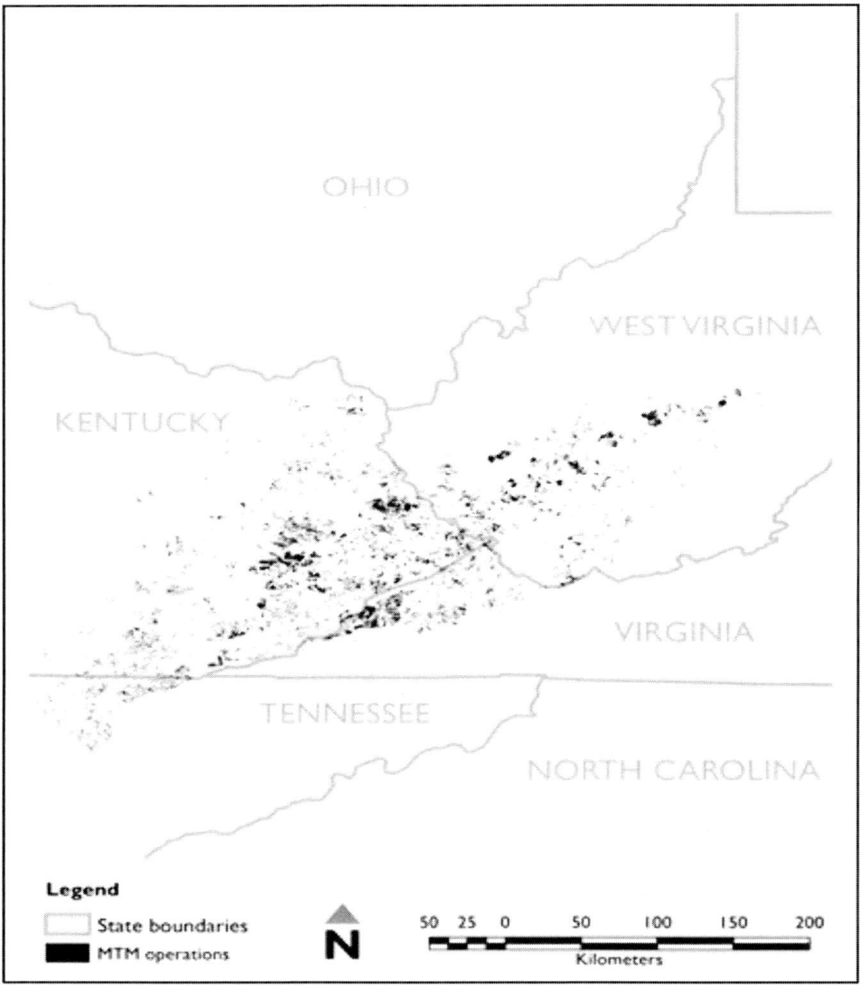

Figure 1. Mountaintop removal mining (MTM) operations of the Central Appalachian coalfields. The EPA [3] estimates that 6.7% of this predominantly forested region has been impacted by MTM and approximately 4000 km of headwater streams have been buried under valley fills. MTM boundaries remote sensed from aerial photography by SkyTruth [5].

Significant progress has been made in understanding the impacts of MTM/VF on downstream chemistry and ecological conditions. A recent review by Griffith *et al.* [6] provides a state-of-understanding about the physiochemical impacts of this practice and recent work by Lindberg *et al.* [7] and Bernhardt *et al.* [8] show cumulative and regional impacts and significant biological impairment downstream of mountaintop mines and valley fills. By placing minespoil in the path of streams, contact time is forced between runoff and overburden chemistry potentially altering stream chemistry (e.g., alkaline or acid mine drainage). Notwithstanding the role hydrology plays on stream chemistry, aquatic ecology, and overall aquatic ecosystem health, and despite the scale, extent, and potential for continued development, the effects MTM/VF on hydrology is poorly understood. Much of our limited knowledge about the hydrologic impacts of surface mining is from only a handful of traditional surface mining (*i.e.*, strip, auger) studies whose numbers pale in comparison to other land use studies (e.g., forest harvesting). The number of MTM studies is considerably less making generalization and inference difficult between watersheds. While in certain respects MTM is similar to traditional surface mining, MTM generally differs in terms of its spatial extent and prolific use of valley fills, although valley fills are sometimes using in conjunction with contour mining While valley fills have been described as acting like an unconsolidated aquifer that modulate runoff [9], the minespoil that forms the valley fill also develops preferential flow paths capable of storing and rapidly routing water [10,11,12].

Investigations into contemporary MTM/VF operations have involved hydrologic modeling or quantifying catchment outlet streamflow responses to precipitation inputs. These studies show an overall increase in baseflow in MTM/VF impacted catchment [11,13,14,15] and generally show increases in discharge for larger storm events [11,16]. However, results from these studies are variable, influenced by climate, geography location, and the stage/age of mining activities, underscoring the need for more hydrological studies.

MTM/VF has become an increasingly polarized issue nationally and particularly in the communities in which it is practiced. Local citizens, environmental advocacy groups, and regulators have expressed concerns over the long term impacts of MTM/VF on downstream water quality, public health, and safety. After several devastating floods in southern West Virginia, public concerns were raised about the potential of MTM to exacerbate flooding in coal region communities, which typically abut streams and rivers in narrow valleys due to the region's steep topography. Industry has countered these concerns by emphasizing the economic benefits of the coal industry [17] and, more effectively, citing the absence of conclusive scientific evidence to support the MTM/VF operations' culpability in the alteration of downstream hydrology. Underscoring the critical knowledge gaps associated with MTM is the need for thorough scientific investigations to inform citizens, the energy industry, and policy makers about the potential environmental consequences of this practice.

In short, there is a lack of data to inform our understanding of how hydrologic systems are responding to this drastic and permanent landscape-scale change. Thus, the overall objective of this paper is to aggregate the existing knowledge base of the hydrologic impacts of MTM/VF to identify areas where further scientific study is critically needed. The organization of this paper is as follows: Section 2 provides an overview of the catchment water balance, important processes that govern runoff, and impacts of land cover changes on streamflow to develop a conceptual understanding of potential impacts of MTM/VF on hydrology; Section 3 describes MTM/VF and traditional surface mining operations and reviews hydrologic impacts associated with each practice; and in Section 4 we identify critical hydrologic knowledge gaps based on our literature search to provide direction for future research.

2. OVERVIEW OF CATCHMENT HYDROLOGY

2.1. Water Balance and Catchment Processes

Small catchments (<10 km^2) and hillslopes are the primary experimental unit for many hydrologic studies because inputs and outputs (Figure 2) are relatively easier to measure at smaller scales. How catchments collect, store, and release water is largely a function of climate, morphology (including topography, soils, geology), and land cover. The storage and flow of water through a catchment affects a number of processes including the quantity and timing of runoff, soil erosion and sediment transport, downstream water chemistry and biology, and biogeochemical cycling.

The primary input to headwater catchments in the Central Appalachian region is precipitation primarily in the form of rain and a transient snowpack. Losses are from evaporation and transpiration (evapotranspiration, E_t), streamflow and groundwater. E_t in this region accounts for, on average, 47% of annual precipitation [18]. Canopy interception in this predominantly deciduous forests accounts for 10%–20% of precipitation [19,20,21].

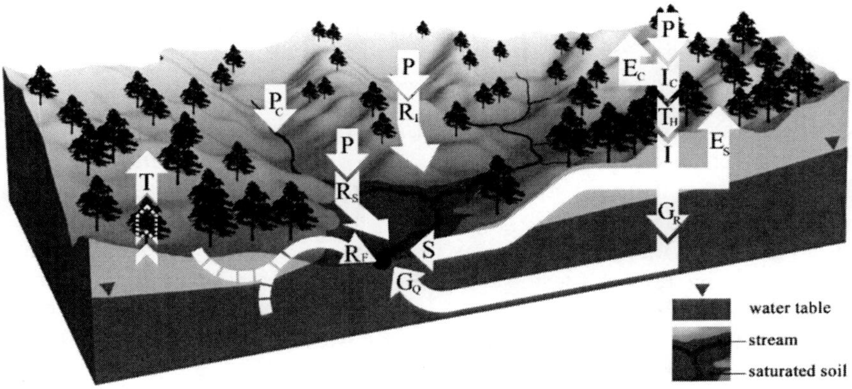

Figure 2. Conceptualization of forest headwater catchment hydrologic processes. Water is added to the catchment as precipitation (**P**), some of which falls directly into the stream channel (**P_C**). Precipitation falling on forested hillslopes is intercepted by the tree canopy (**I_C**) and is either lost to evaporation (**E_C**) or falls to the forest floor as throughfall (**T_H**). On the forest floor, water is infiltrated (**I**) into the soil and is either lost to evaporation (**E_S**) and plant transpiration (**T**) or percolates to the water table and becomes groundwater recharge (**G_R**) which will be stored then discharged to the stream channel (**G_Q**) or to the adjacent riparian area as return flow (**R_F**). Infiltrated water may also take shallow subsurface flow paths as matrix flow or through macropores and soil pipes (**S**). Water not infiltrated into the soil will either be lost to evaporation or runoff to the stream channel as infiltration excess overland flow (**R_I**). Precipitation falling on saturated soil becomes saturation excess overland flow (**R_S**).

Runoff can take a variety surface and subsurface flowpaths from the hillslope to the stream channel that ultimately control the timing and magnitude of runoff. Because of high hydrologic conductivity in soil macropores [21,22], runoff generation in forested catchments is dominated by subsurface flow [23]. The biogeochemical composition of water discharged to the stream channel is a function of how long water remains in the catchment; longer residence times imply greater contact time for biogeochemical transformation [24,25]. Therefore, runoff flowpaths exert important controls on downstream aquatic ecosystems that extend beyond physical hydrology.

2.2. Landuse and Land Cover Changes

The consequences of landuse/land cover changes on streamflow have garnered much attention over the past few decades [26]. While significant research focuses on the impacts of forest harvesting on streamflow (e.g., [27,28,29]) studies also include agriculture (e.g., [30]), and urbanization (e.g., [31,32,33]).

Forest harvesting usually increases runoff by reducing canopy interception and transpiration (e.g., [34]). In addition, soil compaction from heavy machinary and logging roads reduces infiltration thereby increasing surface runoff (e.g., [35]). Logging roads cut into steep hillsides can intercept shallow subsurface flowpaths from upland hillslopes, thereby delivering this water to the stream channel as surface runoff [36]. Because vegetation is usually reestablished after logging operations and the catchment structure is relatively unaltered, the impact of harvesting on streamflow is generally thought to be short-lived, though dense and poorly constructed logging road networks can prolong the impact [37].

Agricultural practices and urbanization can have profound effects of streamflow. The conversion of mature forests to agricultural land changes the amount of water lost through E_t, increasing available water for runoff. Further exposure of bare soils and compaction can lead to rapid routing of runoff to stream channels, often as Hortonian overland flow [38] or through constructed drainage networks [39]. Urbanization generally increases runoff magnitude and volume. Impervious surfaces such as roads, parking lots, and roofs can greatly diminish infiltration resulting in reduced soil moisture, groundwater recharge and increased stormflow [32]. The conversion of forest to lawn or impervious surface can greatly reduce catchment E_t [40]. Runoff is often directly routed to streams through storm sewers and constructed channels. Consequently, urbanized catchments tend to quickly convert rainfall to runoff with flashy, high-magnitude storm hydrographs and limited baseflow compared to forested systems [31,32,33].

Similar to these landuse/land cover changes, surface mining alters the catchment water budget by altering flow paths. Ferrari *et al.* [41] described the impacts of traditional surface mining as being more similar to urbanization than forest harvesting. Confounding better understanding of the impacts of MTM/VF is that this practice causes structural changes in the catchment (*i.e.*, topography, drainage density) in addition to altering the water budget due to loss of forest and soil compaction during reclamation; MTM activities can result in major changes to headwater topography and drainage network but the implication of a wholesale alteration in catchment structure and land cover is not well understood.

3. IMPACTS OF MOUNTAINTOP-REMOVAL MINING AND VALLEY FILL ON HYDROLOGY

3.1. Overview of MTM/VF Operations

MTM/VF is a special form of surface mining adapted to mountainous terrain in which the forest, topsoil, and overlying bedrock is removed using explosives and heavy machinery to gain direct access to deeper coal seams (Figure 3). While MTM/VF is broadly categorized as surface mining and synonymous with traditional strip and contour mining, MTM is distinct in terms of its scale and management of overburden [6]. As much as 300 m of overburden is removed from ridge tops to access underlying coal seams [42]. Because the volume of displaced minespoil precludes it being replaced to the ridge tops, much of this excess material is placed in adjacent valleys burying headwater streams and springs, creating valley fills (Figure 4). Valley fill construction techniques vary; the sorting and placement of spoil material, management of water through or on top, and soil conditions of the valley fill face are often site-specific and vary considerably across the region.

MTM is permitted under the Surface Mining Control and Reclamation Act (SMCRA) of 1977 [43]. Under this act, mining operators are required to restore the topography to approximate original contour (AOC) which states "*...backfilling and grading of the mined area so that the reclaimed area, including any terracing or access roads, closely resembles the general surface configuration of the land prior to mining and blends into and complements the drainage pattern of the surrounding terrain...*", though interpretation and enforcement of this language is left to regulators at the state level [43]. Additionally, variances to the AOC requirement are granted if mine operators propose a post mining land use that would constitute an "improvement" over pre-mining conditions (*i.e.*, industrial, commercial, residential, agriculture, and public land uses). In such cases, mine operators are under no requirement to recreate pre-mining topography, the result of which is flattened ridge-top topography and large valley fill structures.

Figure 3. The process of MTM removes vegetation, soil, and uses explosives and heavy machinery to gain access to deeper coal seams. Overburden is placed in adjacent headwater stream valley creating valley fills. The picture below shows the critical elements of a MTM/ valley fill (VF) operation in a headwater catchment in southern West Virginia (WV): (**a**) Pre-mining forest; (**b**) forest and soils cleared in preparation of blasting; (**c**) Explosives are used to remove overburden geology changing ridgelines (**d**) and altering topography; (**e**) Overburden material is placed in stream valleys creating valley fills (**f**) in lifts altering catchment storage; (**g**) post-mining landscape is reclaimed under the Surface Mining Control and Reclamation Act (SMCRA). Post-mining vegetation is often herbaceous.

The primary objective of reclamation since the passage of SMCRA has been on slope and soil stability to prevent soil erosion from the mine surface [44]. To achieve this, mine soils are heavily compacted using heavy machinery and fast growing (often exotic) herbaceous cover is seeded to quickly establish a vegetative surface. A consequence of the emphasis on slope stability has been the loss of natural tree succession on mine surfaces [45]. The Forestry Reclamation Approach (FRA) advocated by the Appalachian Region Reforestation Initiative (ARRI) has shown that loose dumped spoil is an effective growing medium for Appalachian hardwood tree species [46]; while this method of surface mine reclamation has become more commonplace, it does not represent the predominant reclamation practice on surface mines in the region.

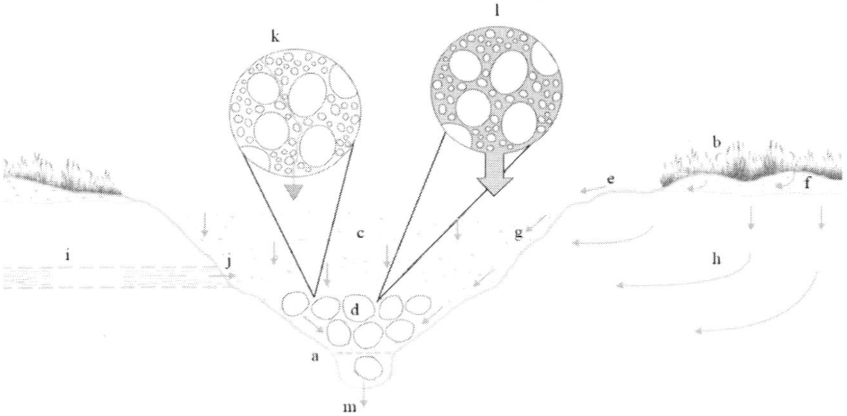

Figure 4. Conceptual model of runoff flow paths in a MTM/VF headwater catchment: (**a**) original stream channel; (**b**) MTM surface reclaimed with herbaceous cover; (**c**) valley fill material and lifts; (**d**) larger overburden material sort to fill bottom; (**e**) rapid Hortonian overland flow; (**f**) infiltrating water hits less permeable mine pavement facilitating horizintal flow; (**g**) percolation of water through valley fill and along original valley slope; (**h**) groundwater flow; (**i**) subsurface flow paths along coal seems and abandoned deep mines; (**j**) groundwater contributions to fill from (**i**); (**k**) steady state matrix flow through tortorous flow paths sustains streamflow exiting valley fill (**m**) during non-storm periods; (**l**) runoff through connected saturated pores and large voids during storm conditions rapidly routes water through valley fill; (**m**) runoff exiting valley fill to surface channel.

Because MTM/VF operations create two distinct landforms, each with unique physical and hydrologic impacts, the following sections discuss the impacts of traditional surface mining and the individual and combined hydrologic impacts of the MTM/VF (Table 1).

3.2. Traditional Surface Mining

Surface mining for coal in the Appalachians began in the mid-20th century and continues today. Much of the mining in the early part of this period constituted small contour and strip mine operations that disturbed areas less than 400 hectares. In contrast, MTM in the coalfields of southern West Virginia, southwestern Virginia, and eastern Kentucky, and Tennessee in the 1980s disturbed areas on the scale of thousands of hectares [64]. Prior to the passage of SMCRA in 1977, the methods and

Table 1. Hydrologic impact studies of surface mining and MTM/VF in the eastern US.

Reference [a]	Year [b]	Location	Mining Method	# of catchments	Spatial Scale [c]	Time scale/ study duration	Key hydrologic observations
[47]	1965	Indiana	pre-SMCRA surface mining	2	HW to MS: 55.4 & 111.4 km^2	3 months	Show that spoil piles act as reservoirs for groundwater storage.
[48]	1966	Indiana	pre-SMCRA surface mining	2	HW: 55 & 111 km^2	12 months	Surface mine impacted watersheds maintained flows while forested catchments went dry during extreme draught conditions.
[49]	1970	Kentucky	pre-SMCRA strip mining	3	HW: 0.67–2.2 km^2	4 years	Greater flow variability, less storage, larger peakflows in mined catchments.
[50,51]	1972	Kentucky	pre-SMCRA strip mining	4	HW: 0.66–1.8 km^2	Single storm event	During a major storm event, surface mined watersheds had smaller peak discharges than adjacent, undisturbed catchments.
[52]	1972	Kentucky	pre-SMCRA surface and underground mining	1	BA: 2442 km^2	25 years	Streamflow is maintained during extended dry periods in mined catchments after streams in unmined catchments have ceased flowing, which the authors attributed to storage of water in underground mines, spoil piles, and strip pits.
[53]	1981	Kentucky	pre-SMCRA strip mining	6	HW: 0.70–1.5 km^2	4 years	Storm flow volumes were unchanged but peak flows increased by 36% in mined catchments. This effect was most prominent during smaller events; higher magnitude peaks were unaffected.
[54]	1987	Pennsylvania	post-SMCRA surface mining	-	PS: 78 Infiltration tests on 5 reclaimed surface mines	30 minutes tests on reclaimed mines 1–4 years old	Infiltration rates on newly reclaimed mine soil were an order of magnitude lower than undisturbed soils. Infiltration rate recovered through time the extent of which appeared to be controlled by overburden lithology.
[9]	1989	Tennessee	pre- and post-SMCRA contour mining, small scale MTM operations	5	HW: 1.74–11.2 km^2	8 years	Total flow increased in mine-impacted catchments, particularly during low flow conditions. Authors attributed this to decreased ET losses from deforestation and the storage and slow release of runoff from mine spoil.
[55]	1989	West Virginia	pre-SMCRA strip and underground mining	5	HW: 4.66–20.07 km^2	3–5 years	Synoptic discharge measurements revealed the importance of underground mining in rerouting subsurface water. A rainfall-runoff model indicated increased ET losses and decreased runoff in surface mined catchments.

diligence of surface mine reclamation varied greatly. While some operations loosely regraded disturbed areas and planted hardwood trees that ultimately developed into healthy forests [71], many areas were left untouched after extraction causing prolific water chemistry, erosion, and sedimentation problems downstream. It was this disparate state of reclamation that prompted the passage of SMCRA, a key objective of which was creating stable landforms to prevent erosion and stream sedimentation [44]. To achieve this goal, regulators emphasized heavily grading minespoils to achieve AOC and the use of quick growing herbaceous grasses and legumes to prevent soil erosion. This methodology resulted in heavily compacted soils often lacking organic matter and competitive groundcovers that prevented natural succession, growth, and survival of native trees species [45,72]. Natural succession from adjacent forest land has occurred, but the overall basal area and species diversity lagged behind natural forests, even after decades [73]. Thus, the predominant state of reclaimed surface mines across Appalachia is grasslands with heavily compacted soils with limited to no tree growth.

The post-SMCRA reclaimed surface mine is characterized by herbaceous grasses and groundcovers growing on heavily compacted mines soils [73]. Common grasses seeded for revegetation include fescues (*Festuca spp* L.), redtop (*Agrostis alba*L.), and perennial rye grass (*Lolium perenne* L.). Crownvetch (*Coronilla varia* L.) and Chinese bushclover (*Lespedeza cuneata*), both legumes, are also commonly used in mine reclamation due to their nitrogen fixing potential [74,75]. While these species have been successful in quickly establishing vegetative cover on disturbed soils, they lack the structural complexity of a mature deciduous tree canopy. Consequently, rainfall interception and storage capacities in grasslands are less than forested canopies, though the rates of evaporation from each are similar [76]. Losses to E_t are generally less for grasslands compared to forests [77]; therefore, a larger proportion of precipitation is converted to runoff [9,57].

During the mining process (Figure 3), soils are removed, stockpiled, and then replaced during reclamation, though variances may be granted that excuse this requirement [78]. The reconstructed post-mining soil structure is drastically altered from pre-mining condition [78,79]. Minesoils [80] are heavily compacted soils with high bulk density and low porosity near the soil surface [72,81,82]. But due to the presence of large rock fragments [83], porosity may increase with depth [80,84]. Excessive surface compaction limits root penetration and growth [78,85] and lowers the available water holding capacity [82,86]. While the conditions described above are nearly ubiquitous for reclaimed surface mines in the region, variation in soil structure and properties result from different machinery and techniques used for reclamation [79,85] as well as the pre-mining overburden lithology [54].

Infiltration rates of minesoil can be an order of magnitude lower than undisturbed forest soils [54,57,65] that cause the initiation of infiltration excess overland flow which dominates storm runoff in mined areas [54,57]. The dominance of surface pathways in runoff generation in mined catchments is further evidenced by the increase in total suspended solids downstream of mining activities [87] that originate from the mine surface during the period immediately following reclamation [88]. However, Ritter and Gardner [57] observed variability in the infiltration rate of mine soils through time; some minesoils maintain low infiltration rates with minimal recovery, while other minesoils return to near pre-mining rates in as little as four years. Jorgensen and Gardner [54] attribute this variability to overburden lithology which ultimately controls the mineralogy and grain size during the redistribution of soils onto the mine surface and initial weathering. Surprisingly, there is little observed change in bulk density of shallow minesoil [54,89,90]. This dissonance is explained by Guebert and Gardner [62], who show the increase in infiltration rate is due to the development of an extensive macropore structure. As the macropore structure develops, subsurface flow paths will become more significant which causes a reduction in the peak discharge during storm events.

While impact of surface mining and reclamation on streamflow has been explored for nearly as long as the practice of surface mining in Appalachia, studies are relatively few in number. Early investigations on the impact of surface mining used a paired catchment approach (e.g., [49]) where change through time was measured in a pre-SMCRA mined catchment and a reference catchment. Collier *et al.* [49] investigated changes in the Beaver Creek Basin of Kentucky from 1955 to 1966 and found that a surface mined catchment had dampened peak stormflow and greater baseflow compared to the undisturbed catchment. They were unable to link the modulated runoff response to surface mining due to an inadequate pre-mining calibration period, but other early studies observed similar effects [47,48,52]. Curtis [50] observed a marked increase in the peak flow volume in pre-SMCRA surface mined catchments in West Virginia, but later noted that such an increase only occurred during active mining and may be ameliorated by reclamation [91], particularly the use of sediment retention ponds [51]. In Kentucky, Bryan and Hewlett [53] observed a 36% increases in peak flow in surface mined catchments but this effect was limited to the summer season; peak discharge in winter and spring months were unchanged and possibly reduced. Citing the difference in magnitude between winter in summer peak flows, Bryan and Hewlett [53] concluded that the increases in summer peak flows from surface mining does not represent a serious flood risk.

In a study of the impact of surface mining on headwaters of the New River in Tennessee from 1972 to 1985, Dickens *et al.*[9] elaborated on the mechanisms that control runoff modulation observed in prior studies. Significant baseflow increases (10×) in mined catchments was a result of infiltration, storage, and slow release of water stored in minespoil. Using monitoring wells, total spoil storage was estimated to be 4×10^6 m^3 (44% of annual catchment yield) in Indian Branch, a catchment mined in the 1950s–1960s. However, the authors observed differences in water storage in spoil banks due to mining and reclamation practices. Minespoil in a catchment reclaimed under Tennessee's partial backfill reclamation standards [80], stored 1×10^5 m^3 of water (11% of annual catchment yield). Contrary to earlier studies, Dickens *et al.* [9] observed a increases in the total catchment water output which was attributed to reduction in E_t from deforestation and water storage in minespoil.

More recent studies of surface mining following the implementation of SMCRA observed different results than the aforementioned studies. In a study of three headwater catchments undergoing surface mining in Ohio, Bonta, *et al.* [59] observed increases in peak stormflow from undisturbed to reclaimed conditions with no consistent pattern in baseflow response to mining. In western Maryland, Negley and Eshleman [65] observed significantly different runoff responses between a forested and mined headwater catchment at the storm event scale. The mined catchment exhibited higher storm runoff coefficients (2.5×), greater total storm runoff (3×), and higher hourly peak runoff rates (2×). In spite of the large storm response, total annual runoff did not significantly differ between the two catchments. The authors attribute this to the heavily compacted soils in the mined catchment; infiltration rates on the reclaimed mine surface were two orders of magnitude lower than those in the forested catchment. This led to an infiltration excess overland flow driven storm response in the mined catchment and poorly sustained baseflows during the winter and spring due to insufficient subsurface flow contributions. Negley and Eshleman [65] also analyzed storm responses using unit hydrograph theory [92]. Surprisingly, the unit hydrographs for the mined and forested catchments were remarkably similar. The authors attributed this similarity to differences in catchment sizes and slopes (the mined catchment was an order of magnitude larger and significantly flatter) that offset the differences in runoff processes caused by landcover change. The authors stress the importance of selecting catchments with similar size, shape, and physical characteristics because confounding variables may obfuscate or exacerbate effects of disturbance.

Few studies attempt to assess the impacts of surface mining at the basin scale (>100 km^2). McCormick *et al.* [66] explored runoff responses in the mined George's Creek basin (188 km^2) and unmined Savage River basin (127 km^2) in Maryland. Results showed that George's Creek had higher peak runoff and shorter lag times that were attributed to landuse. However, George's Creek only produced two thirds of the total stormflow volume of Savage River which the authors attribute to infiltration of subsurface flow into abandoned underground mines and a large, subsurface inter-basin diversion that draws water from George's Creek. Thus, assessing surface impacts can be complicated by legacy subsurface mining. Ferrari *et al.*[41] modeled runoff responses in the George's Creek basin under increasing mining scenarios. Results show that runoff magnitude increases linearly with increasing mining disturbance, a trend that more closely resembled urbanization than deforestation from forest harvesting. The authors call into question the efficacy of modern reclamation practices in returning mined areas to the hydrologic regime that existed prior to disturbance.

3.3. The Valley Fill

Few studies explore the role that valley fills play on controlling water storage and release. Much of our understanding of contemporary valley fill behavior comes from studies of unconsolidated spoil piles associated with pre-SMCRA surface mining operations. While limited inference about valley fill hydrology can be made, the scale and construction methods of contemporary valley fills drastically differ from spoil piles from early surface mining operations. The size of modern valley fills varies, but the largest have volumes of over 150 million m^3 and are >3 km in length [93]. In the period 1985–2001 the US EPA found that average valley fill area was increasing through time in the southern coalfields of West Virginia [93].

Multiple methods exist for constructing valley fills but the predominant technique used in the rugged topography of Appalachia is the durable rock fill technique where spoil is end-dumped from the mine surface in lifts (Figure 3). Valley fills are required to be composed of at least 80% durable rock (rock that will not slake in water or degrade to soil material) so that fine material that could prevent water movement in the underdrain is minimized. Other valley fill construction techniques require that an underdrain be built prior to fill material placement but this regulation is waived for durable rock fill methodology because it is assumed that spoil will naturally segregate during dumping so that fine spoil material stays at higher elevations on the valley fill and large rock and boulders fall to the valley floor. In a study by the Office of Surface Mining and the Kentucky Division of Mine Reclamation and Enforcement, over half of 44 valley fills studied were constructed with less than 80% durable material and that gravity formed underdrains are often poorly formed or even nonexistent [94].

The storage and slow release of water from surface minespoil described by Dickens *et al.* [9] has been shown to apply to valley fills, as multiple studies have observed longer flow durations and augmented baseflow downstream of valley fills [11,15]. However, this conceptualization of minespoil as a storage reservoir may not capture the complexity of water movement in valley fills. Caruccio *et al.* [58] and Caruccio and Geidel [10] described flow through minespoil as pseudokarst [87], a term more often associated with glacial or pyroclastic sediments that have multiple, rapid responding flow paths [88]. Hawkins and Aljoe [95] showed that pseudokarst was a dominant flow path in minespoil during transient, unsteady conditions but during steady state conditions, flow was dominated by matrix flow.

The most complete picture of valley fill hydrology comes from a series of investigations of water movement and storage in a large minespoil area at the Star Fire Mine in eastern Kentucky [12,61]. Using groundwater monitoring wells, dye tracers, measured discharge from valley fill outflows, and structural and topography maps, the authors present a conceptual model of minespoil hydrology with distinct but interconnected saturated zones. Water stored on the former mining bench is slow moving but eventually drains towards two surrounding valley fills where water movement is rapid [12]. Recharge to the valley fills occur from streams, adjacent bedrock and coal aquifers, and surface water that infiltrates into the minespoil from the bedrock-spoil interface [12]. While hydraulic conductivity within the minespoil varied, there was no discernible difference between the spoil interior and the valley fills. Therefore, the discrepancy between water movement in the spoil interior and valley fills was a function of topographic gradients and continued recharge to the valley fills and not differences in the spoil material itself. The authors conclude that movement of water within the spoil body is mostly a function of gradients created from recharge and discharge interactions and the subsurface topography created by the impermeable pavement below the lowest mined coal and drainage patterns in the valleys prior to the start of mining.

3.4. Mountaintop Mining and Valley Fill—A Two Part System

Assessing the hydrologic impacts of MTM/VF is difficult because it is a two part system, each with potentially contradictory effects on the storage and movement of water. While post-SMCRA reclaimed surface mines generally produce rapid, higher magnitude runoff response to storm events, valley fills appear to act as storage reservoirs that dampen storm responses and sustain baseflow but the physical processes controlling runoff generation in these systems is not clear.

The U.S. Geological Survey's (USGS) study of extensively mined (0.5 km^2; 44% MTM/VF), partially mined (5.7 km^2; 40% MTM/VF), and forested (1.4 km^2; no MTM/VF) subcatchments of Ballard Fork in southern West Virginia offers the most complete picture of hydrologic impacts of MTM. Messinger and Paybins [14] investigated differences in water balance components, namely precipitation and daily and monthly mean flow, for three watersheds from 1999 to 2001. Total unit flow in the extensively mined catchment was nearly twice that of the partially mined and forested catchments. The greatest flow differences between the catchments occurred during low flow when the forested catchment went dry compared to sustained year round streamflow in the extensively mined catchment. This corroborates the findings of Wiley *et al.* [15] who show that 90% flow duration was 6×–7× greater downstream of valley fills. High flows were similar between the extensively mined and forested catchments. Messinger and Paybins [14] attribute the increased runoff responses to a reduction in E_t in the heavily mined catchment due to the removal of soil and vegetation from mining operations.

Messinger [11] examined storm responses in this same catchments and study period and found varying outlet responses to different storm intensities. Peak unit runoff for storms where rainfall exceeded 25 mm hr^{-1} was greater in the extensively mined catchment than the forested and partially mined catchments. This relationship was reversed during smaller storms; the extensively mined catchment showed a smaller peak compared to the other two catchments. For storm events with sufficient intensity (>6 mm hr^{-1}), hydrographs from the extensively mined catchment showed a distinct double peak, which Messinger [11] attributed to infiltration excess overland flow causing the first peak and delayed discharges from valley fills contributing to the second peak, though hydrograph separations were not conducted in this study. Total unit flows in the extensively mined catchment were generally twice that of the forested catchment where the greatest differences in flow among the three catchments occurred during the recession limb of the storm hydrograph. Messinger [11] noted that the largest storm event during the study period only produced a return interval of 1.1 years in the forested catchment and that rainfall runoff relations might be different during extreme events.

Wiley and Brogan [15,16] examined peak discharges in six small catchments in the headwaters of the Clear Fork River in West Virginia for a single, large storm event on 6–7 July 2001. Peak discharge was indirectly calculated for six catchments using the slope-area method [91]. Three of these catchments were undisturbed and three had varying degrees of MTM/VF development. Flood recurrence intervals were calculated for the storm event for each catchment. The three undisturbed catchments had recurrence intervals between 10 and 25 years. The disturbed catchments showed greater variability with return intervals ranging from <2 years to >100 years. Variability is likely due to differing extents of valley fill development within each watershed; the lowest return interval occurred in a watershed with one large, reclaimed valley fill while the largest occurred in a catchment with active MTM and an un-reclaimed valley fill. Understanding how the stage of mining and reclamation affects runoff response is important to mitigating the flood risk to communities downstream of MTM/VF reclamation. If flooding is indeed more likely to occur during the mining process, then additional steps are necessary to protect downstream communities during this period.

Due to the dearth of gaged headwater catchments in the MTM region, several studies used hydrologic models to explore the impacts of MTM. In response to extreme flooding in May and July of 2001, the Governor of West Virginia created the Flood Advisory Technical Taskforce (FATT) to investigate the possible impacts of logging and surface mining. Using Natural Resource Conservation Service curve number approach, the study found that surface mining (including MTM) and harvesting increased peak flows between 3% and 21% but the significance of this additional input was lessened in the furthest downstream reaches in the modeled catchments [63]. McCormick and Eshleman [68] calculated curve numbers for surface mined and reclaimed catchments in western Maryland using rainfall-runoff data and found that their curve numbers were greater than if estimated from prevailing engineering methods. Therefore, modeled runoff in mined catchments in the FATT [63] study likely underestimates the magnitude of discharge. This reinforces the need for empirical studies of catchment hydrology. Predicting runoff responses to larger precipitation events will require hydrologic models tested and applied in the coalfields region. Phillips [64] examined runoff and surface and subsurface flow detention using hydrologic models that considered differences in runoff producing conditions in mine and unmined catchments in eastern Kentucky. Results from this study show that runoff production was likely to increase in MTM-impacted catchments compared to unmined catchments but with considerable variability attributed to local geologic, topographic, and pedologic conditions, and to differences in the stage and method of valley fill construction and mine reclamation. Zégre *et al.* [70] modeled the hydrologic response times of four storms in a MTM/VF-impacted headwater catchment in West Virginia and observed steep response curves indicating a rapid translation of rainfall to runoff, large baseflow ratios, and little variability between events but the absence of an unmined reference catchment makes placing these results in context to other catchments difficult.

Few studies have addressed the hydrologic impact of MTM at larger scales. Long term studies of the Tug Fork watershed (4040 km^2) in West Virginia, Kentucky, and Virginia from 1947 to 1978 [96] and sub-watersheds (~225 km^2) of the Russell Fork River in Virginia from 1927 to 1980 [97] show general trends in increased flood magnitudes and increased baseflows but the extent of MTM in those basins during the respective study periods is unclear and likely limited. More recently, Zégre*et al.* [69] characterized MTM-driven land cover changes and annual streamflow from 1994 to 2010 in the 1011 km^2 Big Coal River watershed in southern West Virginia. Despite that land cover changes show a general conversion of forest to mineland, significant streamflow changes were not detected. Zégre *et al.* [70] expanded the study in the Big Coal River by quantifying MTM/VF-driven land cover changes over the lifetime of this practice (1973–2010) and evaluated for changes across the hydrologic regime at multiple temporal and spatial scales. Significant decreases in maximum streamflow and variability and significant increases in baseflow ratio were detected since the start of MTM/VF in the 1970's. These changes were attributed to the influence of valley fills and to unaccounted-for water from deep abondoned underground coal mines [70]. Increased baseflow from surface and underground mine drainage raises significant water quality concerns for these receiving rivers; the degree to which mineral and metal laden discharge from mining-impacted catchments aggregate downstream will have major impacts on water quality (see Lindberg *et al.* [7]). Additionally, altered flow regimes will impact the mid- to long-term geomorphic evolution of these rivers. This is particularly relevant in the Central Appalachian coalfields where communities are often located directly adjacent to streams and rivers.

A review of existing literature highlights significant variability in the hydrologic responses of MTM/VF-impacted watersheds at the headwater and downstream scales. Despite mixed results, it is evident that individually and collectively, mountaintop mines and valley fills exert important controls on catchment hydrology, namely in runoff altering flow paths.Figure 4 shows a conceptual model for runoff flowpaths in this two-part system. Our hypothesis, based in part on the work by Hawkins and Aljoe [95] for surface mine minespoils, is that flow within the valley fill is controlled by matrix flow during non-storm conditions and by large preferentail flowpaths during storm conditions. Elevated baseflow downstream of valley fills is the result of water flowing through torturous flowpaths through small pores (matrix flow), while the larger, air-filled pores remain hydrologically inactive. This results in transient storage and release of water overtime, sustainaing baseflows compared to ephemeral nature of natural headwater streams. Prolonged forced contact time with overburden geology placed in the fill results in elevated solutes downstream of valley fills. During storm conditions, small pores and voids created by large rock fragments are saturated, activating rapid runoff flowpaths, resulting in increased peakflows downstream of valley fills.

4. KNOWLEDGE GAPS AND FUTURE RESEARCH DIRECTIONS

Existing research on the hydrologic impacts of MTM has documented a range of potential impacts to the storm hydrograph and seasonal flow regimes but also reveals considerable variability in hydrologic responses to storm events, extent of disturbance, and stage and method of reclamation. Currently we lack the data to understand the cause of this variability. *What are the dominant runoff generation processes in MTM-impacted catchments? How do these processes change with increasing disturbance from MTM? How do these processes change with differing reclamation techniques? How do contemporary valley fills store, route, and release water? What variability exists within forested catchments in the Central Appalachian coalfields? What are the effects of legacy disturbances such as timber harvesting and underground coal mining?*

These critical questions remain unanswered and little progress can be made in understanding and quantifying the hydrologic impacts of MTM until the volume and type of data necessary to understand the variability observed in the existing literature is collected. The following section expands on the knowledge gaps identified by Zegre *et al.* [70] and our review of the literature.

4.1. Streamflow Generation Processes in MTM-Impacted Catchments

At present, all investigations into the hydrologic impacts of MTM have been measured at the catchment outlet or have relied on hydrologic models in gaged and ungaged catchments. The limitations of conducting controlled scientific investigations in drastically disturbed areas (see Bonta [98]) have made these approaches appropriate and valuable information about the variability associated with MTM has been garnered from this work. But the inherent constraints of black box studies limit our understanding of streamflow generation processes. Significant data is required to quantify variability and inform models to extend information across the MTM region. Differences in runoff responses in pre-SMCRA unconsolidated spoil banks and the post-SMCRA heavily compacted surface mine is well documented in existing literature. However, the dominant streamflow generation processes in MTM catchments where both compacted surface mines and large spoil piles are not yet known.

In order to understand the full range of hydrologic consequences of this practice, future research needs to focus on catchment processes that control the storage, transport, and flowpaths of water. Geochemical and isotopic approaches should be incorporated into hydrometric studies to discern geographic sources of runoff in addition to its magnitude and duration. How these processes change in response to differing climatic inputs, extents of disturbance, reclamation techniques, and alteration to catchment structure and organization will provide insight into the variability observed in the hydrologic studies to date. Isotopic and geochemical tracers have been applied in catchments disturbed by urbanization (e.g., [31,99]), but have seen little use in catchments impacted by surface mining (e.g., [100]). This will ultimately require cooperation between industrial landowners, researchers, and agencies as quantifying catchment wide streamflow generation processes will require access to all reaches of the catchment, not just the outlet. This has been an obstacle to past research in thesesystems [70] and must be bridged in order to understand and ameliorate the environmental problems associated with this mining practice.

4.2. Hydrology of Non-MTM Catchments in the Central Appalachian Coalfields

As Wiley and Brogan [16] demonstrated, different hydrologic responses to large storm events are seen in adjacent, forested, similarly sized catchments. Wiley and Brogan [16] attribute the large variability to heterogeneous and patchy precipitation patterns within the study region, though the complex and heavily fractured geology of the Central Appalachian coalfields likely exerts some control on the observed runoff variability in adjacent catchments.

The hydrologic impact of MTM operations in the context of the heterogonous landscape of the region remains uncertain. Is variability normalized by the landscape scale disturbance of MTM or does the displacement of mountain ridges augment preexisting hydrologic differences caused by topography, geology, and legacy land disturbance?

Surprisingly little is known about the hydrologic processes responsible for movement and storage of water in the context of multiple episodes of land disturbance in the Central Appalachian coalfields. While the MTM region lies between long-term forested experimental catchments at the Fernow Experimental Forest (northern WV) and the Coweeta Hydrologic Laboratory (NC), the information gained from those sites is limited due to differing climatology, geology and historical and current landuse in the Central Appalachian coalfields. Work in the greater Central Appalachian region has shown that stormflow is dominated by subsurface flow [101].

In the Central Appalachian coalfields, groundwater movement is predominantly controlled by a complex network of stress relief fractures in hillslopes and valley bottoms [102,103]. However, a long history of underground coal mining throughout much of this region has drastically altered the structure of the subsurface system [104]. Subsidence associated with abandoned underground mines creates additional fractures which can increase hydrologic connectivity between the surface and subsurface as well as between water-bearing subsurface geologic units [89,105]. Consequently, underground mines and associated subsidence fractures can become major conduits for subsurface water movement. Headwater drainage networks downdip (at the lower end of the sloped coal seam) and below the mined coal beds can receive significant amounts of water while streams underlain by underground mines lose water, especially during baseflow conditions [55]. At the headwater scale, substantial volumes of water can be transferred between basins, increasing the complexity of assessing hydrologic change related to surface and subsurface mining [56]. While Borchers *et al.* [56] examined the combined effects of deep mining and surface strip mining, no study we are aware of examines the interactions between contemporary MTM/VF operations and legacy deep mining. A thorough examination of the spatial and temporal variability of individual catchments due to heterogeneity in catchment characteristics and legacy disturbances is necessary to understand the impacts of this practice across the landscape. A robust hydrologic monitoring network comprised of numerous MTM-impacted and unmined catchments will be necessary to address this variability.

4.3. Valley Fill Hydrology

How do valley fills store and release water? Wunsch *et al.* [12] began to address this question of their study of a large spoil pile from a MTM operation in eastern Kentucky. Their research provides valuable insight into the complexity of valley fill hydrology but uncertainty remains regarding the processes involved in the movement and storage of water in valley fills. Little is known about the geographic sources of water that supply valley fills; the spatial distribution and residence time of water within valley fills; or how water is released during lowflow and storm conditions (Figure 4). Heterogeneity in the surrounding geology coupled with legacy land disturbance (*i.e.*, underground mining) and the multitude of different valley fill construction techniques creates additional complexity. Insights into these uncertainties extend beyond physical hydrology as valley fills are particularly important in terms of downstream water chemistry; overburden placed in drainage pathways forces contact time between runoff and unweathered rock. Numerous studies document increased concentrations of dissolved solutes that degrade aquatic ecosystems downstream of MTM/VF (e.g., [7,104,105]). While the work of Hawkins and Aljoe [95] and Wunsch *et al.* [12] are starting point for understanding, future research should address valley fills that differ in size, construction and reclamation techniques, age, and geologic and topographic settings to represent the range of conditions present across the region. In addition, isotopic and geochemical tracers should also be used to discern the origin and flowpaths of water in valley fills (e.g., [100]).

4.4. MTM/VF Reclamation Techniques

The Forestry Reclamation Approach has gained traction as a potentially viable reclamation technique amongst industry and regulators [46]. In the FRA, minespoil and in some cases topsoil or amended topsoil, is dumped and loosely graded to create a minimally compacted growing medium for high value native hardwood tree seedlings [106]. This technique has shown to be an effective method for establishing forests in on mine lands [44], but its effect on catchment hydrology is not fully understood. Taylor *et al.* [67] describe plot-scale runoff of loose dumped soil to have low runoff ratios, small peak discharges, and long flow durations. Due to the high infiltration capacities of loose dumped minespoil [107], the broad application of the FRA to large surface mines would likely result in the establishment of subsurface flowpaths in reclaimed mined areas. However, given the reduced slopes on reclaimed MTM surfaces, residence time of water stored in the loose dumped spoil profile may be longer than in forested hillslopes of the coalfields region that are generally thought to have thin soils with little storage capacity [108]. Therefore, storm hydrographs from FRA-reclaimed hillslopes might produce more damped hydrographs than pre-mining forests. Hillslope and catchment scale studies of bare and vegetated loose dumped mine reclamation operations are needed to understand the effects of this reclamation practice.

4.5. Thresholds

Disturbance thresholds to detect measurable changes in streamflow have not been established for MTM or traditional surface mining. Forest harvesting studies suggest that 15%–20% of a catchment needs to be disturbed to detect changes in streamflow [34,109]. Bernhardt *et al.* [8] showed that as little as 2% of contributing area impacted by mining results in significant losses of intolerant macroinvertebrate taxa and at >5%, biological impairment occurs. Messinger [11] observed large differences in runoff responses in a heavily MTM-impacted catchment (44% MTM/VF) and moderately MTM-impacted catchment (12% MTM/VF) across storm events indicating that extent of mining and catchment size are likely important factors influencing outlet responses. However, the question of How much mining does it take to alter hydrology? remains unanswered.

Runoff thresholds for precipitation depth, duration, and intensity, and valley fill/catchment moisture conditions would also aid in understanding the hydrologic response of MTM-impacted catchments (Figure 4). Messinger [11] observed a precipitation intensity threshold of 25 mm hr^{-1} that dictated whether unit peak flow was greater in a MTM-impacted catchment (>25 mm hr^{-1}) or the forested catchment (<25 mm hr^{-1}). Understanding the conditions and the type of storms likely to produce flash flooding in MTM-impacted (as well as forested) catchments is critical to limiting the destructive events that have ravaged communities in the Central Appalachian coalfields.

4.6. Multiple Spatial and Temporal Scales

Investigations into the hydrologic impacts of MTM predominantly occur at the headwater scale. While this scale is an appropriate experimental unit and represents the scale of disturbance, it is also necessary to understand the downstream consequences of this practice. Zégre *et al.* [70] provides one of the first multiple spatial and temporal scale studies showing scale-invariant dampening of hydrologic variability, decreases in maximum flows, and increases in baseflow ratio thought to be related to valley fill control and unaccounted-for deep mine drainage. However more research is necessary to understand if this trend is consistent across watersheds in the coalfields region. Flow regime changes at headwater and downstream scales has important flood, geomorphology, and water quality implications to communities dependent on ecosystems derived from mountainous watersheds.

To the extent of our knowledge and extensive literature search, longitudinal studies of catchments undergoing MTM-related disturbance do not exist. Such studies are necessary to understand how hydrology changes in response to ongoing mining and reclamation. In addition, long-term (e.g., decades) studies are needed to assess how reclaimed landscapes, including morphology and vegetation, and hydrologic processes evolve over time. For example, Ritter and Gardner [57] observed differences of the infiltration rate recovery of surface mines in Central Pennsylvania during a decade long study. This elicits questions about the hydrologic evolution of valley fills and reclaimed MTM surfaces, and at a broader scale, the stability of the new topographic structure imposed on MTM/VF-impacted catchments. Surface mines reclaimed under the protocols of the FRA also warrant long-term study as these dynamic landscapes will rapidly change from their initial condition. Long-term studies will also capture infrequent large storm events which are critical to elucidating the role of MTM/VF in flood generation in the Central Appalachian coalfields.

4.7. Altered Topography

MTM/VF displaces as much as 300 m of overburden from Appalachian ridge tops to access deeper coal seams and therefore represents a substantial change to the size, shape, organization, slope, and drainage network of the catchments where it occurs. To date, quantifications of the change to catchment topography from MTM/VF operations remains scarce and, to the authors' knowledge, no study connects this structural reorganization to hydrologic change.

The importance of catchment topography and drainage structure in controlling hydrologic response has been a focus in contemporary literature. McGlynn *et al.* [110] investigated the role of catchment size in hillslope and riparian runoff dynamics. McGuire *et al.* [111] explored the dominant physical controls on the residence time of catchment water. Among their findings were the importance of the headwater riparian zone and flow path gradient in generating runoff in headwater catchments. Given MTM/VF's potential to drastically reduce catchment slope (especially in instances where variances to AOC are granted) and bury headwater riparian areas, framing hydrologic change in the Central Appalachian coalfields as a result of "landcover disturbance" may be insufficient. The residence time and rate of runoff in MTM/VF-impacted catchments will, at least in part, be a function of the new topographic structure created by the coal mining and reclamation process.

This issue is confounded by the addition of drainage structures such as rock-lined channels from the mine surface to the valley bottom or sediment retention ponds at the base of valley fills. Therefore, discerning the hydrologic effect of topographic reorganization from other impacts of MTM will be difficult. This process must begin with a quantification of the extent and magnitude of topographic and drainage network change in MTM/VF-impacted areas. The proliferation of high resolution LiDAR-based topographic models of Central Appalachian coalfields will be critical in this step.

5. CONCLUDING REMARKS

MTM represents a dramatic and likely permanent landscape-scale disturbance with important local and regional impacts. Despite its scale, extent, and potential of continued use, key knowledge gaps regarding hydrologic consequences of this practice exist. Water storage and movement in these disturbed landscapes exert important controls on the well documented downstream water quality issues associated with MTM/VF. The culpability of surface mining operations in extreme flooding events in the Appalachian coalfields is still being debated amongst citizens, industry, and regulators. Thus, hydrologic studies of the impacts of MTM/VF play a critical role in elucidating the consequences of this mining practice.

Most investigations into the hydrologic impacts of surface mining involve traditional contour and strip mining techniques that occur at smaller scales and do not consider the role that valley fills have on hydrology, presenting additional uncertainties. Studies to date demonstrate that since the enactment of SMCRA, heavily compacted mine surfaces have decreased infiltration capacity and tree growth and consequently produce flashier, higher magnitude runoff responses to storm events. While minespoil studies provide insight into the potential hydrologic controls of valley fills, no studies specifically address valley fill hydrology. Previous studies dealing with hydrologic impacts of MTM operations have been successful in establishing a range of possible responses of this practice, but these studies underscore catchment heterogeneity and response variability limiting understanding of the downstream consequences and warranting further investigation through controlled experiments.

MTM's expected proliferation in the coming decades coupled with the adjacency of communities to streams and rivers in the Central Appalachian coalfields makes understanding the hydrologic consequences of this practice necessary. The observed hydrologic effects of MTM/VF present landowners, policy makers, engineers, and water resource managers with many difficult questions about how to manage the altered landscapes created by this mining progress. Such progress will ultimately depend on expanding the number of hydrologic studies in the Central Appalachian coalfields region and concentrating research efforts on the physical processes that control hydrologic response in these disturbed systems.

ACKNOWLEDGMENTS

Funding for this research was provided in part by NSF EAR Award 1042683; Oak Ridge Associated University (ORAU) Ralph E. Powe Junior Faculty Enhancement Award; West Virginia University Senate Research Award; and Hatch Act Formula Fund. We are grateful to Jane Lefevre of West Virginia University Davis College for graphical assistance on Figure 4. West Virginia Agricultural and Forestry Experimental Station Scientific Article No. 3204.

CONFLICTS OF INTEREST

The authors declare no conflict of interest.

REFERENCES

1. World Coal Association. Available online: http://www.worldcoal.org (accessed on 1 June 2013).
2. Sayler, K.L. *Land Cover Trends: Central Appalachians*; US Geological Survey: Washington, DC, USA, 2008.
3. US Environmental Protection Agency (EPA). *The Effects of Mountaintop Mines and Valley Fills on Aquatic Ecosystems of the Central Appalachian Coalfields*; US EPA: Washington, DC, USA, 2011; p. 153.
4. US Energy Information Administration (EIA). *Annual Energy Outlook 2013: Early Release Overview*; DOE/EAI-0383ER(2013); U.S. Department of Energy: Washington, DC, USA, 2012; pp. 1–16.
5. SkyTruth Mountaintop Removal Mining, Part 1: Measuring the Extent of Mountaintop Removal in Appalachia. Available online: http://blog.skytruth.org/2009/12/measuring-mountaintop-removal-mining-in.html (accessed on 5 March 2013).
6. Griffith, M.B.; Norton, S.B.; Alexander, L.C.; Pollard, A.I.; LeDuc, S.D. The effects of mountaintop mines and valley fills on the physicochemical quality of stream ecosystems in the central Appalachians: A review. *Sci. Total Environ.* **2012**, *417–418*, 1–12.

7. Lindberg, T.T.; Bernhardt, E.S.; Bier, R.; Helton, A.M.; Merola, R.B.; Vengosh, A.; Di Giulio, R.T. Cumulative impacts of mountaintop mining on an Appalachian watershed. *Proc. Natl. Acad. Sci. USA* **2011**, *108*, 20929–20934.

8. Bernhardt, E.S.; Lutz, B.D.; King, R.S.; Fay, J.P.; Carter, C.E.; Helton, A.M.; Campagna, D.; Amos, J. How Many Mountains Can We Mine? Assessing the regional degradation of central appalachian rivers by surface coal mining.*Environ. Sci. Technol.* **2012**, *46*, 8115–8122.

9. Dickens, P.S.; Minear, R.A.; Tschantz, B.A. Hydrologic alteration of mountain watersheds from surface mining. *J. Water Pollut. Control Fed.* **1989**, *61*, 1249–1260.

10. Caruccio, F.T.; Geidel, G.; Williams, R. Induced Alkaline Recharge Zones to Mitigate Acidic Seeps. In Proceedings of the National Symposium on Surface Mining, Hydrology, Sedimentology and Reclamation, Lexington, KY, USA, 2 December 1984.

11. Messinger, T. *Comparison of Storm Response of Streams in Small, Unmined and Valley-Filled Watersheds, 1999–2001, Ballard Fork, West Virginia*; US Geological Survey: Charleston, WV, USA, 2003; p. 22.

12. Wunsch, D.R.; Dinger, J.S.; Graham, C.D.R. Predicting ground-water movement in large mine spoil areas in the Appalachian Plateau. *Int. J. Coal Geol.* **1999**, *41*, 73–106.

13. Green, J.; Passmore, M.; Childers, H. *A Survey of the Condition of Streams in the Primary Region of Mountaintop Mining/Valley Fill Coal Mining*; US Environmental Protection Agency: Philadelphia, PA, USA, 2000.

14. Messinger, T.; Paybins, K.S. *Relations between Precipitation and Daily and Monthly Mean Flows in Gaged, Unmined and Valley-Filled Watersheds, Ballard Fork, West Virginia, 1999–2001*; WRi No.2003–4113; US Geological Survey: Charleston, WV, USA, 2003; p. 51.

15. Wiley, J.B.; Evaldi, R.D.; Eychaner, J.H.; Chambers, D.B. *Reconnaissance of Stream Geomorphology, Low Streamflow, and Stream Temperature in the Mountaintop Coal-Mining Region, Southern West Virginia, 1999–2000*; US Geological Survey: Charleston, WV, USA, 2001; pp. 1–38.

16. Wiley, J.B.; Brogan, F.D. *Comparison of Peak Discharges Among Sites with and without Valley Fills for the July 8–9 2001, Flood in the Headwaters of Clear Fork, Coal River Basin, Mountaintop Coal-*

Mining Region, Southern West Virginia; U.S. Department of the Interior, U.S. Geological Survey: Charleston, WV, USA, 2003; pp. 1–16.

17. Higginbotham, A.; Witt, T.S.; Childs, R.; Kent, C.; Pardue, E.; Risch, C. *The West Virginia Coal Economy: 2008*; Bureau of Business and Economic Research, West Virginia University and Center for Business and Economic Research, Marshall University: Huntington, WV, USA, 2008.

18. Adams, M.B.; Edwards, P.J.; Ford, W.M.; Schuler, T.M.; Thomas-Van Gundy, M.; Wood, F. *Fernow Experimental Forest: Research History and Opportunities*; USDA Forest Service: Washington, DC, USA, 2012; p. 26.

19. Carlyle-Moses, D.E.; Price, A.G. An evaluation of the Gash interception model in a northern hardwood stand. *J. Hydrol.* **1999**, *214*, 103–110.

20. Helvey, J.D.; Patric, J.H. Canopy and litter interception of rainfall by hardwoods of eastern United States. *Water Resour. Res.* **1965**, *1*, 193–206.

21. Price, K.; Jackson, C.R.; Parker, A.J. Variation of surficial soil hydraulic properties across land uses in the southern Blue Ridge Mountains, North Carolina, USA. *J. Hydrol.* **2010**, *383*, 256–268.

22. Moore, I.D.; Burch, G.J.; Wallbrink, P.J. Preferential flow and hydraulic conductivity of forest soils. *Soil Sci. Soc. Am. J.***1985**, *50*, 876–881.

23. Bonell, M. Progress in the understanding of runoff generation dynamics in forests. *J. Hydrol.* **1993**, *150*, 217–275.

24. Burns, D.A.; McDonnell, J.J.; Hooper, R.P.; Peters, N.E.; Freer, J.E.; Kendall, C.; Beven, K. Quantifying contributions to storm runoff through end-member mixing analysis and hydrologic measurements at the Panola Mountain Research Watershed (Georgia, USA). *Hydrol. Process.* **2001**, *15*, 1903–1924.

25. Hornberger, G.M.; Scanlon, T.M.; Raffensperger, J.P. Modelling transport of dissolved silica in a forested headwater catchment: The effect of hydrological and chemical time scales on hysteresis in the concentration-discharge relationship. *Hydrol. Process.* **2001**, *15*, 2029–2038.

26. Eshleman, K.N. Hydrological Consequences of Land Use Change: A Review of the State-of-Science. In *Ecosystems and Land Use Change*; AGU: Washington, DC, 2004; pp. 13–29.

27. Hornbeck, J.W.; Pierce, R.S.; Federer, C.A. Streamflow changes after forest clearing in New England. *Water Resour. Res.* **1970**, *6*, 1124–1132.

28. Jones, J.A.; Grant, G.E. Peak flow responses to clear-cutting and roads in small and large basins, western Cascades, Oregon. *Water Resour. Res.* **1996**, *32*, 959–974.

29. Thomas, R.B.; Megahan, W.F. Peak flow responses to clear-cutting and roads in small and large basins, western Cascades, Oregon: A second opinion. *Water Resour. Res.* **1998**, *34*, 3393–3403.

30. Potter, K.W. Hydrological impacts of changing land management practices in a moderate-sized agricultural catchment. *Water Resour. Res.* **1991**, *27*, 845–855.

31. Gremillion, P.; Gonyeau, A.; Wanielista, M. Application of alternative hydrograph separation models to detect changes in flow paths in a watershed undergoing urban development. *Hydrol. Process.* **2000**, *14*, 1485–1501.

32. O'Driscoll, M.; Clinton, S.; Jefferson, A.; Manda, A.; McMillan, S. Urbanization effects on watershed hydrology and in-stream processes in the southern United States. *Water* **2012**, *2*.

33. Rose, S.; Peters, N.E. Effects of urbanization on streamflow in the Atlanta area (Georgia, USA): A comparative hydrological approach. *Hydrol. Process.* **2001**, *15*, 1441–1457.

34. Bosch, J.M.; Hewlett, J.D. A review of catchment experiments to determine the effect of vegetation changes on water yield and evapotranspiration. *J. Hydrol.* **1982**, *55*, 3–23.

35. Wemple, B.C.; Jones, J.A.; Grant, G.E. Channel network extension by logging roads in two basins, western Cascades, Oregon. *J. Am. Water Resour. Assoc.* **1996**, *32*, 1195–1207.

36. Jones, J.A.; Swanson, F.J.; Wemple, B.C.; Snyder, K.U. Effects of roads on hydrology, geomorphology, and disturbance patches in stream networks. *Conserv. Biol.* **2000**, *14*, 76–85.

37. Eisenbies, M.H.; Aust, W.M.; Burger, J.A.; Adams, M.B. Forest operations, extreme flooding events, and considerations for hydrologic modeling in the Appalachians—A review. *For. Ecol. Manag.* **2007**, *242*.

38. Sidle, R.C.; Ziegler, A.D.; Negishi, J.N.; Nik, A.R.; Siew, R.; Turkelboom, F. Erosion processes in steep terrain—Truths, myths, and uncertainties related to forest management in Southeast Asia. *For. Ecol. Manag.* **2006**, *224*, 199–225.

39. Skaggs, R.W.; Brevé, M.A.; Gilliam, J.W. Hydrologic and water quality impacts of agricultural drainage. *Crit. Rev. Environ. Sci. Technol.* **1994**, *24*, 1–32.

40. Dow, C.L.; DeWalle, D.R. Trends in evaporation and Bowen Ratio on urbanizing watersheds in eastern United States. *Water Resour. Res.* **2000**, *36*, 1835–1843.

41. Ferrari, J.R.; Lookingbill, T.R.; McCormick, B.; Townsend, P.A.; Eshleman, K.N. Surface mining and reclamation effects on flood response of watersheds in the central Appalachian Plateau region. *Water Resour. Res.* **2009**, *45*.

42. Peng, S.S. Mountaintop removal controversy slows West Virginia coal mining. *Min. Eng.* **2000**, *52*, 53–58.

43. US Congress. Surface Mining Control and Reclamation Act. US Congress, Washington, DC, USA, 1977.

44. Angel, P.; Graves, D.; Barton, C.; Warner, R.C.; Conrad, P.; Sweigard, R.C.; Agouridis, C. Surface mine reforestation research: Evaluation of tree response to low compaction reclamation techniques. In Proceedings of the 7th International Conference on Acid Rock Drainage (ICARD); American Society of Mining and Reclamation (ASMR): St. Louis, MO, USA, 2006; pp. 45–58.

45. Graves, D.H.; Ringe, J.M.; Pelkki, M.H.; Sweigard, R.J.; Warner, R. High value tree reclamation research. In *Environmental Issues and Management of Waste in Energy and Mineral Production*; CRC Press: Boca Raton, FL, USA, 2000; pp. 413–421.

46. Zipper, C.; Burger, J.; Skousen, J.; Angel, P.; Barton, C.; Davis, V.; Franklin, J. Restoring forests and associated ecosystem services on appalachian coal surface mines. *Environ. Manag.* **2011**, *47*, 751–765.

47. Traux, C.N. Water storage potential of surface coal mined lands. *Min. Congr. J.* **1965**, *51*, 40–46.

48. Agnew, A.F. A quarter to zero—Surface mining and water supplies. *Min. Congr. J.* **1966**, *52*, 29, 32–34, 38–40.

49. Collier, C.R.; Pickering, R.J.; Musser, J.J. *Influences of Strip Mining on the Hydrologic Environment of Parts of the Beaver Creek Basin, Kentucky, 1955–66*; US Geological Survey: Lexington, KY, USA, 1970.

50. Curtis, W.R. Strip-mining increases flood potential of mountain watersheds. In *National Symposium of Watersheds in Transition*;

American Water Resources Association: Fort Collins, CO, USA, 1972; pp. 357–360.

51. Curtis, W.R. *Surface Mining and the Flood of April 1977*; US Forest Service Research Note NE-248: Berea, KY, USA, 1977; pp. 1–4.

52. Grubb, H.F.; Ryder, P.O. *Effect of Coal Mining on the Water Resources of the Tradewater River Basin, Kentucky*; US Geological Survey: Lexington, KY, USA, 1972; pp. 1–91.

53. Bryan, B.A.; Hewlett, J.D. Effect of surface mining on storm flow and peak flow from six small basins in eastern Kentucky. *JAWRA J. Am. Water Resour. Assoc.* **1981**, *17*, 290–299.

54. Jorgensen, D.W.; Gardner, T.W. Infiltration capacity of disturbed soils: Temporal changes and lithologic controls.*JAWRA JAWRA J. Am. Water Resour. Assoc.* **1987**, *23*, 1161–1172.

55. Puente, C.; Atkins, J.T. *Simulation of Rainfall-Runoff Response in Mined and Unmined Watersheds in Coal Areas of West Virginia*; US Geological Survey: Charleston, WV, USA, 1989; pp. 1–56.

56. Borchers, J.W.; Ehlke, T.A.; Mathes, M.V.; Downs, S.C. *The Effects of Coal Mining on the Hydrologic Environmnet of Selected Stream Basins in Southern West Virginia*; U.S. Geological Survey: Charleston, WV, USA, 1991.

57. Ritter, J.B.; Gardner, T.W. Hydrologic evolution of drainage basins disturbed by surface mining, central Pennsylvania.*Geol. Soc. Am. Bull.* **1993**, *105*, 101–115.

58. Caruccio, F.T.; Geidel, G. *Status Report: Long-Term Effects of Alkaline Trenches and Funnels at the Mercer Site*; West Virginia Surface Mine Drainage Task Force Symposium: Morgantown, WV, USA; 4; April; 1995.

59. Bonta, J.V.; Amerman, C.R.; Harlukowicz, T.J.; Dick, W.A. Impact of surface coal mining on three Ohio watersheds—Surface water hydrology. *JAWRA J. Am. Water Resour. Assoc.* **1997**, *33*, 907–917.

60. Wunsch, D.; Dinger, J.; Taylor, P. *Design, Construction, and Monitoring of the Ground-Water Resources of a Large Spoil Area: Star Fire Tract, Eastern Kentucky*; Kentucky Geological Survey: Lexington, KY, USA, 1992; pp. 1–16.

61. Wunsch, D.; Dinger, J.; Taylor, P.; Carey, D.; Graham, C. *Hydrogeology, hydrogeochemistry, and spoil settlement at a large mine-spoil area in eastern Kentucky: Star Fire Tract*; Kentucky Geological Survey: Lexington, KY, USA, 1996; pp. 1–55.

62. Guebert, M.D.; Gardner, T.W. Macropore flow on a reclaimed surface mine: Infiltration and hillslope hydrology.*Geomorphology* **2001**, *39*, 151–169.

63. Flood Advisory Taskforce. *Runoff Analyses of Seng, Scrabble, and Sycamore Creeks, Part 1*; WV Department of Environmental Protection: Charlseton, WV, USA, 2002.

64. Phillips, J.D. Impacts of surface mine valley fills on headwater floods in eastern Kentucky. *Environ. Geol.* **2004**, *45*, 367–380.

65. Negley, T.L.; Eshleman, K.N. Comparison of stormflow responses of surface-mined and forested watersheds in the Appalachian Mountains, USA. *Hydrol. Process.* **2006**, *20*, 3467–3483.

66. McCormick, B.C.; Eshleman, K.N.; Griffith, J.L.; Townsend, P.A. Detection of flooding responses at the river basin scale enhanced by land use change. *Water Resour. Res.* **2009**, *45*, W08401.

67. Taylor, T.J.; Agouridis, C.T.; Warner, R.C.; Barton, C.D.; Angel, P.N. Hydrologic characteristics of Appalachian loose-dumped spoil in the Cumberland Plateau of eastern Kentucky. *Hydrol. Process.* **2009**, *23*, 3372–3381.

68. McCormick, B.C.; Eshleman, K.N. Assessing hydrologic change in surface-mined watersheds using the curve number method. *J. Hydrol. Eng.* **2011**, *16*, 575–584.

69. Zégre, N.P.; Maxwell, A.; Lamont, S. Characterizing streamflow response of a mountaintop-mined watershed to changing land use. *Appl. Geogr.* **2013**, *39*, 5–15.

70. Zégre, N.; Miller, A.J.; Maxwell, A.; Lamont, S. Multi-scale analysis of hydrology in a mountaintop mineimpacted watershed. *J. Am. Water Resour. Assoc.* in press.

71. Rodrigue, J.A.; Burger, J.A. Forest productivity and woody species diversity on pre-SMCRA mined land. In *National Meeting for the American Society of Mine Reclamation*; American Society of Mine Reclamation: Tampa, FL, USA, 2000; pp. 205–225.

72. Bussler, B.H.; Byrnes, W.R.; Pope, P.E.; Chaney, W.R. Properties of minesoil reclaimed for forest land use. *Soil Sci. Soc. Am. J.* **1984**, *48*, 178–184.

73. Holl, K.D. Long-term vegetation recovery on reclaimed coal surface mines in the eastern USA. *J. Appl. Ecol.* **2002**, *39*, 960–970.

74. Barnhisel, R.I.; Hower, J.M. Coal surface mine reclamation in the eastern United States: The revegetation of disturbed lands to hayland/pasture or cropland. In *Advances in Agronomy*; Academic Press: San Diego, CA, USA, 1997.

75. Bradshaw, A.D. The reclamation of derelict land and the ecology of ecosystems. In *Restoration Ecology: A Synthetic Approach to Ecological Research*; Jordan, W.R., Gilpin, M.E., Aber, J.D., Eds.; Cambridge Unversity Press: Cambridge, UK, 1990; pp. 53–75.

76. Kelliher, F.M.; Leuning, R.; Schulze, E.D. Evaporation and canopy characteristics of coniferous forests and grasslands.*Oecologia* **1993**, *95*, 153–163.

77. Webb, W.; Szarek, S.; Lauenroth, W.; Kinerson, R.; Milton, S. Primary productivity and water use in native forest, grassland, and desert ecosystems. *Ecology* **1978**, *59*, 1239–1247.

78. Bell, J.C.; Cunningham, R.L.; Anthony, C.T. Morphological characteristics of reconstructed prime farmland soils in western Pennsylvania. *J. Environ. Qual.* **1994**, *23*, 515–520.

79. Indorante, S.J.; Jansen, I.J.; Boast, C.W. Surface mining and reclamation: Initial changes in soil character. *J.Soil Water Conserv.* **1981**, *36*, 347–351.

80. Ciolkosz, E.J.; Cronce, R.C.; Cunningham, R.L.; Petersen, G.W. Characteristics, genesis, and classification of Pennsylvania minesoils. *Soil Science* **1985**, *139*, 232–238.

81. Chong, S.K.; Cowsert, P.T. Infiltration in reclaimed mined land ameliorated with deep tillage treatments. *Soil Tillage Res.* **1997**, *44*, 255–264.

82. Silburn, D.M.; Crow, F.R. Soil properties of surface mined land. *Am. Soc. Agric. Eng.* **1984**, *27*, 827–832.

83. Power, J.F.; Ries, R.E.; Sandoval, F.M. Reclamation of coal-mined land in the northern great plains. *J. Soil Water Conserv.* **1978**, *33*, 69–74.

84. Potter, K.; Carter, F.; Doll, E. Physical properties of constructed and undisturbed soils. *Soil Sci. Soc. Am. J.* **1988**, *52*, 1435–1438.

85. McSweeney, K.; Jansen, I.J. Soil structure and associatedrooting behavior in minesoils. *Soil Sci. Soc. Am. J.* **1984**, *48*, 607–612.

86. Pedersen, T.A.; Rogowski, A.S.; Pennock, R. Physical characteristics of some minesoils. *Soil Sci. Soc. Am. J.* **1980**, *44*, 321–328.

87. Bonta, J.V. Impact of coal surface mining and reclamation on suspended sediment in three Ohio watersheds. *JAWRA J. Am. Water Resour. Assoc.* **2000**, *36*, 869–887.

88. Fox, J.F. Identification of sediment sources in forested watersheds with surface coal mining disturbance using carbon and nitrogen isotopes. *JAWRA J. Am. Water Resour. Assoc.* **2009**, *45*, 1273–1289.

89. Lemieux, C.R. Infiltration Characteristics and Hydrologic Modeling of Disturbed Land, Moshannon, Pennsylvania. Master's Thesis, Pennsylvania State University, University Park, PA, USA, 1987.

90. Ritter, J.B. Surface Hydrology of Drainage Basins Disturbed by Surface Mining and Reclamation, Central Pennsylvania. Ph.D Thesis, Pennsylvania State University, University Park, PA, USA, 1990.

91. Curtis, W.R. Surface mining and the hydrologic balance. *Min. Congr. J.* **1979**, *7*, 35–40.

92. Sherman, L.K. Streamflow from rainfall by the unit-graph method. *Eng. News Record* **1932**, *108*, 501–505.

93. US Environmental Protection Agency (EPA). *Draft Programmatic Environmental Impact Statement —Mountaintop Mining/Valley Fills in Appalachia*; US EPA-Region 3: Philadelphia, PA, USA, 2003.

94. Michael, P.R.; Superfesky, M.J.; Uranowski., L.J. Challenges to Applying Geomorphic and Stream Reclamation Methodologies to Mountaintop Mining and Excess Spoil Fill Construction in Steep-Slope Topography (e.g. Central Appalachia). In Proceedings of the 2010 National Meeting of the American Society of Mining and Reclamation, Pittsburgh, PA, USA, 5–11 June 2010.

95. Hawkins, J.W.; Dunn, M. Hydrologic characteristics of a 35-year-old underground mine pool. *Mine Water Environ.***2007**, *26*, 150–159.

96. Hirsh, R.M.; Scott, A.G.; Wyant, T. *Investiagation of trends in flooding in the Tug Fork basin of Kentucky, Virginia, and West Virginia*; US Geological Survey: Alexandria, VA, USA, 1982; pp. 1–41.

97. Larson, J.D.; Powell, J.D. *Hydrology and Effects of Mining in the Upper Russell Fork Basin, Buchanan and Dickenson Counties,*

Virginia; US Geological Survey: Lexiongton, KY, USA, 1986; pp. 1–62.

98. Bonta, J.V. Challenges in conducting hydrologic and water quality research in drastically disturbed watersheds. *J. Soil Water Conserv.* **2005**, *60*, 121–133.

99. Meriano, M.; Howard, K.W.F.; Eyles, N. The role of midsummer urban aquifer recharge in stormflow generation using isotopic and chemical hydrograph separation techniques. *J. Hydrol.* **2011**, *396*, 82–93.

100. Zegre, N.; McGuire, K. Characterizing the Hydrologic Impacts of Mountaintop Mining Using Stable Isotopes. EOS Trans. Am. Geophys. Union, 2011. *Fall Meeting 2011*, abstract #H31A-1120.

101. DeWalle, D.R.; Swistock, B.R.; Sharpe, W.E. Three-component tracer model for stormflow on a small Appalachian forested catchment. *J. Hydrol.* **1988**, *104*, 301–310.

102. Kipp, J.A.; Dinger, J.S. Stress-Relief Fracture Control of Ground-Water Movement in the Appalachian Plateaus, In Proceedings of the Fourth Annual Eastern Regional Ground Water Conference, Burlington, VT, USA, 14–16 July 1987; National Water Welll Association: Burlington, VT, 1987; pp. 423–438.

103. Wyrick, G.G.; Borchers, J.W. *Hydrologic Effects of Stress-Relief Fracturing an Appalachian Valley [West Virginia]*; US Geological Survey: Washington, DC, USA, 1981; pp. 1–51.

104. Merriam, E.R.; Petty, J.T.; Merovich, G.T.; Fulton, J.B.; Strager, M.P. Additive effects of mining and residential development on stream conditions in a central Appalachian watershed. *J. North Am. Benthol. Soc.* **2011**, *30*, 399–418.

105. Petty, J.T.; Fulton, J.B.; Strager, M.P.; Merovich, G.T.; Stiles, J.M.; Ziemkiewicz, P.F. Landscape indicators and thresholds of stream ecological impairment in an intensively mined Appalachian watershed. *J. North Am. Benthol. Soc.***2010**, *29*, 1292–1309.

106. Burger, J.; Graves, D.; Angel, P.; Davis, V.; Zipper, C. The forestry reclamation approach; Appalachian Regional Reforestation Initiative. *Forest Reclamation Advisory.* **2005**, *2*.

107. Rogowski, A.S.; Jacoby, E.L. Monitoring water movement through strip mine spoil profiles. *Trans. ASAE* **1979**, *22*, 104–109.

108. Ehlke, T.E.; Runner, G.S.; Downs, S.C. Hydrology of Area 9, Eastern Coal Province, West Virginia. US Geological Survey, Charleston, WV, USA, 1982.

109. Stednick, J.D. Monitoring the effects of timber harvest on annual water yield. *J. Hydrol.* **1996**, *176*, 79–95.

110. McGlynn, B.L.; McDonnell, J.J.; Seibert, J.; Kendall, C. Scale effects on headwater catchment runoff timing, flow sources, and groundwater-streamflow relations. *Water Resour. Res.* **2004**, *40*, W07504.

111. McGuire, K.J.; McDonnell, J.J. Hydrological connectivity of hillslopes and streams: Characteristic time scales and nonlinearities. *Water Resour. Res.* **2010**, *46*, W10543.

Index